NMR Spectroscopy of Polymers

NMR Spectroscopy of Polymers

Edited by

ROGER N. IBBETT

Courtaulds Research and Technology
Coventry

BLACKIE ACADEMIC & PROFESSIONAL
An Imprint of Chapman & Hall
London · Glasgow · New York · Tokyo · Melbourne · Madras

Published by Blackie Academic & Professional, an imprint of Chapman & Hall, Wester Cleddens Road, Bishopbriggs, Glasgow G64 2NZ

Chapman & Hall, 2–6 Boundary Row, London SE1 8HN, UK

Blackie Academic & Professional, Wester Cleddens Road, Bishopbriggs, Glasgow G64 2NZ, UK

Chapman & Hall Inc., 29 West 35th Street, New York NY 10001, USA

Chapman & Hall Japan, Thomson Publishing Japan, Hirakawacho Nemoto Building, 6F, 1-7-11 Hirakawa-cho, Chiyoda-ku, Tokyo 102, Japan

DA Book (Aust.) Pty Ltd, 648 Whitehorse Road, Mitcham 3132, Victoria, Australia

Chapman & Hall India, R. Seshadri, 32 Second Main Road, CIT East, Madras 600 035, India

First edition 1993

Typeset in 10/12pt Times by Thomson Press (India) Ltd., New Delhi
Printed in Great Britain by St. Edmundsbury Press, Bury St. Edmunds, Suffolk.

ISBN 0 7514 0005 X

A catalogue record for this book is available from the British Library

Library of Congress Cataloging-in-Publication data

NMR spectroscopy of polymers / edited by Roger N. Ibbett. -- 1st ed.
p. cm.
Includes bibliographical references and index.
ISBN 0-7514-0005-X
1. Polymers--Spectra. 2. Nuclear magnetic resonance spectroscopy.
I. Ibbett, Roger N., 1958–
QC463.P5N57 1993
547.7'046--dc20 93-17124
CIP

Printed on acid-free text paper, manufactured in accordance with ANSI/NISO Z39.48-1992 (Permanence of Paper).

Contributors

Professor J.C. Bevington The Polymer Centre, School of Physics and Materials, University of Lancaster, Bailrigg, Lancaster LA1 4YA, UK

Professor P.T. Callaghan Department of Physics and Biophysics, Faculty of Science, Massey University, Palmerston North, New Zealand

Dr J.R. Ebdon The Polymer Centre, School of Physics and Materials, University of Lancaster, Bailrigg, Lancaster LA1 4YA, UK

Dr. F. Heatley Manchester Polymer Centre, Department of Chemistry, University of Manchester, Oxford Road, Manchester M13 9PL, UK

Dr I.R. Herbert Courtaulds Research and Technology, PO Box 111, Lockhurst Lane, Coventry CV6 5RS, UK

Dr O.W. Howarth Department of Chemistry, Centre for Nuclear Magnetic Resonance, University of Warwick, Coventry CV4 7AL, UK

Dr T.N. Huckerby The Polymer Centre, School of Physics and Materials, University of Lancaster, Bailrigg, Lancaster LA1 4YA, UK

Dr R.N. Ibbett Courtaulds Research and Technology, PO Box 111, Lockhurst Lane, Coventry CV6 5RS, UK

Dr A.M. Kenwright Department of Chemistry, University of Durham, South Road, Durham DH1 3LE, UK

Dr F. Lauprêtre Laboratoire de Physico-Chimie Structurale et Macromoléculaire associé au CNRS, ESPCI, 10 rue Vauquelin, 75231 Paris cedex 05, France

Dr D.M. Rice Varion, 3120 Hansen Way, Palo Alto, California 94304–1030, USA

Mr B.J. Say Department of Chemistry, University of Durham, South Road, Durham DH1 3LE, UK

Dr A.E. Tonelli Department of Textile Engineering, Chemistry and Science, College of Textiles, North Carolina State University, PO Box 8301, Raleigh, NC 27695–8301, USA

Contents

Editorial introduction

R.N. IBBETT

This book provides a source of information on all major aspects of NMR spectroscopy of synthetic polymers. It represents a deliberate attempt to pull together the numerous strands of the subject in a single comprehensive volume, designed to be readable at every scientific level. It is intended that the book will be of use to the vast majority of polymer scientists and NMR spectroscopists alike.

Readers new to NMR will find extensive information within the book on the available techniques, allowing full exploration of the many polymer science applications. Readers already established within a branch of NMR will find the book an excellent guide to the practical study of polymers and the interpretation of experimental data. Readers who have specialised in polymer NMR will find the book a valuable dictionary of proven methodologies, as well as a guide to the very latest developments in the subject.

Workers from all of the main branches of polymer NMR have been invited to contribute. Each chapter therefore contains information relating to a particular investigative topic, indentified mainly on the basis of technique. The book is loosely divided between solution and solid-state domains, although the numerous interconnections confirm that these two domains are parts of the same continuum. Basic principles are explained within each chapter, combined with discussions of experimental theory and applications. Examples of polymer investigations are covered generously and in many chapters there are discussions of the most recent theoretical and experimental developments. Exciting advances are being made in both solution and solid-state polymer NMR and this book provides an unparalleled opportunity to establish the direction of research across the whole field. The extensive reference information can be used as a direct route forward into the varied polymer and NMR oriented research literature.

Throughout the book numerous references are made to the available range of general purpose NMR spectroscopy texts. The reader should turn to these for a full introduction to solution and solid-state techniques and their application to chemical science. Alternatively, texts are available which treat NMR as one of the range of spectroscopies of relevance to polymer characterisation. The reader will also be made aware of some excellent works on specific aspects of polymer NMR, including microstructural determination and high-resolution solid-state methods. In a number of cases these would form a natural extension to chapters contained within this book, and are therefore recommended as further reading.

Many scientists will be aware of solution-state NMR as a powerful technique for structural determination, as used in synthetic chemistry. Alternatively, many will have used it in an analytical context for the determination of chemical compositions, utilising the correspondence of spectral areas to molar proportions. To organic chemists a polymer molecule may seem inordinately uninteresting, being the endless repetition of a very simple structure, and the use of NMR might seem limited to analytical applications. NMR spectra of polymers in solution were acquired shortly after the development of the earliest commercial spectrometers, and it was indeed confirmed that the molecular structures of the repeat units could be identified. However, it was soon realised that the asymmetric nature of many vinyl polymer units gave rise to more than the expected number of resonances, reflecting subtle differences in chemical environments along polymer chains. This aspect of polymer NMR rapidly developed into what is now the most powerful means of determining average chain stereochemical sequence distributions. This has had enormous ramifications concerning the understanding of polymer synthesis and structure–property relations. The monomer sequence distributions in copolymers are determined using similar NMR methods, which can then be related to monomer reactivities. Ingenious ways have been found for overcoming effects associated with drifting feed compositions, and turning such effects to experimental advantage. A comparison between the NMR average historical view and the instantaneous kinetic view of polymer formation often provides a fruitful course of study in its own right.

The resolving power of NMR spectroscopy is directly proportional to magnetic field strength, and the introduction of superconducting magnets has dramatically enhanced the level of stereochemical and sequence information that can be accessed. In this respect the study of polymers is no different from any other type of chemical NMR, in that technical advances have revolutionised the ease of acquisition and content of spectroscopic data. In recent years the ensemble of multi-pulse techniques has been applied to polymer problems and many of these techniques are now in the polymer spectroscopist's standard repertoire. It has become possible to identify added ingredients down to very low levels, as well as the assortment of impurities that may be present in industrial polymer samples. Hence, in its more routine role NMR is invaluable to the synthetic polymer scientist or the process chemist. The search for minor architectural structures and end-groups has also developed in its own right and has advanced to the point where such information is of direct use in establishing polymer reaction mechanisms and kinetics. Without such progress it would be impossible to develop new types of polymers using synthetic routes of ever-increasing sophistication.

The spectral domain provides immediately accessible chemical information, which can be often be interpreted quickly and efficiently. In many cases the task can be performed by the owner of the sample, thereby increasing their confidence in the final outcome. The same is not true for those NMR experiments which probe the time domain, especially those that deal with NMR relaxation. This second tier of information is contained within the pre-Fourier transformed free-induction-decay, in spectral intensities and in linewidths. Its value relies on the fact that the form of the NMR response is uniquely sensitive to local molecular motion, exchange and diffusion processes. The dynamic properties of polymers in solution are therefore accessible. Despite the convoluted nature of the data and the more involved technical procedures, some impressive theories have been developed. These allow subtle details of polymer chain motions to be deduced. Such theories are gradually being introduced into the language of polymer science, although they sometimes retain an aura of mystery. It is hoped that this book will help those students intent on improving their understanding of this field.

The knowledge that polymer segments rapidly exchange between conformational states in solution explains why spectral lines are often tolerably narrow. They can be narrowed further by modest increases in temperature, so the spectroscopist has little excuse for claiming that polymer spectra will be uninformative. Conformational theories make the approximation that polymer segments occupy a few favourable discrete states, and that the bulk chain properties can be related to the weighted populations of these. The fact that not all conformers are equally favoured provides a basis for understanding the reasons behind tacticity-induced chemical shifts, and indeed NMR shifts can be predicted using the same weighted conformational populations. Conversely, this raises the exciting possibility of determining polymer solution properties from NMR spectral data, or of explaining conformational arrangements in polymeric solids.

In recent years impressive advances have been made in the field of high-resolution solid-state NMR. This has progressed from the status of a research technique, fraught with difficulties, to that of a routinely applicable method. The combination of *magic angle spinning, dipolar decoupling* and *cross-polarisation* into a complete methodology is one of the success stories of NMR. Whilst it is fair to say that the production of spectra still takes a degree of commitment and skill, the technique has revolutionised the study of polymers. Not only does it provide high chemical resolution, but it can also give key insights into polymer chain packing and morphology. Those who have compared solution and solid-state spectra of polymers might dispute the claims of chemical resolution, but in fact it is often physical heterogeneity rather than instrumental effects that leads to spectral broadening. Rather than limiting the information content, these physical influences can often be interpreted in terms of chain arrangements. The chemical information content is usually quite adequate for the study of curing and solid-state polymer reactions, with the important bonus that spectral intensities are sensitive to solid-state dynamics.

It might be thought that high-resolution solid-state NMR would find most application in the study of intractable polymers. It has been used extensively in such areas, but has probably made equal impact as a tool for determining solid-state polymer dynamics. For example, as a thermoset polymer cures the chains lose flexibility and this can be manifested in relative changes in relaxation behaviour of reacted and unreacted species. Hence, NMR takes on the role of a molecular scale tool for mechanical analysis, with the additional chemical dimension. A variety of pulse sequences are available which select resonances based on local or regional motion, and a number of these are utilised for routine spectral editing. NMR has also helped with the design of sophisticated models for solid polymer dynamics, and this book shows how such models can apply to rubbery solids as well as to solutions. Different models apply to non-rubbery polymers, where motion is far less extensive, and NMR has also played a key role in their development.

There is a point when the NMR characteristics of a bulk polymer have to be treated as non-liquid-like, that is below the glass transition temperature. The NMR spectrum will then be dominated by static effects, such as the orientation dependence of chemical shift and the dipolar interaction. The earliest forms of solid-state polymer NMR were developed in the knowledge that these solid-state effects would be present, and would complicate the resulting data interpretations. Proton NMR studies of polymers have an impressive pedigree, and have relied on applications of a distinct solid-state theory. But it is only relatively recently that aspects of this theory have been refined to the point where experimental observations can be understood more fully. In particular, the process of spin diffusion is much better appreciated. Proton broad-line NMR does not necessarily require the use of large magnets and despite the theoretical uncertainties it has been widely accepted within the realms of rapid analysis. In its simplest form it has been used to quantify liquid and solid ingredients, for example, in plasticised polymers.

An additional more complex type of static magnetic interaction is experienced by nuclei which have a spin quantum number greater than one half. Far from ruling out the studies of such species, application of quadrupolar techniques has added a new dimension to the characterisation of polymers. This is because the quadrupolar interaction is remarkably sensitive to order, orientation and local motion. With the necessary synthetic skills quadrupolar atoms such as deuterium can be inserted into a polymer at a chosen segmental site and can then be persuaded to report on their surroundings. The spectrometer is tuned to the specific nuclear frequency and the data are collected without unwanted responses from the rest of the sample. It is only the need for labelling that has restricted the more routine use of deuterium NMR for the study of polymers. Despite this, it has developed into an uniquely powerful research tool.

A different array of instrumentation is required if one is to gain access to spatially resolved NMR information. NMR imaging is most widely known as a medical tool, where the resonances of fluid species such as water are relatively easy to observe. Solid resonances decay very quickly and hence the imaging of bulk polymeric materials places great demands on equipment. Some very elegant techniques have been developed and it is now true to say that useful images of polymeric solids can be obtained. In addition to true solids, a wealth of spatial information can also be obtained from polymers in solution or molten states. It is possible to generate diffusional or velocity maps of flowing polymers by NMR, and these techniques have recently been extended to very high shear fields. Theories of polymer reptation can now be directly tested against NMR data.

The purpose of this introduction has been to set out some of the important themes developed within the chapters of this book. Each author has written their contribution in the knowledge that the subject must be viewed as a series of complementary disciplines, each building on and reinforced by the others. It is probably not an exaggeration to say that NMR can provide information on almost every aspect of polymer character, and this introduction has tried to emphasise the richness of the accessible data. The chemical shift is merely the starting point.

Although NMR is very powerful it has recognised limitations, for example in signal-to-noise, absolute spatial resolution, and insensitivity to long-range effects. It is always foolhardy to use a narrow approach and the numerous references in this book to other polymer methods bear testimony to this. With this in mind NMR probably remains the most versatile technique for the study of polymers, in any state.

1 Introduction to NMR and its use in the study of polymer stereochemistry

F. HEATLEY

1.1 Introduction

NMR is now a powerful, mature analytical tool which has widespread applications in all areas of synthetic chemistry, and polymer science is no exception. High-resolution solution-state NMR is a versatile technique which not only provides accurate qualitative and quantitative information on the chemical structure of a polymeric material, but is also capable of providing detailed information on certain aspects of chain structure which is not accessible by any other technique. The purpose of this chapter is to introduce the physical basis of the NMR phenomenon, to outline the effects that determine the form of the NMR spectra of liquid-state samples, to describe the experimental techniques of solution-state NMR of polymers, and to introduce those aspects of the structure of synthetic polymers that are specially amenable to NMR analysis.

For those wishing to study the theory and practice of high-resolution NMR in more detail, numerous texts and reviews are available [1–13], while other discussions of the high-resolution NMR of polymers may be found in [14–18]. A comprehensive survey of the application of NMR to polymers appears in the *Nuclear Magnetic Resonance* volume in the *Specialist Periodical Report* series published annually by the Royal Society of Chemistry.

1.2 Basic principles of NMR

As a consequence of non-zero spin angular momentum, characterised by a nuclear spin quantum number I, many nuclei possess a magnetic moment, μ. The magnetic moment is quantised according to the usual scheme for the quantisation of angular momentum, i.e. the magnitude of the magnetic moment is given by

$$|\mu| = \gamma \frac{h}{2\pi} \sqrt{I(I+1)} \tag{1.1}$$

and the allowed components of μ in a given direction (e.g. a magnetic field) are given by

$$\mu_z = \gamma \frac{h}{2\pi} I_z \tag{1.2}$$

where γ is the magnetogyric ratio of the nucleus and I_z is the azimuthal quantum number taking $(2I + 1)$ values from $+I$ to $-I$. The most important nuclei in analytical NMR (^1H, ^{13}C, ^{15}N, ^{19}F, ^{29}Si, ^{31}P) all have $I = \frac{1}{2}$, giving two possible orientations (or 'spin states'), conventionally denoted $\alpha(I_z = +\frac{1}{2})$ and $\beta(I_z = -\frac{1}{2})$, relative to a reference direction. In NMR, the nuclei are subject to a strong magnetic field, conventionally denoted B_0, which defines the reference direction, as illustrated in Figure 1.1(a). The magnetic field has two effects on the nuclear magnetic moments. The first effect is to cause the moments to precess about the direction of B_0 as shown in Figure 1.1(a). The precession or resonance frequency ν_0 (in Hz) is given by

$$\nu_0 = \gamma B_0 / 2\pi \tag{1.3}$$

The second effect of B_0 is to lift the degeneracy of the spin states, as shown in Figure 1.1(b). The energy of a particular state is

$$E = \gamma h B_0 I_z / 2\pi \tag{1.4}$$

The energy difference between α and β, ΔE, is therefore given by

$$\Delta E = \gamma h B_0 / 2\pi \tag{1.5}$$

At thermal equilibrium, the precession phases of the nuclei are random but there is an excess of spins in the lower energy state, so the bulk or total nuclear magnetic moment, denoted M_0, lies along B_0, as shown in Figure 1.1(c). To

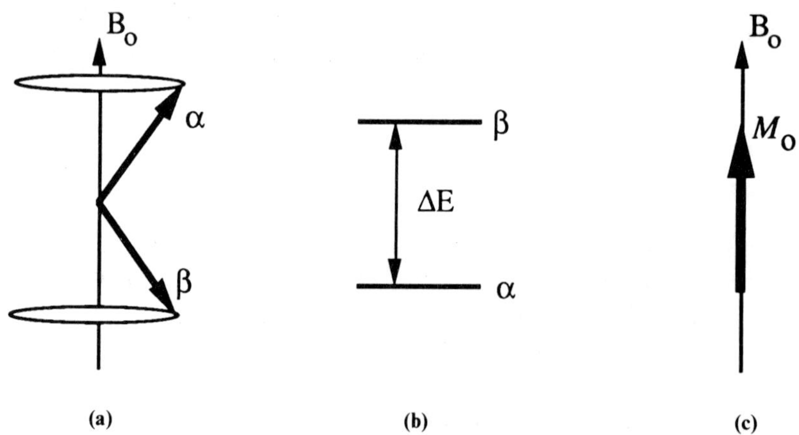

Figure 1.1 (a) The two spin states α and β, of a nucleus with $I = \frac{1}{2}$. (b) The energy levels of a nucleus with $I = \frac{1}{2}$. (c) The total nuclear magnetisation at thermal equilibrium (M_0) in a magnetic field B_0.

obtain an NMR spectrum, the nuclei are subjected to a second small magnetic field, denoted B_1, which is oriented perpendicular to B_0 and which oscillates at a frequency at or near ν_0. However, the field B_1 may be applied in two distinct ways, either using the so-called continuous-wave (CW) technique or using the pulsed Fourier transform (FT) technique.

In CW NMR, historically the first technique used for high-resolution NMR, the B_1 field is applied continuously and weakly while its frequency is swept through a range encompassing ν_0. In simple terms, the generation of a spectrum may be envisaged as the induction of $\alpha \leftrightarrow \beta$ transitions by the B_1 field when its oscillation frequency ν matches the transition frequency, i.e. when

$$\nu = \Delta E/h = \nu_0. \tag{1.6}$$

Thus transitions leading to a detectable absorption of energy occur at a frequency corresponding to the resonance frequency. In more sophisticated terms, the B_1 field when applied at the resonance frequency, slightly rotates the net magnetisation M_0 away from the B_0 direction thus inducing a component transverse to B_0 which precesses about B_0 at the frequency ν_0. The NMR

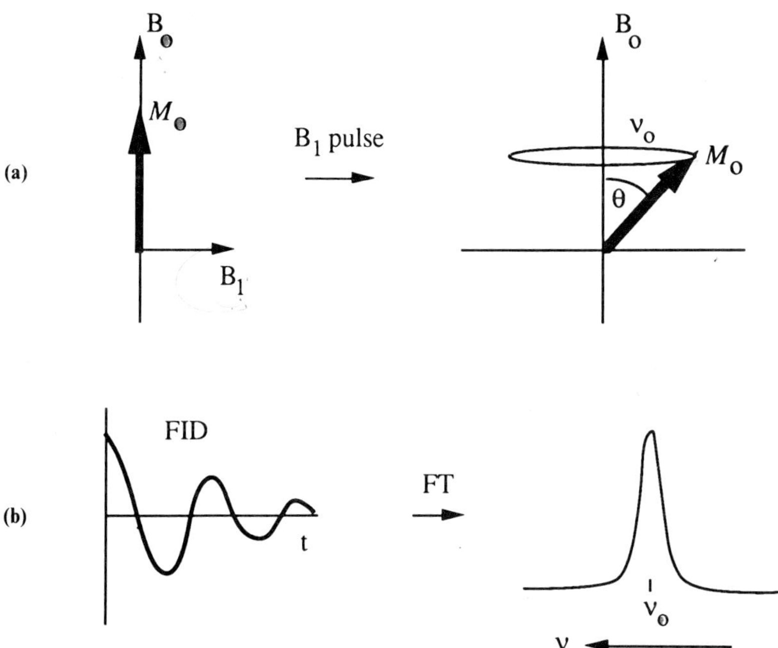

Figure 1.2 (a) The action of a pulsed oscillating magnetic field B_1 on the total nuclear magnetisation. θ is the flip angle and ν_0 is the precessional frequency. (b) The time-based free induction decay (FID) resulting from detection of the precessing transverse component of the nuclear magnetisation after a pulse, and its Fourier transform to give the frequency sweep spectrum.

signal is detected by the voltage induced by the precessing transverse component in a coil oriented with its axis perpendicular to B_0.

In the FT NMR technique, illustrated in Figure 1.2(a), the B_1 field is applied in the form of a short, intense pulse of the order of microseconds in duration at a fixed frequency equal to or near the resonance frequency. The effect of the pulse is to rotate the net magnetisation through a large angle, typically 90°, determined by the magnitude of B_1 and the pulse duration, so generating a large nuclear magnetisation component transverse to B_0. After the pulse, the rotated magnetisation precesses about B_0 at its resonance frequency while at the same time relaxing back to the equilibrium state of Figure 1.1(c). The NMR signal is detected by the voltage induced in a coil by the precessing and decaying transverse component. This signal, an exponentially damped oscillation known as the free induction decay (FID), is illustrated in Figure 1.2(b). The FID lasts for a time of the order of a second and is recorded digitally by computer. Fourier transformation of the FID (a function of time) then yields the NMR spectrum (a function of frequency) consisting of a peak at the resonance frequency with linewidth inversely related to the FID decay time constant. The spectrum is identical to that obtained from a slow sweep CW experiment.

The value of ν_0 depends on two factors, the nucleus (via γ) and B_0. Typical values of B_0 in present-day spectrometers range from about 2 to 14 T; values of ν_0 for common nuclei in a field of 7.046 T are given in Table 1.1. The frequencies lie in the radio-frequency (RF) region, and different nuclei have widely different frequencies. NMR spectrometers are therefore operated so that the spectrum of only one particular nucleus is observed at a time. Many spectrometers, however, may be tuned over a range of frequencies to allow multi-nuclear operation. For the analysis of polymers, by far the most important nuclei are the ubiquitous ^{1}H and ^{13}C isotopes, although ^{15}N, ^{19}F, ^{29}Si and ^{31}P NMR may find a role in specialised circumstances. Also with the provision of a second (or even third) RF oscillator, it is possible to irradiate two or more types of nuclei simultaneously as in spin-decoupling and two-dimensional NMR outlined below.

Table 1.1 Resonance frequencies in a magnetic field of 7.046 T; all nuclei have $I = \frac{1}{2}$ except ^{2}H which has $I = 1$

Nucleus	Natural abundance (%)	Frequency (MHz)
^{1}H	99.98	300.00
^{2}H	0.015	46.05
^{13}C	1.11	75.43
^{15}N	0.37	30.40
^{19}F	100	282.23
^{29}Si	4.7	59.60
^{31}P	100	121.44

Following this brief basic description, one advantage of FT NMR over CW NMR becomes apparent; the sensitivity of FT NMR is much greater because the magnitude of the transverse magnetisation detected is larger. Other advantages are described in the following sections.

1.3 The form of a liquid-state NMR spectrum

The value of NMR as an analytical tool arises because the actual resonance frequency of a nucleus is modified by several factors depending on the molecular environment. For nuclei with $I = \frac{1}{2}$, the factors in order of decreasing magnitude are dipole–dipole coupling, chemical shifts, and spin–spin coupling. For nuclei with $I \geqslant 1$, such as ^{2}H, ^{14}N and ^{35}Cl, a fourth effect, quadrupolar coupling, may also be significant, even dominant (see chapter 8). In isotropic solutions, dipole–dipole coupling (the direct interaction of nuclear magnetic moments) and quadrupolar coupling (the interaction of a nuclear quadrupole moment with an electric field gradient) average to zero, so these are not considered further here. However, they are most important factors in the NMR spectra of solids, as discussed in chapters 5, 7 and 8, and in magnetic relaxation, as discussed in chapters 4 and 6.

1.3.1 Chemical shifts

From the basic outline above, it would appear that every nucleus of a particular species would give the same resonance frequency. In fact, it is found that the frequency depends on the chemical environment of the nucleus, and so varies according to such structural factors as bond hybridisation, the nature of neighbouring atoms, delocalisation, stereochemistry and hydrogen bonding. It is this sensitivity to structure which gives NMR its analytical powers. The frequency depends on the chemical environment because the surrounding electrons, in response to the applied field B_0, generate a small induced field opposed to B_0 thus shielding or screening the nucleus from B_0. The shielding field is proportional to B_0, and the effect is therefore quantified by a shielding constant, σ, according to

$$B' = B_0(1 - \sigma) \tag{1.7}$$

where B' is the reduced field at the nucleus. From equation (1.3), the resonance frequency thus becomes

$$\nu_0 = \gamma B_0(1 - \sigma)/2\pi \tag{1.8}$$

The magnitude of σ, and hence ν_0, depends on the electronic response, i.e. on the chemical structure. The variation of ν_0 with structure is known as the chemical shift. For isotropic liquids, differences in ν_0 are much greater than the resonance linewidth, so distinct resonances are observed for chemi-

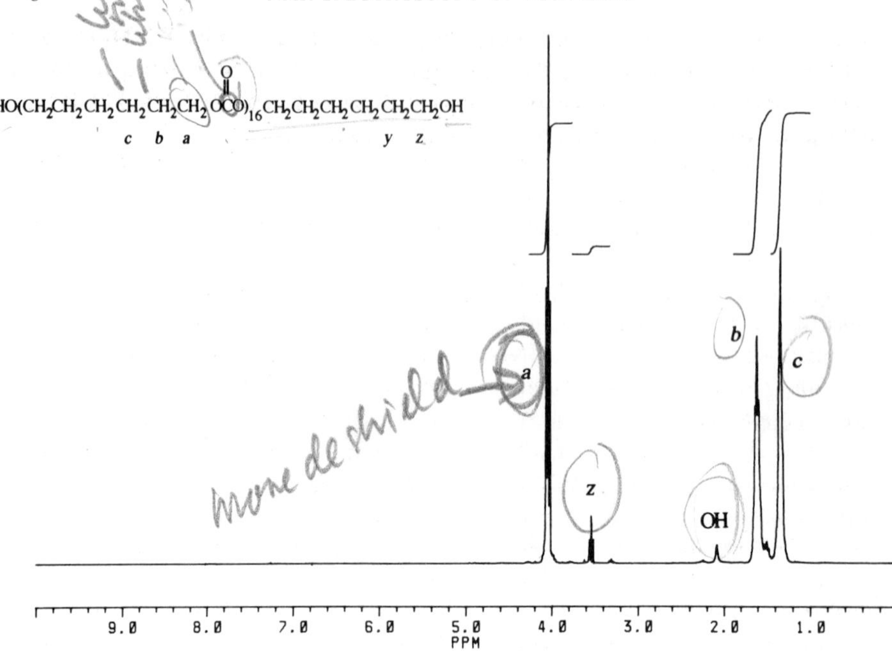

Figure 1.3 The 300 MHz ^{1}H spectrum of low molecular weight poly(hexamethylene carbonate) in $CDCl_3$.

cally different nuclei. A simple polymer example, the ^{1}H spectrum of a low molecular weight sample of poly(hexamethylene carbonate), is shown in Figure 1.3. The principal feature of this spectrum is the three intense groups of resonances corresponding to the three chemically distinct methylene groups in the interior repeat units. (The multiplet splitting of each group arises from spin–spin coupling which is described below.) A further manifestation of the chemical shift effect which is particularly relevant to polymers is the observation of groups of less intense peaks attributable to hydroxy end-groups as indicated.

At this point, it is essential to note that the 'frequency' scale of Figure 1.3 is not expressed in Hertz, as the discussion above would indicate, but is expressed using dimensionless units of 'parts per million' (ppm). This convention is used because the actual resonance frequency of a particular nucleus depends on the value of B_0 in the instrument used, and frequency comparisons are of no value. Instead, the position of a resonance is measured relative to a reference compound which for ^{1}H and ^{13}C NMR is defined to be tetramethylsilane ($Si(CH_3)_4$, TMS). On the ppm scale, the chemical shift, δ_X, of a sample peak X is defined as

$$\delta_X = \left[\frac{(\nu_X - \nu_{ref})}{\nu_{ref}}\right] \times 10^6 \tag{1.9}$$

where ν_X and ν_{ref} are the resonance frequencies of peak X and the reference, respectively. δ_X is dimensionless and independent of B_0. Nuclei less shielded than the reference have positive δ_X, nuclei more shielded than the reference have negative δ_X.

From examination of the spectra of compounds of known structure, correlations between chemical shift and structure have been developed, particularly for ^{13}C NMR [12, 13]. Several useful collections of reference spectra have also been assembled [19–23]. Simple guidelines which may be helpful in following the examples in this book are summarised in Table 1.2. In a B_0 field of 7.046 T, ^{1}H chemical shifts cover a range of some 4 kHz, and ^{13}C chemical shifts cover a range of some 17 kHz. Since the linewidths are about the same, the chemical discriminatory power of ^{13}C NMR is some fourfold greater than ^{1}H.

Practically all samples of chemical interest contain nuclei of different chemical shift, i.e. several different resonance frequencies are observable for a particular type of nucleus. In CW NMR, the magnitude of B_1 is very small and only resonances within 1 Hz or so of the frequency of B_1 are stimulated. Each resonance is excited individually when the frequency of B_1 reaches the resonance frequency of that particular nucleus, and it may take several minutes to scan a complete spectrum sufficiently slowly to avoid distortion. In FT NMR, however, the B_1 pulse perturbs nuclei whose resonance frequencies are offset from the applied frequency as well as nuclei which are on-resonance. Since the excitation bandwidth of a pulse is approximately equal to the reciprocal of the pulse length, typically 10 μs, normal pulses in

Table 1.2 Typical chemical shift ranges for broad classes of ^{1}H and ^{13}C nuclei

Functional group	Chemical shift (ppm)
(a) ^{1}H	
H–CR$_1$R$_2$R$_3$	1–2.5
(R$_1$, R$_2$, R$_3$ = H or C)	
H–CR$_1$R$_2$R$_3$	3–5
(R$_1$ = N, O or Hal, R$_2$, R$_3$ = H or C)	
H–C=C	4.5–6.5
Aromatic	6.5–8
–CHO	8–10
R–OH	1–14[a]
(b) ^{13}C	
CR$_1$R$_2$R$_3$R$_4$	10–50
(R$_1$, R$_2$, R$_3$, R$_4$ = H or C)	
CR$_1$R$_2$R$_3$R$_4$	40–90
(R$_1$ = N or O, R$_2$, R$_3$, R$_4$ = H or C)	
Olefinic or aromatic	90–150
R$_1$COOR$_2$	160–190
R$_1$COR$_2$	190–220
(R$_1$, R$_2$ = H or C)	

[a] Variable depending on the degree of hydrogen bonding.

FT NMR excite all nuclei of a given species simultaneously. Therefore, the FID is a complex interferogram consisting of the superposition of a number of damped oscillations with resonance frequency and relaxation time characteristic of each chemically distinct nuclear environment. Fourier transformation separates the components in the frequency domain. The same chemical shift information is obtained as in CW NMR but in a timescale of seconds rather than minutes. FT NMR is therefore a much more efficient method for the application of time-averaging techniques.

1.3.2 Spin–spin (scalar) coupling

In Figure 1.3, it is seen that the resonance peak attributed to the a-CH_2 protons is split into a well resolved triplet, while the resonances attributed to the b and c CH_2 protons are each split into a broadened quintet. The magnitudes of the splittings are all about 6.5 Hz. This fine structure is termed spin–spin or scalar coupling and arises from an interaction between neighbouring nuclear magnetic moments transmitted by electrons through intermediate bonds. The effect is thus only intramolecular. The magnitude of the splitting is termed the spin–spin coupling constant and is conventionally denoted by the symbol J. In simple terms, the splitting may be envisaged as a dependence of the resonance frequency of one nucleus on the spin states of its neighbours, the term 'neighbours' in this context meaning magnetic nuclei up to three bonds distant. Thus the a CH_2 protons are split into three lines with relative intensities 1:2:1 (a 'triplet') because the two adjacent b protons may adopt one of the following four combined spin states.

$$\alpha\alpha \qquad (\alpha\beta \text{ or } \beta\alpha) \qquad \beta\beta$$

The $\alpha\beta$ and $\beta\alpha$ states are degenerate so have the same effect on the resonance frequency. The b and c protons are each split into five lines with relative intensities 1:4:6:4:1 (a 'quintet') because each has four neighbouring protons with the following possible combined spin states:

$\alpha\alpha\alpha\alpha$	$\alpha\alpha\alpha\beta$	$\alpha\alpha\beta\beta$	$\alpha\beta\beta\beta$	$\beta\beta\beta\beta$
	$\alpha\alpha\beta\alpha$	$\alpha\beta\alpha\beta$	$\beta\alpha\beta\beta$	
	$\alpha\beta\alpha\alpha$	$\beta\alpha\alpha\beta$	$\beta\beta\alpha\beta$	
	$\beta\alpha\alpha\alpha$	$\alpha\beta\beta\alpha$	$\beta\beta\beta\alpha$	
		$\beta\alpha\beta\alpha$		
		$\beta\beta\alpha\alpha$		

By analogy with these patterns, it is readily seen that a group of nuclei coupled to a single $I = \frac{1}{2}$ neighbour is split into two lines of equal intensity (a 'doublet') while a group coupled to three identical $I = \frac{1}{2}$ neighbours is split into four lines with relative intensities 1:3:3:1 (a 'quartet'). In general a nucleus coupled to n neighbours of spin quantum number I is split into $(2nI + 1)$ lines.

Figure 1.3 also illustrates two other important aspects of spin–spin coupling. First, no splittings arise from interactions between nuclei of the same chemical shift, e.g. in Figure 1.3 there is no splitting from coupling between the protons within each CH_2 group. Thus a molecule such as acetone or benzene in which all protons are identical gives a single line NMR spectrum. Second, there is no coupling involving carbon and oxygen because the overwhelmingly abundant isotopes of those atoms, ^{12}C and ^{16}O, have no magnetic moment and so are invisible to NMR except insofar as they influence chemical shifts.

It should be noted that the simple patterns described above, termed 'first-order' coupling patterns, are observed only if the difference between the resonance frequencies of the coupled nuclei is much greater than the spin–spin coupling constant. If this condition is not met, deviations in line frequency and intensity from the simple models appear. However, such spectra can be fully analysed using a quantum mechanical description of the spin energy levels [1–4, 24].

A further illustration of spin–spin coupling is shown in the natural abundance ^{13}C spectrum of poly (isobutylene) in Figure 1.4. This spectrum illustrates a simple polymer ^{13}C spectrum with heteronuclear coupling to protons. Because the abundance of ^{13}C is only 1.1%, the probability of two ^{13}C nuclei occurring within coupling distance of each other is very small, and the only coupling effects are those between ^{13}C and neighbouring protons. The methyl carbon is split into a quartet by coupling to three attached protons and the CH_2 carbon is split into a triplet by coupling to two attached protons, the

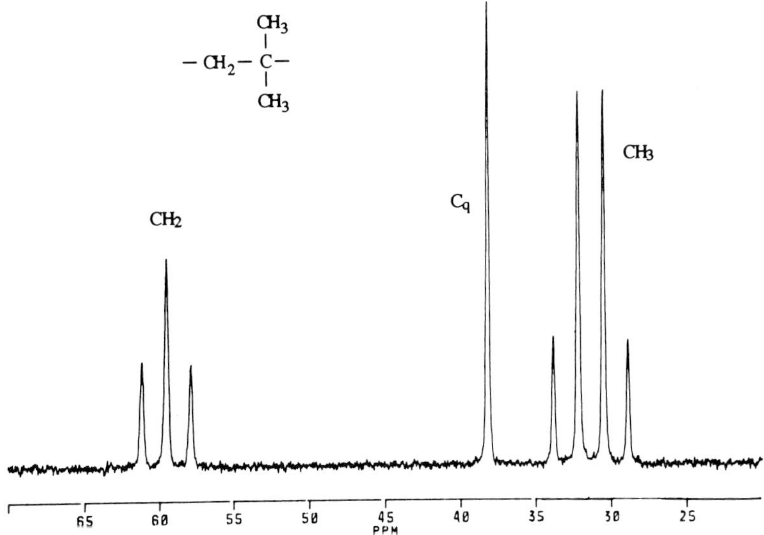

Figure 1.4 The ^1H-coupled 75.5 MHz ^{13}C spectrum of high molecular weight poly(isobutylene) in $CDCl_3$.

Table 1.3 Typical values of ^{1}H–^{1}H and ^{1}H–^{13}C coupling constants

Coupling	J(Hz)
(a) ^{1}H–^{1}H coupling	
H–C–H (sp^3)	9–15
H–C–H (sp^2)	0–3
H–C–C–H (sp^3, *gauche*)	2–4
H–C–C–H (sp^3, *trans*)	10–12
H–C–C–H (sp^3, average)	3–10
H–C=C–H (olefinic, *cis*)	6–16
H–C=C–H (olefinic, *trans*)	14–22
H–C=C–H (aromatic, *ortho*)	5–8
(b) ^{1}H–^{13}C coupling	
H–C$R_1R_2R_3$ (R_1, R_2, R_3 = H or C)	120–130
H–C$R_1R_2R_3$ (R_1 = N or O, R_2, R_3 = H or C)	140–150
$\underline{\text{H}}$–$\underline{\text{C}}$=C	150–170
$\underline{\text{H}}$–$\underline{\text{C}}$=O	180–220
$\underline{\text{C}}$–C–$\underline{\text{H}}$	0–6
$\underline{\text{C}}$–C–C–$\underline{\text{H}}$	1–10

one-bond C–H coupling constant in each case being about 125 Hz. The quaternary carbon gives a singlet. All lines are broadened by unresolved multiple two-bond and three-bond coupling of the order of a few Hertz to other protons.

Typical coupling constants in ^{1}H and ^{13}C spectroscopy are summarised in Table 1.3. Note that vicinal coupling constants are strongly dependent on the dihedral angle between the coupled nuclei. For H–H coupling in an H–C–C–H fragment, the coupling constant is 10–12 Hz for a *trans* orientation and 2–4 Hz for a *gauche* orientation. The values of vicinal coupling constants may therefore be used to investigate polymer conformations (see chapter 4).

1.3.3 Intensities in NMR spectra

In NMR, there is no absolute intensity scale equivalent for example to the optical density scale in IR or UV spectroscopy. Plotted NMR intensities are arbitrary and are not related directly to concentration. However, *relative* numbers of nuclei in a given sample are easily obtained because the total intensity of a group of peaks is directly proportional to the total number of contributing nuclei in the sample, provided certain conditions are fulfilled with regard to time allowed for relaxation after a pulse and (in proton-decoupled spectra of other nuclei) possible differential nuclear Overhauser enhancements; these factors are described in more detail below. The relative intensity of a group of peaks is obtained in practice from its computer-generated integral, as exemplified in Figure 1.3.

1.3.4 Spin-decoupling

From the two examples in Figures 1.3 and 1.4, it is apparent that spin–spin coupling provides valuable information on the number of neighbours within coupling distance of a particular nucleus at the price of greater spectral complexity and loss of sensitivity due to distribution of the signal intensity over a number of lines. In complex ^{13}C spectra in particular, the large magnitude of one-bond C–H coupling constants generally leads to a confusing overlap of adjacent multiplets. However, spin–spin coupling may easily be eliminated using the technique of spin-decoupling. This is a double-resonance experiment in which the effect in the spectrum of a nucleus A due to coupling with another nucleus X is removed by simultaneously irradiating X at its resonance frequency while observing A. In homonuclear decoupling, A and X are the same isotope, normally ^1H, and the decoupling irradiation is applied rela-

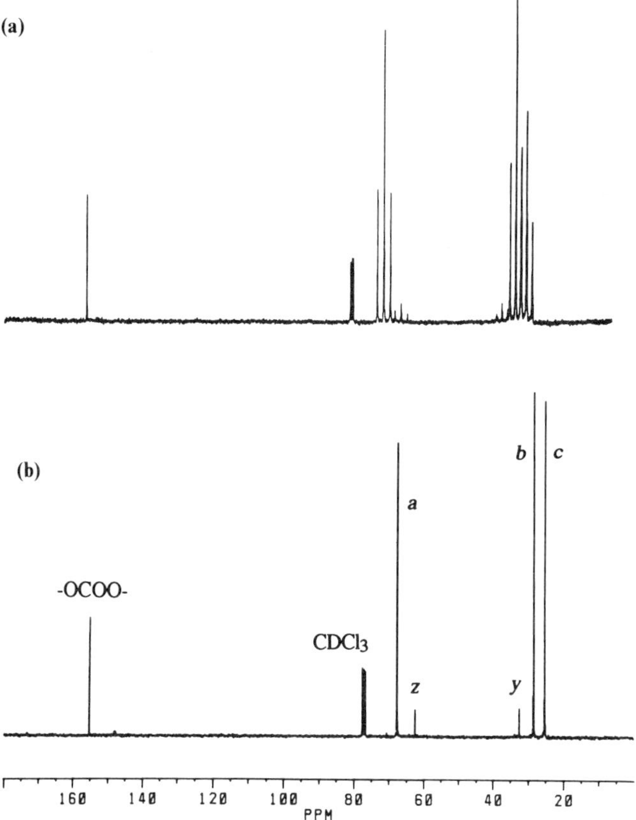

Figure 1.5 75.5 MHz ^{13}C spectra of poly(hexamethylene carbonate) in $CDCl_3$. (a) ^1H-coupled spectrum; (b) spectrum with continuous broadband ^1H-decoupling. For peak labels, see Figure 1.3. The intensity scale in (a) is approximately tenfold greater than that in (b).

tively weakly in order to decouple only one type of proton. In heteronuclear decoupling, commonly encountered in the form of ^1H-decoupled ^{13}C spectra, high-power broadband decoupling at the proton frequency is used in order to decouple all protons. Figure 1.5 compares coupled and decoupled ^{13}C spectra of poly(hexamethylene carbonate). In the coupled spectrum in Figure 1.5(a), the three types of CH_2 carbon give triplets due to one-bond C–H coupling, and the C=O carbon gives a singlet. In the decoupled spectrum, all carbons are reduced to singlets with a consequent simplification of the spectrum and vast improvement in signal-to-noise ratio. Also, in the decoupled spectrum, peaks from hydroxy end-groups are more clearly resolved.

1.4 Nuclear magnetic relaxation

1.4.1 Basic principles

After perturbation by an applied oscillating magnetic field, either CW or pulsed, the bulk nuclear magnetisation returns inevitably to its equilibrium state by natural relaxation processes. After a 90° pulse, the longitudinal component of the bulk magnetisation parallel to B_0, denoted M_z, is initially zero and grows to its equilibrium value M_0, while the transverse component, denoted M_x, is initially equal to M_0 and decays to its equilibrium value of zero. The longitudinal and transverse relaxation processes, however, do not necessarily occur at the same rate because of their different effects on the total nuclear spin energy. A change in M_z requires a change in the spin state populations, i.e. an exchange of energy with the thermal energy bath of the surrounding molecules. For this reason, the longitudinal relaxation process is often referred to as 'spin–lattice' relaxation. In contrast, a change in M_x does not require a change in the total spin energy, but merely a dephasing of individual nuclear magnetic moments. This process is also termed 'spin–spin' relaxation. Normally in liquids, both relaxation processes are exponential. Conventionally, relaxation of M_z is characterised by the longitudinal or spin–lattice relaxation time, T_1, and relaxation of M_x by the transverse or spin–spin relaxation time, T_2. After an arbitrary perturbation, the time dependences of M_z and M_x are given by

$$M_z(t) = M_0 + [M_z(0) - M_0]\exp(-t/T_1) \tag{1.10}$$

$$M_x(t) = M_x(0)\exp(-t/T_2) \tag{1.11}$$

where $M_z(0)$ and $M_x(0)$ are the initial values.

Both longitudinal and transverse relaxation are stimulated by time-dependent perturbations acting on the nuclei, such as dipole–dipole coupling. Usually the time-dependence arises from molecular motion, and measurements of relaxation times are a powerful method of studying polymer motion in both solution and the solid state. Detailed discussions of such applications, as well as methods of measuring relaxation times, may be found in chapters 4, 6 and 7. For the present purposes, an understanding of magnetic relaxation is important only insofar as it affects the use of NMR for the characterisation of polymer structure.

1.4.2 Practical implications of relaxation

In general terms, T_2 influences linewidths of an NMR spectrum whereas T_1 influences relative intensities.

The influence of T_2 on linewidth arises because T_2 is the decay time constant of the FID and the inverse relationship between time and frequency leads to the following equation linking T_2 and the linewidth-at-half-height, $\Delta\nu_{1/2}$ (in Hz):

$$\Delta\nu_{1/2} = \frac{1}{\pi T_2} \tag{1.12}$$

It is clearly advantageous to maximise T_2, which can be achieved by using lower concentrations, less viscous solvents and higher temperatures. However, in practice, there are frequently other contributions to the linewidth, such as inhomogeneities in B_0 or unresolved chemical shifts and coupling constants which are not affected by changes in sample conditions. These other factors are often dominant in polymer NMR, and usually the FID has decayed well before the longitudinal component has recovered.

The influence of T_1 on intensities arises because in FT NMR, the signal intensity is proportional to the longitudinal magnetisation before the pulse. If insufficient time between pulses is allowed for complete relaxation, the intensity will be less than maximum. Also, if different nuclei relax at different rates, as frequently occurs, the reduction in intensity will vary and the relative intensities will not reflect the relative amounts of each type of nucleus. This differential relaxation effect can be ameliorated to some extent by using a flip angle of less than 90°. For quantitative accuracy, it is necessary to ensure that all nuclei of interest are allowed to relax between pulses. Anticipating chapter 4 slightly, the value of T_1 depends on two factors, the mobility of the segment containing the nucleus in question and the proximity of neighbouring magnetic nuclei. For most synthetic polymers in solution at normal temperatures, T_1 increases with increasing mobility, so that nuclei situated at the ends of chains or in small side-groups (see chapter 3) generally have longer relaxation times than interior backbone nuclei. T_1 increases with increasing distance between the nucleus and the nearest magnetic nuclei, so that methine protons have longer T_1 values than similar methylene protons. In ^{1}H NMR, values of T_1 typically range from 0.1 to 5 s, depending on polymer structure and temperature. A pulse interval of 10 s is usually sufficient for quantitative accuracy. In ^{13}C NMR, the relaxation times of protonated carbons are similar to those of the protons but quaternary carbons usually

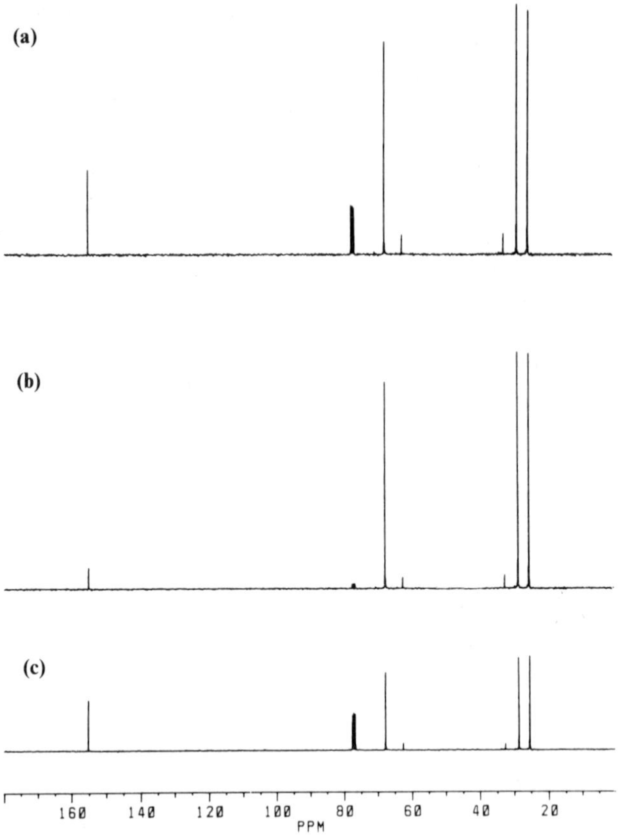

Figure 1.6 ^1H-decoupled 75.5 MHz ^{13}C spectra of poly(hexamethylene carbonate) in $CDCl_3$. (a) Continuous ^1H irradiation, pulse interval 30 s; (b) continuous ^1H irradiation, pulse interval 3 s; (c) NOE-suppressed, pulse interval 30 s. All spectra are plotted with the same intensity scale and were obtained using the same number of acquisitions.

have considerably longer relaxation times, perhaps by a factor of ten or more, and a pulse interval of up to 30 s may be required if accurate intensities are desired. Figures 1.6(a), (b) demonstrate the effect of the pulse interval on the intensities of the peaks in the ^{13}C spectrum of poly(hexamethylene carbonate). The relative intensities of the interior CH_2 peaks are essentially independent of the pulse interval, but the carbonyl peak is considerably attenuated when the pulse interval is short. The terminal CH_2OH peak is also reduced, although not to the same extent.

1.4.3 The nuclear Overhauser effect in ^{13}C NMR

Comparison of integrated peak intensities in Figures 1.5(a), (b) shows that the peaks in the decoupled spectrum are in fact about 2.5 times greater than those in the coupled spectrum. This enhancement arises as a result of magnetisation transfer from the decoupled protons to the ^{13}C nuclei and is known as the nuclear Overhauser effect (NOE). Fundamentally, the NOE is a consequence of mutual relaxation of ^{1}H and ^{13}C nuclei by the direct interaction of their magnetic dipole moments, and is closely related to the longitudinal relaxation process. The origin and theory of the NOE is explained more fully in chapter 4. For the present purpose, it is only necessary to observe that like T_1, the NOE depends on motional and structural factors and for ^{13}C spectra with ^{1}H decoupling may vary from a minimum of 15% enhancement for relatively slow motion to a maximum of 200% enhancement for relatively rapid motion. For most polymer systems, the NOE is essentially the same for protonated backbone carbon nuclei but more mobile end-groups and side-groups may have a larger NOE. The NOE is reduced if the carbon experiences relaxation by a mechanism other than dipole–dipole interaction with protons. In practice, this occurs for unsaturated quaternary carbons in B_0 fields of about 5 T or greater, i.e. ^{1}H resonance frequencies of 200 MHz or greater. The competing relaxation mechanism in this case is the anisotropic chemical shift mechanism. The effect is seen in Figure 1.5(a) where the carbonyl peak, although undergoing complete relaxation between pulses, nevertheless remains less than 50% of the CH_2 peaks. The NOE may be suppressed using a gated decoupling technique (Figure 1.7) in which the decoupling field is switched on only during the second or so during which the FID is acquired and is off during the interval allowed for restoration of

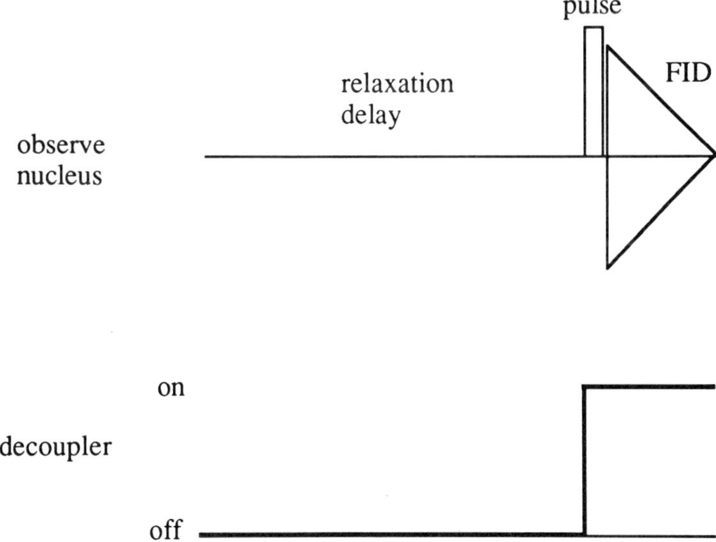

Figure 1.7 The pulse sequence used to observe a decoupled spectrum without the nuclear Overhauser enhancement (inverse gated decoupling).

equilibrium. The decoupling effect is instantaneous but the NOE, being in essence a T_1-related phenomenon, takes a time of the order of T_1 to be established. In the long delay, NOE-suppressed spectrum of poly(hexamethylene carbonate) in Figure 1.6(c), all interior peaks are the same (integrated) intensity per contributing nucleus.

1.5 Experimental practice in high-resolution solution-state NMR of polymers

1.5.1 The spectrometer

A block diagram of a high-resolution multi-nuclear FT instrument is shown in Figure 1.8. The polarising B_0 field is provided either by an electro- or permanent magnet (for ^1H frequencies $\leqslant$ 100 MHz) or by a superconducting magnet (for ^1H frequencies from 200 to 600 MHz). The sample, contained in a precision glass sample tube normally of either 5 or 10 mm diameter, is held in the B_0 field in a so-called probe which also supports the coil used to apply the oscillating B_1 fields. The application of pulses, the digitisation of the FID, and data processing are carried out under the control of a computer. For high-resolution NMR, it is necessary that the B_0 field be both stable and homogeneous. The stability is provided by means of a field-frequency lock device in which B_0 is 'locked' to the ^2H signal from a deuterated solvent produced by CW irradiation at the ^2H frequency. Any fluctuations in B_0 are

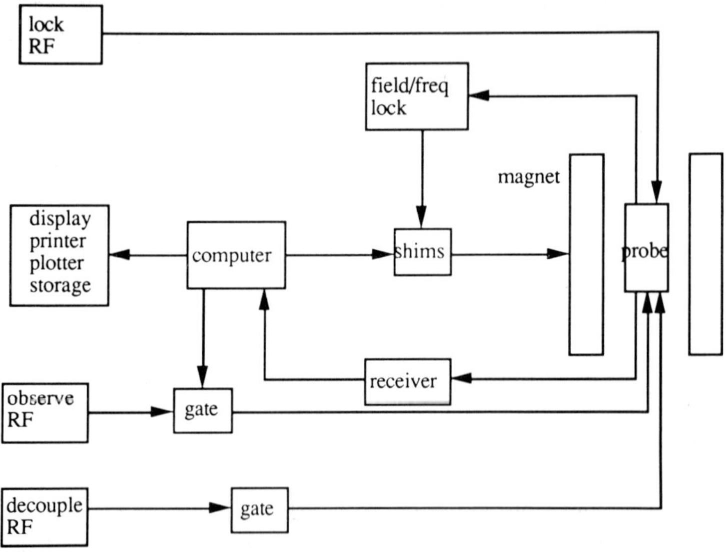

Figure 1.8 A block diagram of the elements of an FT NMR spectrometer.

automatically corrected in order to maintain the solvent in resonance. This mechanism is essential for spectrometers employing electromagnets since these are extremely susceptible to small variations in temperature, but it is not absolutely necessary for systems with superconducting magnets. The lock signal, however, provides a convenient means of optimising the homogeneity of B_0 in order to minimise the linewidth. The homogeneity is adjusted by means of auxiliary 'shim' coils designed to produce magnetic fields which, superimposed on the main B_0 field, correct the inhomogeneities in B_0. Because the possible resolution of liquid-state NMR is extremely high (< 1 Hz in several hundred MHz), the homogeneity adjustment is necessary for each sample. If a deuterated solvent is not used, the homogeneity must be adjusted by repeated observation of the ^{1}H or ^{13}C spectrum.

1.5.2 The sample

The purposes for which a high-resolution NMR spectrum is run differ widely and only general guidance about the sample conditions can be given. However, to achieve high-resolution without recourse to the high power RF and 'magic angle spinning' techniques described in chapter 5, there is a general requirement that the sample be in liquid form, normally in solution. A perdeuterated solvent is usually employed; common solvents such as chloroform, acetone, benzene, dimethyl sulphoxide, methanol and water are readily available in perdeuterated form. In addition to providing a ^{2}H NMR signal for field-frequency locking and homogeneity adjustment described above, the use of a perdeuterated solvent also serves to substantially reduce the strong signal from the solvent which could otherwise obscure weaker solute peaks. This consideration is particularly important in proton NMR because of the limited range of chemical shifts. The strong solvent signal also affects the accuracy of digitisation of the FID, since in FT NMR, the solute and solvent signals are excited simultaneously and the input range of the analogue-to-digital converter (ADC) must be set to cover the largest signal. The solute signal may then be reduced below the digital threshold. The use of a perdeuterated solvent is not so important in ^{13}C NMR for several reasons. First, the range of chemical shifts is larger than in ^{1}H NMR so the probability of overlap is less. Second, solvent nuclei have much longer relaxation times than polymer solutes and so the solvent intensity is reduced by incomplete relaxation. Third, the signals in ^{13}C NMR are much weaker than in ^{1}H NMR and the input to the ADC is dominated by noise rather than the FID. Under these conditions, the acquisition of a weak signal in the presence of a strong signal is achievable.

In many cases, polymers are insoluble at normal temperatures but may be soluble at elevated temperatures. Commercial spectrometers are normally capable of operation up to temperatures of 150 or 200°C. In extreme cases of insolubility, such as for some fluoropolymers [25], the spectrometer may be operated at even higher temperatures using custom-built apparatus to investigate the samples in the molten state.

Many polymer systems of interest do not dissolve to give true solutions but rather swell to give gels, for example because of chemical crosslinks or because of physical entanglements in very high molecular weight materials. The restricted mobility of such samples leads to considerable line broadening which, in ^{1}H NMR, is often large enough to obscure chemical shift information. However, in ^{13}C NMR, the broadening is less significant because of the larger chemical shift range and because a part of the broadening, that arising from incompletely averaged ^{13}C–^{1}H dipole–dipole coupling, is removed by proton spin-decoupling. Useful ^{13}C spectra may therefore be obtained from gels (see chapter 4).

The amount of sample required varies greatly depending on the nucleus studied, the mobility of the chain, the complexity of the spectrum, the purpose of the investigation and the experimental time available. For a ^{1}H spectrum, where the major constituent peaks are of interest, for example in determining the repeat unit structure or the composition of a copolymer, a concentration of 5–20 mg cm^{-3} is sufficient. The ^{1}H spectrum of such a sample requires only a few minutes for acquisition. A ^{13}C spectrum for the same purpose requires 50–500 mg cm^{-3} and may take an hour or more for acquisition. If the interest of the investigation lies in minor features, such as end-groups or branching, the acquisition time must be correspondingly increased, perhaps to several hours. This is also the timescale for more sophisticated experiments such as relaxation measurements, DEPT spectra and two-dimensional spectra.

1.6 Advanced Fourier transform techniques

The pulses applied in FT NMR are phase coherent in two ways. First, the B_1 field is phase coherent over the whole sample, and second, successive pulses may be applied with a defined phase relationship. Consequently, it is possible to manipulate the nuclear magnetisations by a sequence of pulses to produce different overall responses. A wide variety of pulse techniques has been developed which have greatly increased the power and versatility of NMR as an analytical method. These methods are conveniently divided into two classes, one-dimensional techniques in which a spectrum is recorded as a function of one frequency as in standard NMR, and two-dimensional techniques in which the spectrum is recorded as a function of two frequencies. Here, attention is focused on the basic types of experiment, and on the form of the spectra and the type of information obtained, rather than on the theory and operation of the techniques. Methods for determining relaxation times are described in chapters 4, 6 and 7, so are not dealt with here.

1.6.1 One-dimensional techniques

1.6.1.1 Spectrum editing: APT and DEPT. As described above, ^{13}C spectra are routinely acquired with broadband ^{1}H decoupling thus losing so-called multiplicity information, i.e. whether a carbon is primary, secondary, tertiary or quaternary. This information can be regained while retaining the advantages of ^{1}H decoupling by using one of several techniques of which the 'attached proton test' (APT) [26–28] and 'distortionless enhancement by polarisation transfer' (DEPT) [29–31] techniques are the most useful. DEPT is the more powerful and is the method of choice. In both cases, a series of pulses is

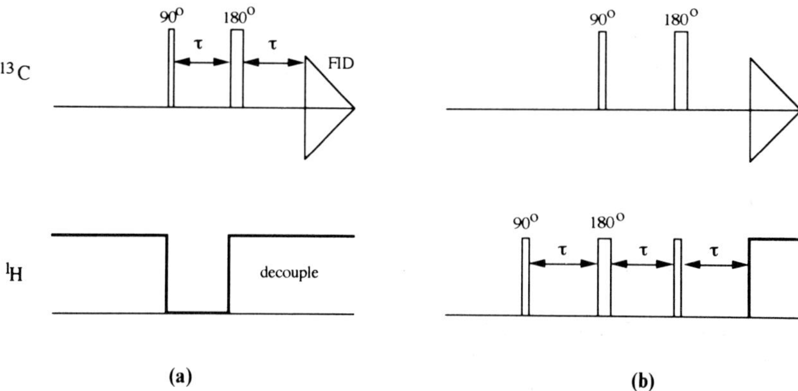

Figure 1.9 (a) The APT pulse sequence. Signals from carbons of different multiplicity vary with the variable pulse interval τ according to Figure 1.10(a). (b) The DEPT pulse sequence. The final ^{1}H pulse is of variable flip angle θ. Signals from carbons of different multiplicity vary with θ according to Figure 1.10(b).

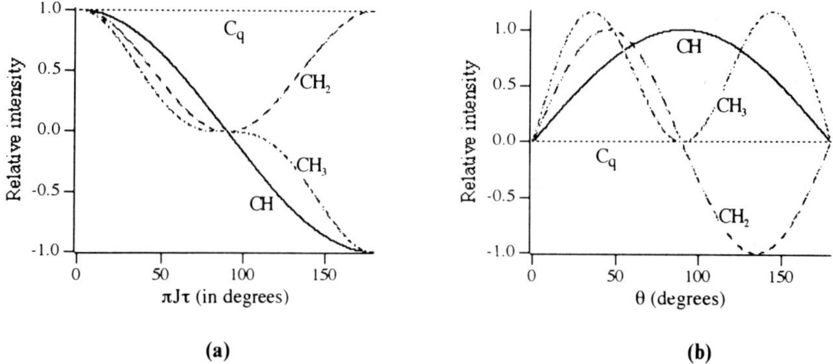

Figure 1.10 (a) The variation of the signals for carbons of different multiplicity with τ in the APT sequence; (b) the variation of the signals for carbons of different multiplicity with θ in the DEPT sequence.

applied to both ^1H and ^{13}C nuclei, the final carbon intensities depending on their multiplicity and on the value of one of the sequence variables, the delay τ in APT and the flip angle θ in DEPT. The sequences and the ideal carbon responses are shown in Figures 1.9 and 1.10; the dependence of intensity on multiplicity and the variable is apparent. In practice, spectra deviate slightly from the ideal because of imperfect pulses and because of a range of values of J_{CH}. The APT sequence has the advantage that it is not necessary to have the ability to switch the ^1H RF power during the sequence from a high value required for broadband ^1H pulses to the lower value required for continuous ^1H decoupling. However, the DEPT sequence is more sensitive and more robust to variations in the value of J_{CH}. Also quaternary carbons do not appear in DEPT spectra.

Using suitable combinations of APT or DEPT spectra, it is possible to perform spectral editing whereby subspectra containing signals of only one multiplicity appear. For DEPT spectra, the editing uses spectra with final

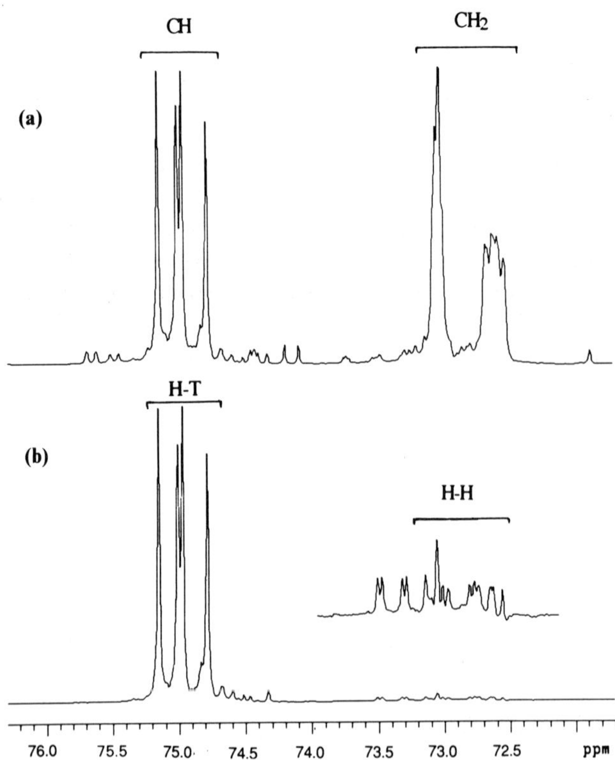

Figure 1.11. 125.5 MHz ^{13}C spectra of the backbone region of anionically polymerised poly(propylene oxide). (a) The normal spectrum; the intense signals from interior CH and CH_2 carbons are indicated; the minor peaks are from end-groups; (b) the CH subspectrum; the labels H–T and H–H indicate CH signals from head-to-tail and head-to-head addition, respectively.

flip angles θ of 45°, 90° and 135°. Denoting these by [45], [90] and [135], the appropriate combinations of (ideal) spectra are

CH subspectrum:	[90]
CH_2 subspectrum:	[45] − [135]
CH_3 subspectrum:	$[45] + [135] - [90]/\sqrt{2}$

In practice, some empirical adjustment to the combination coefficients is necessary to eliminate peaks of the undesired multiplicity. As well as being useful in simplifying complex spectra, subspectral analysis of this type may be very useful in revealing small but structurally important peaks obscured in the standard spectrum by peaks of a different multiplicity. An example is shown in Figure 1.11 which compares the standard spectrum and the CH subspectrum of the backbone carbon region of poly(propylene oxide). Small CH peaks are observed in the CH subspectrum in the region 72.5–73.5 ppm which are obscured by intense CH_2 peaks in the standard spectrum. These minor CH peaks arise from head-to-head units of the structure $-OCH_2CH(CH_3)OCH(CH_3)CH_2-$. The intense CH peaks in the region 74.5–75.5 ppm arise from the predominant head-to-tail units of the structure $-OCH_2-CH(CH_3)OCH_2CH(CH_3)-$ [32, 33]. Other applications of DEPT spectra have been reported [34–36].

1.6.1.2 Solvent suppression. If the use of a protonated solvent is essential, a number of techniques are available to substantially reduce the solvent signal. The principal use of these techniques has been to eliminate the H_2O signal in aqueous solutions of biopolymers, as reviewed in [37]. Applications to synthetic polymers are uncommon, but it is worth mentioning them briefly as part of the NMR armoury.

The simplest technique is pre-saturation in which the solvent is subjected to CW irradiation for a short time at a level sufficient to saturate the solvent without affecting other solute peaks. The saturating field is removed immediately before the observation pulse so that the solvent peak has no opportunity to relax. Several solvent peaks can be saturated if rapid frequency switching can be performed during the saturation period. If proton exchange between solvent and solute occurs, for example exchange of OH and NH with H_2O, the solvent saturation can be transferred to the solute. However, this is rarely a problem in synthetic polymers, and because of its ease of application the pre-saturation method is the preferred method.

A second method makes use of differential longitudinal relaxation times of solvent and solute. Application of a 180° pulse inverts the nuclear magnetisation. During the ensuing relaxation process, the magnetisation necessarily passes through zero, the so-called null point. However, the solvent normally has a longer relaxation time than a polymer solute, and at the solvent null point, the polymer signals will have substantially recovered. Application of an observation 90° pulse at this time will effectively suppress the solvent peak. However, the solute intensities will possibly be distorted by differential relaxation between themselves.

A wide range of other methods for solvent suppression has been developed which may collectively be classed as tailored excitation. These rely on the application of appropriate combinations of pulses to excite protons lying outside a narrow band of frequencies while leaving those within that band (i.e. the solvent) undisturbed. Examples of these are the Redfield pulse [38], the 'jump-return' technique [39] and 'binomial' sequences [40].

1.6.1.3 Resolution enhancement. Because of compositional and stereochemical heterogeneity, polymer spectra often consist of many overlapping, incompletely resolved lines. It is possible to artificially improve the resolution of such spectra, at the price of degradation in the signal-to-noise ratio and lineshape distortion. This is achieved by multiplication of the FID before Fourier transformation by a function which emphasises the intense points at the beginning of the FID and de-emphasises the points containing only spectrometer noise at the end of the FID. Two common functions implemented on many spectrometers are the Lorentz–Gaussian transformation [41] and the sine-bell function [42]. The first of these uses the function

$$\exp\left(\frac{t}{T_{\mathrm{E}}}\right)\exp\left[-\left(\frac{t}{T_{\mathrm{G}}}\right)^{2}\right]$$

where T_{E} and T_{G} are best treated as empirical parameters selected by trial and error. The first exponential counteracts the normal FID decay so T_{E} is approximately equal to the FID decay time constant, and the second exponential converts the envelope to a Gaussian. The FT of a Gaussian FID is a Gaussian peak that has much smaller wings than a Lorentzian function of the same linewidth. T_{E} and T_{G} may be different for different multiplet signals.

The sine-bell method multiplies the FID by half a sine cycle using the function $\sin(\pi t/T)$ where T is the acquisition time of the FID. There are no adjustable parameters, although a variant technique, a shifted sine-bell using the function [43]

$$\sin\left[\pi\left(\frac{t+S}{T+S}\right)\right]$$

where S is an empirical shift parameter, gives some flexibility.

1.6.2 Two-dimensional NMR spectroscopy [6–10, 13, 18]

In one-dimensional NMR, a single FID is acquired digitally as a series of points extending over one time variable. After Fourier transformation, the spectrum consists of a plot of intensity as a function of one frequency. In two-dimensional NMR, a sequence of pulses is applied and a number of FIDs

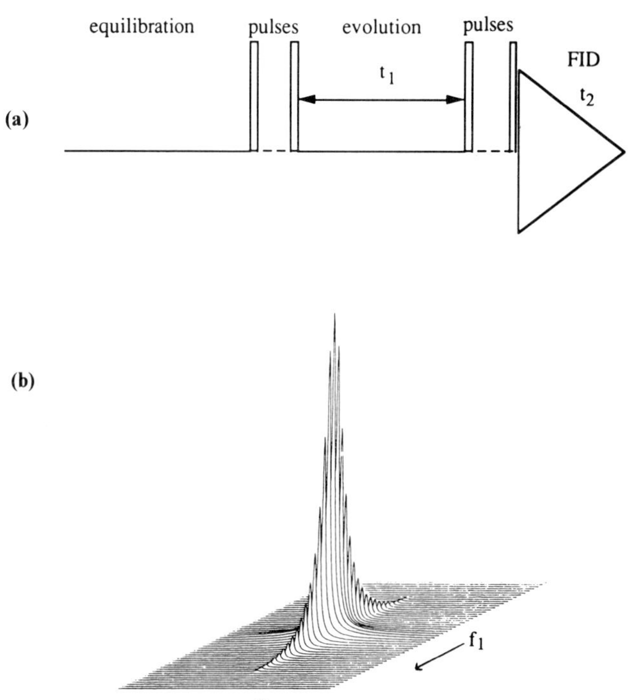

Figure 1.12 (a) Schematic diagram of a pulse sequence for two-dimensional NMR; (b) view of an NMR peak in two dimensions with frequency axes f_1 and f_2.

are acquired for a succession of values of one of the sequence time parameters. The experiment is represented schematically in Figure 1.12(a). Three periods may be distinguished: (a) the preparation period in which the system is allowed to reach a steady initial state; (b) following one or more initial pulses, the evolution period in which the magnetisations oscillate according to chemical shifts and/or coupling constants; (c) following further pulses, the detection period in which the resultant FID is digitised. Conventionally the time variable incremented during the evolution period is denoted t_1 and the time variable during the FID acquisition is denoted t_2. The FID data therefore consists of an array of FIDs, typically 512, for different values of t_1, i.e. an array of points as a function of two times, t_1 and t_2. A double FT over both t_1 and t_2 produces the spectrum consisting of an intensity as a function of two frequencies, conventionally denoted f_1 and f_2, as shown in Figure 1.12(b). The 2-D spectrum is in practice normally plotted as a contour plot. A peak in a 2-D spectrum is characterised by two frequency coordinates, and it is often convenient to think of a 2 D peak as *correlating* these two frequencies. A wide variety of 2-D sequences has been developed [6–13] which differ in the significance of f_1 and f_2 and in the mechanism by which f_1 and f_2 are correlated. Many are variations on a theme, differing for example in refinements affecting their tolerance of instrumental imperfections or mis-setting of parameters. Here only the basic types of 2-D experiment useful for synthetic polymer structural characterisation are described. Examples of the application of 2-D methods to polymer tacticity are described in section 1.8.2. It should be noted that the total evolution time may extend over several tens or hundreds of milliseconds, during which time natural relaxation also occurs. In relatively stiff chains, T_2 may be so short that considerable transverse relaxation takes place during evolution and the final FID intensity is much attenuated.

Heteronuclear correlation (HETCOR). This experiment is used to correlate the signals of two different nuclei X and Y using X–Y spin–spin coupling. In the form normally encountered, X is ^{13}C, Y is ^{1}H, one-bond spin–spin coupling correlations are established, f_1 is the ^{1}H frequency and f_2 is the ^{13}C frequency. In other versions, correlations using two- or three-bond H–C coupling are established. In so-called 'relayed' experiments, correlations of

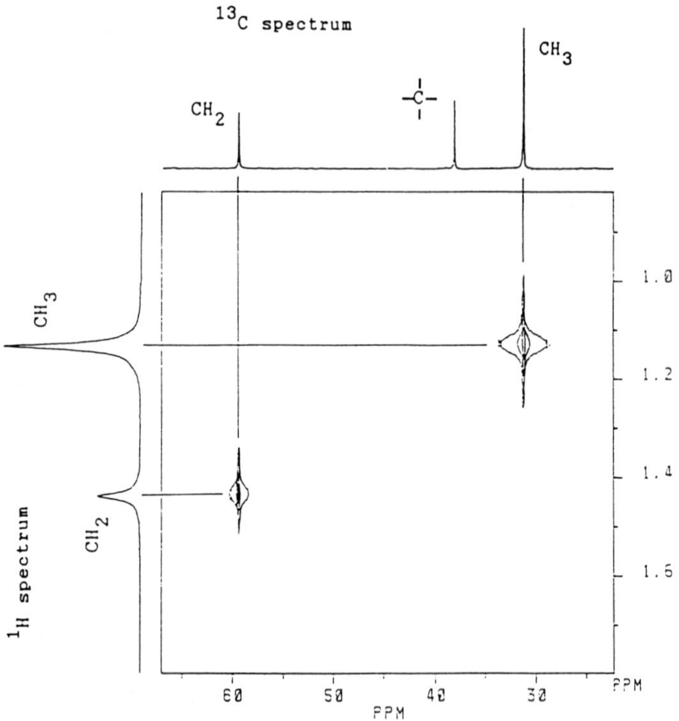

Figure 1.13 The ^{13}C–^{1}H HETCOR 2-D spectrum of poly(isobutylene).

the type H_a–H_b–C are determined, where H_a and H_b are spin-coupled and H_b and C are directly bonded. As an example of the appearance of a basic 2-D contour spectrum, Figure 1.13 shows the one-bond HETCOR experiment applied to poly(isobutylene). The spectrum is shown as the usual contour plot with standard 1-D spectra for reference. Note there is no correlation peak for the unprotonated quaternary carbon.

Homonuclear correlation via spin–spin coupling (COSY). The established acronym 'COSY' stands for COrrelated SpectroscopY. Both f_1 and f_2 represent the frequency of the same nucleus, usually protons, although several applications using ^{19}F NMR have been reported. Correlation is established using homonuclear coupling, so the technique essentially shows the same information in one plot as a series of 1-D homonuclear decoupling experiments. A potentially extremely useful application of this class of 2-D experiment is to investigate ^{13}C–^{13}C spin–spin correlations but two problems are encountered in such work. The first problem is that the technique requires ^{13}C nuclei in two adjacent structural environments. In natural abundance, this probability is only $1.2 \times 10^{-2}\%$ so the experiment is very insensitive and suitable only for concentrated samples rich in carbon content. Secondly, the spin-coupled peaks from ^{13}C–^{13}C pairs are weak compared to the peaks from isolated ^{13}C nuclei and special steps must be taken to suppress the latter. The version used to achieve this suppression is known as INADEQUATE, a technique that has been particularly extensively used in assigning the spectra of polypropylene [44].

Homonuclear correlation via the nuclear Overhauser effect (NOESY). The NOESY experiment correlates peaks by means of the nuclear Overhauser enhancement and so identifies pairs of nuclei which are sufficiently close together in space to relax by their dipole–dipole interaction. This technique is not so applicable in determining stereochemical assignments as those described previously, but may be extremely useful in determining the chain conformation as demonstrated by Mirau *et al.* in a study of the alternating copolymer of styrene and methyl methacrylate [45] (see chapter 4).

J-Resolved spectroscopy. In the *J*-resolved experiment, the f_1 axis is the normal one-dimensional chemical shift frequency axis and f_2 represents the frequency displacements of multiplet peaks which result from spin–spin coupling. The effect is to separate chemical shift effects and spin–spin coupling effects into different dimensions, alleviating peak overlap considerably and making it easier to establish *J* coupling constants of complicated spin systems. Furthermore, by taking a 45° projection of the 2-D spectrum, a spectrum is obtained from which all homonuclear spin–spin coupling is removed.

1.7 A survey of applications of high-resolution solution-state NMR to polymers

The applications of high-resolution NMR to polymers are many and varied, reflecting the wide range of chain structural features accessible by NMR. Many applications are no different from the ubiquitous use of NMR as a qualitative analytical tool in organic structural analysis, but others exploit the capability of NMR in providing information about aspects of molecular architecture that are peculiar to polymers and can be investigated by no other technique. Here the principal types of information are summarised in approximate order of sophistication.

1.7.1 Determination of functional groups and composition of composite systems

At the basic level, the chemical shifts observed in an NMR spectrum characterise to varying degrees of precision, the environment of the nucleus under observation, thus providing data on the chemical structures present in an unknown sample. Simple identification of functional groups may also provide useful data on aspects of the polymerisation process. An example is the polymerisation of butadiene, CH_2=CH–CH=CH_2, which may add to a chain either by 1,2 addition producing a pendant vinyl group, or by 1,4 addition producing an in-chain olefinic bond (Figure 1.14). In the ^{13}C spectrum, the 1,2 addition mode gives olefinic CH_2 and CH resonances of equal intensity at ca. 111 and 140 ppm, respectively, while the 1,4 addition mode gives only olefinic CH resonances at ca. 127 ppm.

Quantitatively, the basic use of NMR is to determine the composition of a mixture in terms of the fraction of each species present from the relative intensities of peaks from each component. The components may be different chains in a polymer blend, different monomer units in a copolymer, or a polymer and an additive, for example in a polymer/plasticiser or polymer/stabiliser system.

1.7.2 Determination of end-groups

The end-groups of a chain are determined by initiation, chain transfer and termination reactions and determination of the end-group structures is

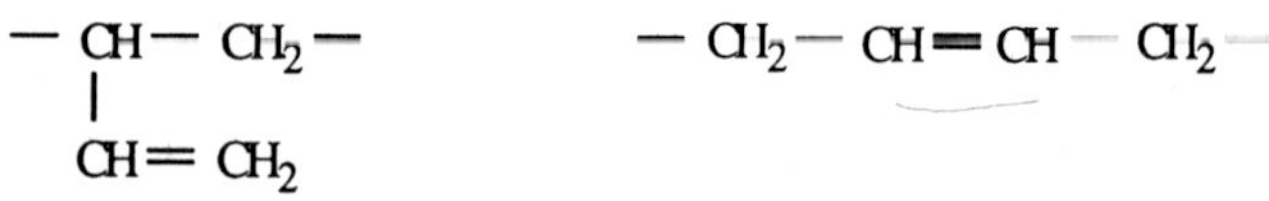

— CH — CH_2 —
CH = CH_2

(a)

— CH_2 — CH = CH — CH_2 —

(b)

Figure 1.14 Addition modes of buta-1,4-diene. (a) 1,2 addition; (b) 1,4 addition.

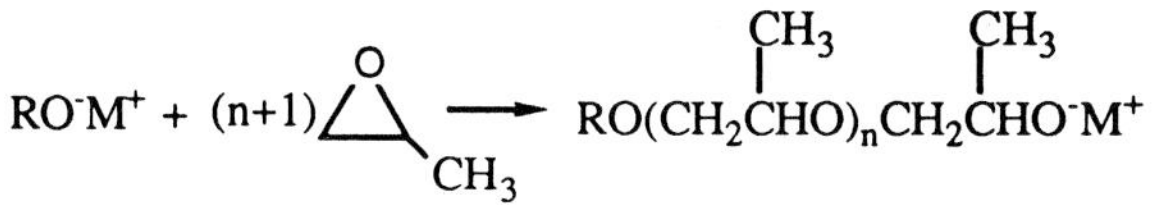

$$RO^-M^+ + (n+1)\,\underset{CH_3}{\overset{O}{\triangle}} \longrightarrow RO(CH_2\overset{CH_3}{\overset{|}{C}}HO)_nCH_2\overset{CH_3}{\overset{|}{C}}HO^-M^+$$

(a)

$$RO^-M^+ + \underset{CH_3}{\overset{O}{\triangle}} \longrightarrow RO^- \cdots H{-}CH_2 \cdots M^{\pm}{-}O \longrightarrow ROH + CH_2{=}CHCH_2O^-M^+$$

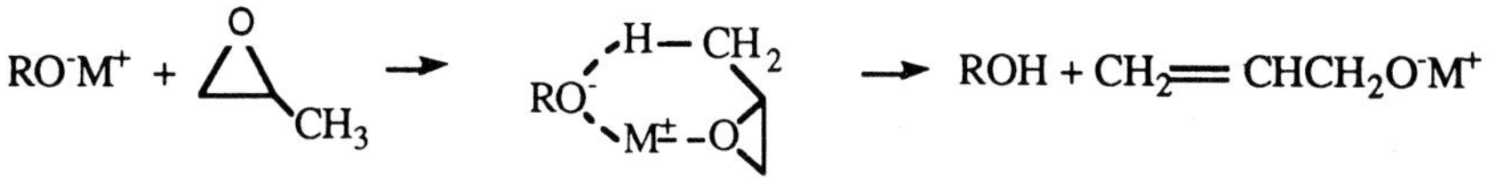

(b)

Figure 1.15 (a) The normal head-to-tail addition propagation mode in the anionically catalysed polymerisation of propylene oxide; (b) the chain transfer reaction generating an allyl alkoxide.

often extremely valuable in studying those processes (see chapter 3). As a simple example here, we mention the anionic polymerisation of propylene oxide. The normal propagation reaction is an addition to the growing chain by ring-opening at the CH_2–O bond producing a secondary alcohol end (Figure 1.15(a)). However, occasionally the propagating alkoxylate anion undergoes chain transfer by abstracting a proton from the monomer methyl, generating an allyl alkoxylate which then initiates a new chain. The allyl end-groups are readily identified by olefinic CH_2 and CH peaks at 117 and 135 ppm, respectively [32].

Quantitatively, the end-group intensity relative to the main-chain intensity gives the number-average degree of polymerisation directly. In practice, because of incomplete resolution of end-group peaks and insufficient sensitivity, this use of NMR is usually restricted to degrees of polymerisation of the order of 100 or less.

1.7.3 Statistical characterisation of the structure of irregular chains

Many polymers have an irregular structure which can be quantitatively characterised only in statistical terms. The irregularity may be stereochemical in origin as in many free-radical homopolymers, or chemical in origin as in a statistical copolymer. The chemical shift of a nucleus is often sensitive not only to the structure of the monomer unit in which the nucleus is located, but also to the structure of nearest and next-nearest monomer units. Thus in irregular polymers, a particular type of nucleus often gives several peaks reflecting the types and probability of occurrence of different neighbours, i.e. the distribution of different sequences of elementary units. This information is not generally available from any other technique. The application of NMR to stereochemically irregular homopolymers is described below, and the application to statistical copolymers is described in chapter 2.

1.8 The observation of polymer stereochemistry (tacticity) by NMR

Polymer stereochemistry is the aspect of polymer structure in which high-resolution NMR made its first significant contribution to polymer science. The use of NMR in this area dates from the pioneering work of Bovey and Tiers [46] who showed that ^{1}H NMR allowed a rapid, unequivocal identification of stereochemically different forms of poly(methyl methacrylate). Subsequently, it became apparent that NMR could not only identify the stereochemistry of stereoregular polymers but could also provide information not available by any other technique on the stereochemical sequence distribution and statistics of stereo-irregular polymers. The application of NMR to polymer stereochemistry continues to develop in two respects. First, improved instrumentation such as higher magnetic fields and new techniques such as 2-D NMR have allowed a more detailed analysis of the stereochemical distribution of existing polymers, and second, determination of the stereochemistry is of considerable importance in characterising the structure of newly synthesised polymers.

This section defines the basic concepts of polymer stereochemistry, and describes how the stereochemistry is manifest in NMR spectra.

The stereochemistry of linear polymers has been recently reviewed in detail [47].

1.8.1 Stereochemistry of vinyl polymers

Historically vinyl polymers were the first to be classified stereochemically and the first to be studied by NMR. In some respects they are also the simplest. The basic concepts of polymer stereochemistry and its influence on NMR spectra are developed first with specific reference to vinyl polymers. The ideas introduced are then extended to other systems.

1.8.1.1 Definitions and notation. Consider a regular vinyl polymer of general structure $[-CH_2-CXY-]_n$. In principle, since the polymer chains on either side are of different lengths, the CXY carbon should be chiral, i.e. resolvable into optically active *R* and *S* enantiomers. These are shown in Figure 1.16 together with their representation as Fischer projections. Except for the very few carbons at or near the chain ends, the difference between the two chain substituents is in fact immaterial, and the group is effectively optically inactive. The configuration is more correctly described as pseudochiral. However, when the *relative* configuration of neighbouring units is considered, stereochemically distinct diastereomers are possible. The two basic stereoregular

Figure 1.16 The enantiomers of a CXY group in a vinyl polymer.

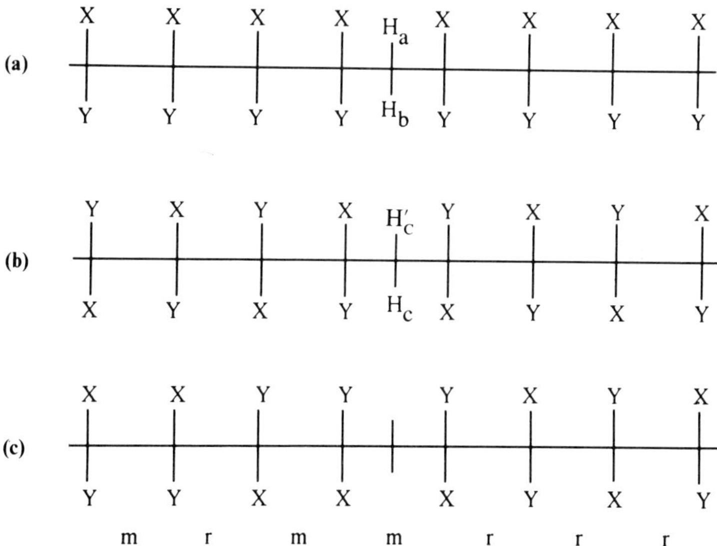

Figure 1.17 The basic tactic structures of a vinyl polymer. (a) Isotactic; (b) syndiotactic; (c) atactic. Methylene protons are equivalent in syndiotactic dyads but not in isotactic dyads.

diastereomers are shown in Figures 1.17(a), (b). In the first, termed *isotactic*, all groups are of the same configuration, whereas in the second, termed *syndiotactic*, the configurations alternate. A third possibility is an irregular structure exemplified by the fragment in Figure 1.17(c); this is termed *atactic*. Examples of all three structures are known; in general, stereoregular polymers are formed by coordination catalysts whereas atactic polymers are formed by uncoordinated catalysts such as free radicals or free ions. Stereoregular polymers are often partially crystalline, and isotactic and syndiotactic diastereomers may have different properties. An example is poly(methyl methacrylate); the glass transition temperatures of the isotactic and syndiotactic forms are 38 and 105°C, respectively. Atactic polymers are normally amorphous.

The stereochemistry of a particular chain is specified completely by the absolute configuration of every pseudochiral carbon. However, since enantiomers have the same properties and are spectroscopically indistinguishable (in a normal achiral environment), it is more convenient to specify the stereochemistry in terms of the relative configuration (or *tacticity*) of adjacent units. The system suggested by Frisch *et al.* [48] is adopted here, although others have used a binary notation [16]. Adjacent units of the same configuration are termed an *m* dyad, while adjacent units of opposite configuration are termed an *r* dyad. This *m*/*r* notation derives from the terms meso and racemic used originally in describing the chirality of molecules with two equivalent chiral centres. Thus an isotactic chain of either chirality would be represented as ...*mmmm*..., and a syndiotactic chain by ...*rrrr*.... The atactic fragment in Figure 1.17(c) would be represented as ...*mrmmrrr*....

1.8.1.2 NMR spectra of stereoregular polymers. We consider first proton spectra. The key distinctive feature is the spectrum of the CH_2 group; in an

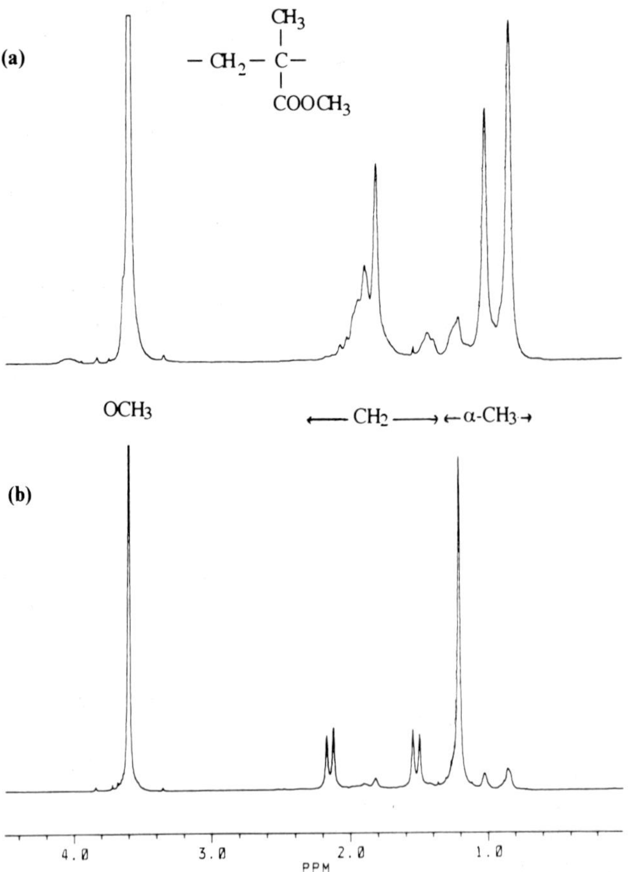

Figure 1.18 300 MHz ^1H spectra of poly(methyl methacrylate) in $CDCl_3$. (a) Predominantly syndiotactic; (b) predominantly isotactic.

isotactic chain, the two CH_2 protons are (in principle) not identical and two CH_2 resonances are observed, whereas in a syndiotactic chain the CH_2 protons are identical by symmetry so only one CH_2 resonance is observed. In Figure 1.17(a), $H_a \not\equiv H_b$ whereas in Figure 1.17(b), $H_c \equiv H_c'$. Following the rules for spin–spin coupling and disregarding coupling to X and Y, the syndiotactic CH_2 spectrum therefore is of the type designated A_2, i.e. a singlet, and the isotactic CH_2 spectrum is of the type designated AX, i.e. two doublets due to geminal coupling (in the first-order limit). These distinctive patterns are apparent in the CH_2 region (1.5–2.5 ppm) of the spectra of poly(methyl methacrylate) shown in Figure 1.18. The sample in Figure 1.18(a), prepared by free-radical initiation, shows essentially a single CH_2 peak at 1.8 ppm indicating a predominantly syndiotactic structure. The sample in Figure 1.18(b), prepared using a Grignard reagent as initiator, shows two doublets at 1.5 and 2.2 ppm with a coupling constant of 14 Hz, indicating an isotactic structure. Other stereoregular polymers which show these characteristic forms are isotactic and syndiotactic poly(propylene) [49] and isotactic poly(isopropyl acrylate) [50]. Note that in polymers where the pseudochiral carbon is –CHX–, there is also vicinal coupling between the CH_2 and CH protons. Each of the CH_2 peaks is further split into a triplet with a typical coupling constant of 6 Hz. It should be noted that although the two CH_2 protons in the isotactic polymer are in principle different, they may fortuitously be experimentally indistinguishable. A case in point is isotactic poly(styrene). At 60 MHz[51], the chemical shift between the CH_2 protons is so small that they give only a single triplet from coupling to the adjacent methine. However, at 220 MHz [52], the chemical shift increases to such an extent that the non-equivalence is apparent, albeit distorted from the first-order pattern by second-order effects.

The ^{13}C spectra of stereoregular polymers show a single sharp line for each chemically distinct carbon because within each type of chain, each monomer residue is identical. However, the chemical shifts for isotactic and syndiotactic chains are not the same. For example, in isotactic poly(propylene), the CH_3, CH and CH_2 carbons occur at 20.0, 27.1 and 44.4 ppm, respectively, whereas in syndiotactic poly(propylene), the corresponding shifts are 18.7, 27.0 and 45.4 ppm [53]. ^{13}C NMR does not yield an unequivocal identification of the tacticity as does the CH_2 proton spectrum, but it does permit different tacticities to be distinguished.

1.8.1.3 NMR spectra of atactic polymers. It might be thought that the spectrum of an atactic polymer, consisting of an irregular sequence of *m* and *r* dyads would simply be a superposition of the spectra of isotactic and syndiotactic polymers with intensities according to the relative proportions of the dyads. In fact the spectra of atactic polymers are frequently considerably more complex than this simple picture. As an example, Figure 1.19 shows the ^{13}C spectrum of free-radical poly(vinyl chloride), $[-CH_2CHCl-]_n$. There are two different types of carbon, CH_2 and CH, but the former is split clearly

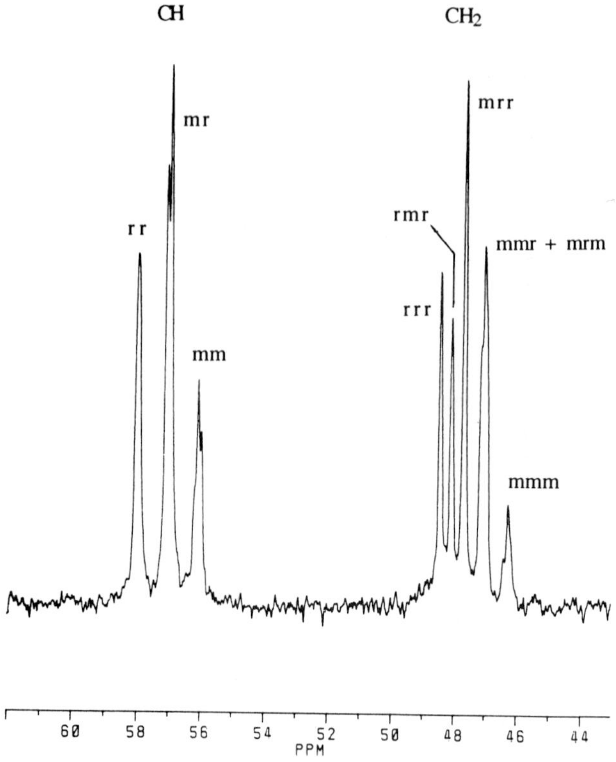

Figure 1.19 75.5 MHz ^{13}C spectrum of atactic poly(vinyl chloride) in *ortho*-dichlorobenzene at 100°C.

into five lines and the latter into three lines with indications of further fine structure. The complexity poses interpretational problems, but at the same time is the source of significant stereochemical information. The abundance of lines fundamentally arises because the chemical shift is sensitive to stereochemistry over sequences longer than a dyad. In order to explain the form of the spectra, it is convenient to consider the spectra of the CXY and CH_2 groups separately.

(*a*) *The spectrum of the CXY group.* At the first level, the chemical shifts of nuclei in the CXY group will depend on the tacticity of its nearest neighbours, i.e. the tacticity of a sequence of three monomer units, termed a *triad*. Since a triad is two successive dyads each of which may be *m* or *r*, there are four distinct triad tacticity environments, *mm*, *mr*, *rm* and *rr*, as shown in Figure 1.20. The *mr* and *rm* configurations are mirror images and therefore indistinguishable by NMR. For chemical shift sensitivity at the triad level, three chemical shifts would therefore be expected for each nucleus. This is

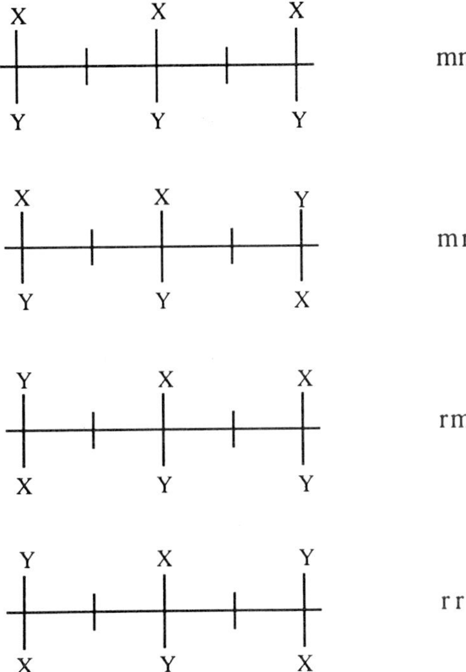

Figure 1.20 Triad configurations in a vinyl polymer.

the origin of the three principal peaks for the CH carbon in PVC in Figure 1.19. A further example of triad splitting is shown in the α-CH_3 region (1–1.5 ppm) in the proton spectra of PMMA in Figure 1.18. Although the two samples were largely stereoregular, each contained some dyads of the other tacticity. In both spectra, three α-CH_3 peaks are observed at the same chemical shifts (0.85, 1.03 and 1.1 ppm) but not of the same relative intensity. The assignment of the three peaks to the appropriate triad is not of course immediately obvious from the CXY spectrum alone. To achieve that objective, the predominant tacticity must be identified. In the case of PMMA, this is known from the form of the CH_2 ^1H spectrum as explained above, and the α-CH_3 peaks can readily be assigned as *mm*, (*mr* + *rm*) and *rr* in order of increasing shielding based on the variation of the relative intensities. The derivation of the PVC ^{13}C triad assignment given in Figure 1.19 is possible via the analysis of the proton CH_2 spectrum, as explained below.

The tacticity sensitivity of the CXY group may extend further than triads, being influenced by the next nearest neighbours also, i.e. a sequence of five monomer units (four dyads), termed a *pentad*. There is a total of 16 pentad enantiomers, but only ten are distinguishable. Extending the *m*/*r* notation to encompass four dyads, the pentads are

mmmm	(*mmrm* + *mrmm*)	*mrrm*
(*mmmr* + *rmmm*)	(*mmrr* + *rrmm*)	(*mrrr* + *rrrm*)
rmmr	(*rmrm* + *mrmr*)	*rrrr*
	(*rmrr* + *rrmr*)	

The partial splitting of the PVC CH triad peaks in Figure 1.19 is due to pentad sensitivity, although not all are resolved [54]. A more complete example is provided by the carbonyl ^{13}C spectrum of free-radical PMMA shown in Figure 1.21. Eight peaks are resolved. The assignment given is based on statistical arguments [55]. Note that the relative intensities of the *mm*-centred group, the (*mr* + *rm*)-centred group and the *rr*-centred group are the same as those of the triad peaks in the α-CH_3 ^{1}H spectrum.

In some cases, notably the ^{13}C spectrum of poly(propylene) [56] and poly(vinyl alcohol) [57], the CXY chemical shifts may show resolvable stereochemical sensitivity over even longer ranges. At the next highest level, a sequence of seven monomer units or *heptad*, there is a total of 64 enantiomers of which 36 are distinguishable.

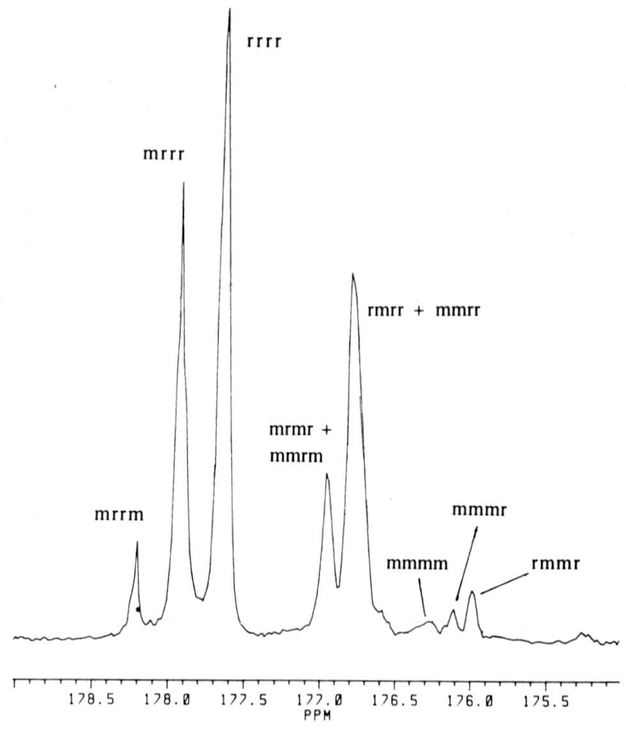

Figure 1.21 75.5 MHz ^{13}C spectrum of the carbonyl carbon in free-radical poly(methylmethacrylate) in $CDCl_3$.

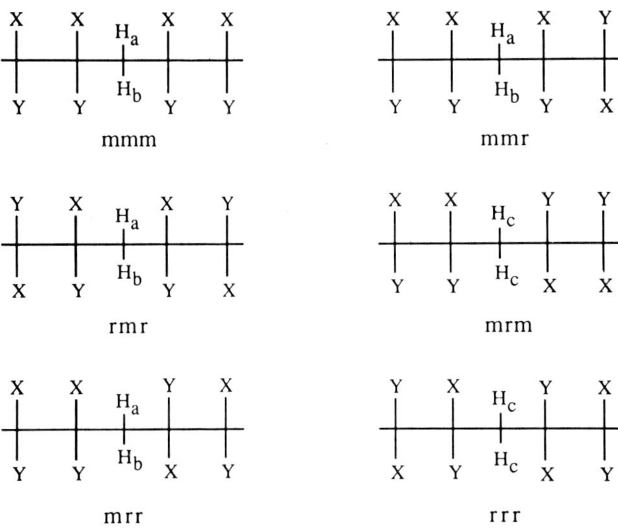

Figure 1.22 Tetrad configurations in a vinyl polymer, illustrating the equivalence or non-equivalence of methylene protons.

(*b*) *The spectrum of the* CH_2 *group.* The form of the spectrum expected for CH_2 groups can be evaluated using the principles applied to the CXY group. If the chemical shift depends on the tacticity of the nearest-neighbour CXY groups, there are two CH_2 environments, namely *m* and *r* dyads as discussed above for regular isotactic and syndiotactic polymers. The ^{13}C spectrum would show two lines and the ^{1}H spectrum would show a total of three chemical shifts comprising one shift for the *r* dyad and two for the *m* dyad.

At the next-nearest neighbour level encompassing a *tetrad* of monomer units, there is a total of eight enantiomers of which six are distinguishable, namely *mmm*, (*mmr* + *rmm*), *rmr*, *mrm*, (*mrr* + *rrm*) and *rrr*, as shown in Figure 1.22. In the ^{13}C spectrum, six lines would therefore be expected. This degree of sensitivity is the origin of the five lines in the ^{13}C spectrum of the CH_2 group in PVC shown in Figure 1.19; two of the tetrads fortuitously overlap. In the ^{1}H spectrum, the protons within the *mrm* and *rrr* tetrads are identical pairs so each gives a spectrum of the type A_2. However, the protons within the other four tetrads are in principle non-equivalent so each gives a spectrum of the type AX (or AB if second-order) with additional splitting by vicinal coupling to X or Y if one is hydrogen. It is rare to find all ten proton chemical shifts resolved; PMMA is one example [58]. A further example more typical of the incompletely resolved ^{1}H spectra of vinyl polymers is shown in Figure 1.23. This is a simulation of the CH_2 proton spectrum of PVC as a composite lineshape from overlap of six tetrads [59]. The chemical shifts of each tetrad were obtained from an analysis of the spectrum of poly(vinyl chloride-α-d_1)

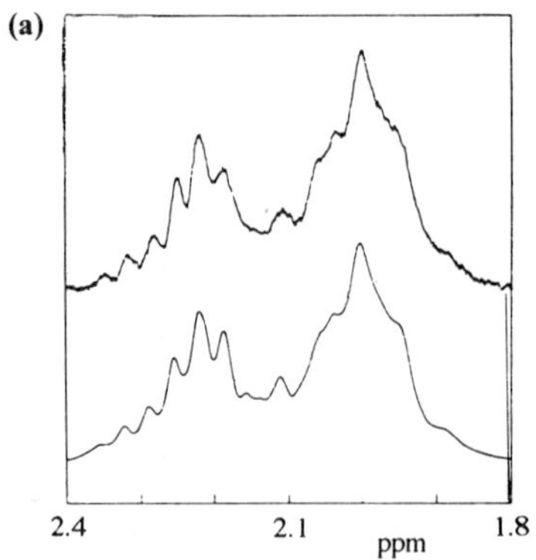

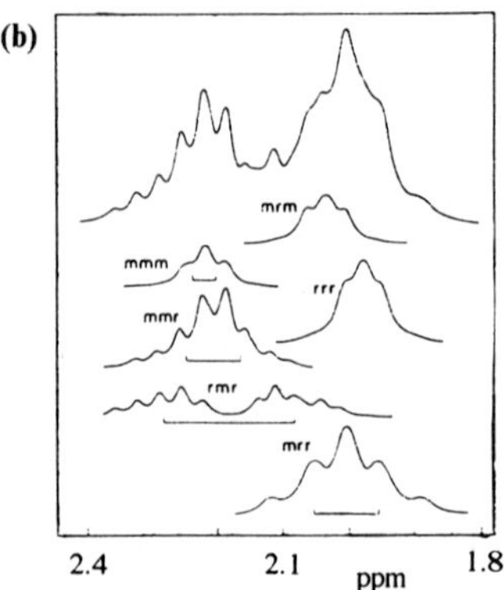

Figure 1.23 (a) Comparison of (upper) the experimental and (lower) the calculated 220 MHz ^1H spectra of the CH_2 protons in free-radical poly(vinyl chloride) in chlorobenzene at 150°C. (b) Decomposition of the calculated spectrum into tetrad components. (Reprinted with permission from [59], as discussed in the text).

in which vicinal coupling is absent. The relative intensities of the tetrad subspectra unequivocally indicate a slight preference for the *r* dyad, which permits the assignment given for the ^{13}C peaks in Figure 1.19.

Extension of CH_2 stereochemical sensitivity to the next highest level entails a sequence of six monomer units, i.e. a *hexad*. There are 32 hexads of which 20 are distinguishable.

1.8.1.4 The experimental observation of stereochemistry. It is worth pointing out some of the characteristics of the relationship between stereochemistry and NMR spectra which arise from the examples described. These practical considerations must be borne in mind when approaching the interpretation of a spectrum in terms of stereochemistry.

1. The range of stereochemical sensitivity, indeed whether different sequences are resolved at all, depends unpredictably on the polymer structure, on the nucleus studied, and on the position of the nucleus in the chain. Thus in the $-C(CH_3)(COOCH_3)-$ group in PMMA, the carbonyl carbon shows well-resolved pentad structure, the α-CH_3 protons show well-resolved triad structure whereas the OCH_3 protons show no tacticity splitting. A second example is polystyrene, a polymer ubiquitous in studies of fundamental properties. Here the backbone CH carbon is effectively completely insensitive to tacticity whereas the CH_2 carbon shows incompletely resolved splitting at the hexad level [60]. The phenyl quaternary carbon shows pentad splitting [61]. A successful measurement of tacticity by NMR is not assured, but if some tacticity sensitivity is present, the number of lines can be understood in terms of the dyad, triad, tetrad, etc. system.
2. When present, tacticity effects do not necessarily vary regularly with

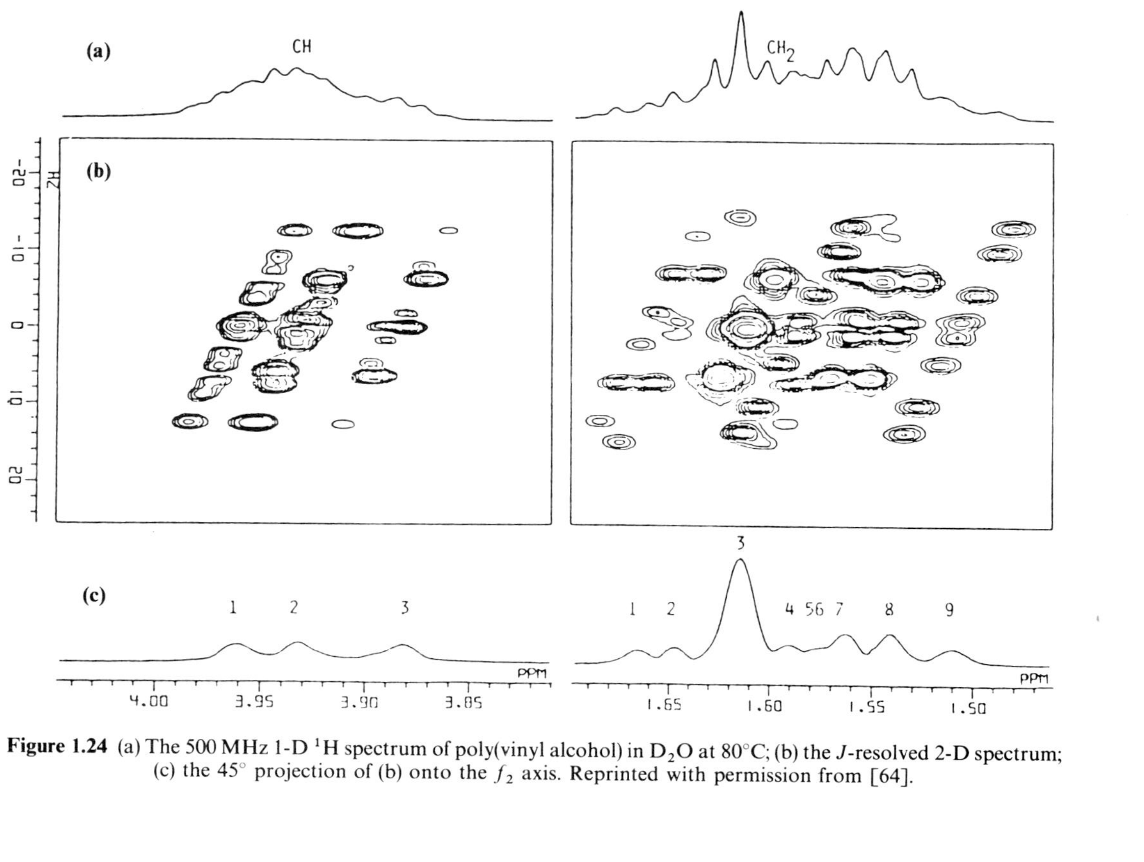

Figure 1.24 (a) The 500 MHz 1-D ^1H spectrum of poly(vinyl alcohol) in D_2O at 80°C; (b) the *J*-resolved 2-D spectrum; (c) the 45° projection of (b) onto the f_2 axis. Reprinted with permission from [64].

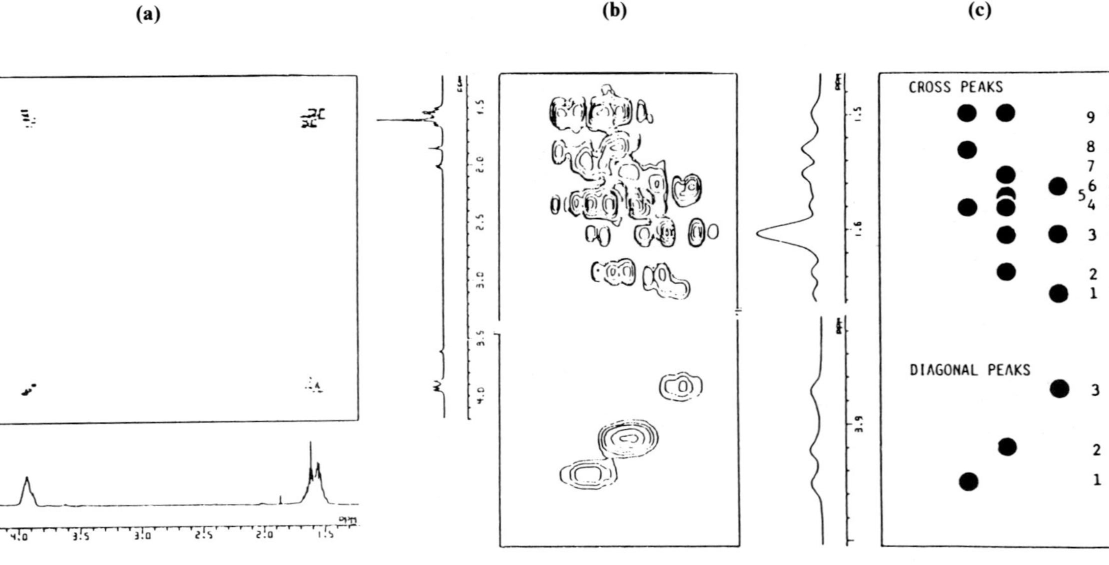

Figure 1.25 (a) The broadband decoupled ^1H COSY spectrum of poly(vinyl alcohol) under the same conditions as Figure 1.24; (b) expansions of the CH diagonal peaks and the CH/CH_2 correlation peaks; (c) schematic representation of the correlation peaks, with assignments given in Table 1.5. Reprinted with permission from [64].

distance. In many cases, tetrad splittings are comparable in magnitude to dyad splittings, e.g. the CH_2 carbon in PVC.

3. The splittings may be solvent-dependent.
4. An absolute determination of tacticity requires the use of proton spectra of the CH_2 group. However, ^{13}C nuclei are usually sensitive to the tacticity over a longer range than protons.
5. Frequently, the linewidth in atactic polymers is determined by an unresolved dispersion of chemical shifts from higher sequence levels. Consequently the linewidth (in Hertz) is proportional to the B_0 field, and an increase in B_0 may not therefore necessarily improve the resolution.
6. The precise dependence of chemical shifts on tacticity cannot be predicted a priori. However, once established, an assignment can often be rationalised semi-empirically in terms of differences in accessible conformations [14] (see chapter 4).

1.8.2 Application of 2-D techniques for the assignment of tacticity-related peaks

2-D methods have been used extensively to assign tacticity-dependent peaks in poly(methyl methacrylate) [58, 62], poly(vinyl chloride) [63] and poly(vinyl alcohol) [64]. The principles underlying such applications are well illustrated by the work on the proton spectrum of atactic poly(vinyl alcohol) [64]. Figure 1.24 shows the 1-D and 2-D *J*-resolved backbone ^1H spectra together with the 45° projection of the 2-D spectrum. The 1-D spectrum shows poorly resolved complex multiplets as a consequence of spin–spin coupling and stereochemical irregularity whereas in the 2-D *J*-resolved spectrum, the spin–spin coupling components are clearly separated in the f_1 (vertical) dimension.

Table 1.4 Connectivity between CH triad peaks and CH_2 tetrad peaks in a vinyl polymer

Triad	Tetrad
	mmm
mm	
	mmr
	rmr
mr	
	mrm
	mrr
rr	
	rrr

Table 1.5 Assignment of the correlation peaks in the broadband decoupled ^{1}H COSY spectrum of poly(vinyl alcohol) in Figure 1.25 (from [64])

	Peak number	Sequence
CH	1	*rr*
	2	*mr*
	3	*mm*
CH_2	1	*mmm*
	2	*rmr*
	3	*mmr*
	4	*mrr*
	5	*rmr*
	6	*mmm*
	7	*mrm*
	8	*rrr*
	9	*mrr*

In the projection, the spin–spin coupling is removed, and each proton gives a single line. In the CH region, three peaks are resolved corresponding to triads. In the CH_2 region, nine peaks are resolved whereas ten would be expected for full resolution of all distinct environments at the tetrad level; evidently two are coincidental. The various CH and CH_2 peaks are correlated by the broadband decoupled COSY spectrum shown in Figure 1.25. The form of the off-diagonal correlation peaks can be understood by means of the triad-tetrad connectivity diagram in Table 1.4. For example the *rr* CH peak is spin-coupled only to the CH_2 protons in either *rrr* or *mrr* tetrads. Hence, if CH peak 1 is the *rr* peak, then the CH_2 peaks 4, 8 and 9 must arise from *rrr* and *mrr*. Similarly, the *mm* CH peak correlates with *mmm* and *mmr* tetrad peaks and the *mr* triad peak correlates with *mmr*, *mrm*, *rmr* and *mrr* tetrad peaks. From arguments such as these, together with other information, the peaks were assigned as in Table 1.5.

1.8.3 Tacticity sequence statistics in vinyl polymers

Since a vinyl monomer CH_2=CXY is achiral, the configuration of the CXY groups, and hence the chain tacticity, is created during the propagation step. For characterising chain stereochemistry, the basic building blocks are *m* and *r* dyads rather than *R* and *S* configurations. A monomer may be added creating either an *m* or *r* dyad although not necessarily with the same probability. The quantitative analysis of tacticity consists of two stages. The first is the definition of the type of statistical information available, and the second is the interpretation of the experimental data in terms of a model for the probabilities of forming *m* or *r* dyads during the propagation. As well as giving a quantitative measure of the average tacticity, investigation of the propagation statistics can give useful information on the polymerisation mechanism. The techniques are extended to the analysis of copolymer propagation in chapter 2.

1.8.3.1 Sequence probabilities. Experimentally, an NMR spectrum in general shows distinct peaks for stereochemically distinguishable sequences of different lengths. The intensity of each peak is proportional to the frequency of occurrence of that sequence. Hence, the intensity of a peak expressed as a fraction of the total intensity of that type of nucleus gives the probability of that sequence. For example, the probability of the *mm* triad in PVC is given by the intensity of the *mm* CH peak in Figure 1.19 expressed as a fraction of the total CH intensity. The probability of a sequence X is conveni-

Table 1.6 Distinguishable probabilities for tacticity sequences in vinyl polymers

Dyads	Triads	Tetrads	Pentads
(m)	(mm)	(mmm)	$(mmmm)$
(r)	(mr)	(mmr)	$(mmmr)$
	(rr)	(mrm)	$(rmmr)$
		(rmr)	$(mmrm)$
		(mrr)	$(mmrr)$
		(rrr)	$(rmrm)$
			$(rmrr)$
			$(mrrm)$
			$(mrrr)$
			$(rrrr)$

Table 1.7 Necessary relationships between tacticity sequence probabilities in vinyl polymers

Tetrad–tetrad

$(mrr) + 2(mrm) = (mmr) + 2(rmr)$

Pentad–pentad

$(mmmr) + 2(rmmr) = (mmrm) + (mmrr)$
$(mrrr) + 2(mrrm) = (rmrr) + (mmrr)$

Dyad–triad

$(m) = (mm) + \frac{1}{2}(mr), \quad (r) = (rr) + \frac{1}{2}(mr)$

Triad–tetrad

$(mm) = (mmm) + \frac{1}{2}(mmr), \quad (rr) = (rrr) + \frac{1}{2}(mrr)$

Tetrad–pentad

$(mmm) = (mmmm) + \frac{1}{2}(mmmr), \quad (mmr) = (mmmr) + 2(rmmr) = (mmrm) + (mmrr)$
$(mrm) = \frac{1}{2}(mmrm) + \frac{1}{2}(mrmr), \quad (mrr) = (mrrr) + 2(mrrm) = (mmrr) + (rmrr)$
$(rmr) = \frac{1}{2}(mrmr) + \frac{1}{2}(rmrr), \quad (rrr) = (rrrr) + \frac{1}{2}(mrrr)$

ently represented by the notation (X), it being understood that if X is unsymmetrical, the quantity (X) also includes the reverse sequence since these are experimentally indistinguishable. The distinct probabilities for sequences up to pentads are summarised in Table 1.6. Because the only possibilities considered are mutually exclusive m and r dyads, the sum of the probabilities in each column must necessarily be unity. Furthermore, since triads are built out of dyads, tetrads out of triads, and so on, there are certain necessary relationships within sequences of a particular order, and between sequences of different orders [48]. These relationships, which are independent of the propagation mechanism, are given in Table 1.7. They are frequently useful in making peak assignments or in determining the probabilities of overlapping peaks. Note that because of the necessary relationships and the normalisation condition, the sequence probabilities are not all independent; of the dyads, only one is independent, of the triads only two, of the tetrads only four, and of the pentads only six. Sometimes it is useful to express the tacticity in terms of the average length of runs of m or r dyads, N_m and N_r. These are determined from dyad and triad probabilities according to

$$N_m = \frac{2(m)}{(mr)} \tag{1.13}$$

$$N_r = \frac{2(r)}{(mr)} \tag{1.14}$$

1.8.3.2 Statistical propagation models. In general, the probability of forming an m or r dyad will depend on the tacticity of previous dyads, i.e. the sequence statistics will be described by a Markov model of some order. Here, only the most common cases of zero- and first-order models are considered. For zero-order Markov statistics (more usually termed Bernoullian statistics), the probability of forming a dyad is independent of the previous dyad. Hence, the sequence probabilities can be expressed simply in terms of two propagation probabilities, the probability of forming an m dyad, denoted P_m, and the probability of forming an r dyad, denoted P_r. Necessarily

$$P_m + P_r = 1 \tag{1.15}$$

so only one probability is independent. The sequence probabilities for this model are given in Table 1.8, expressed in terms of P_m. The value of P_m is found from the probability of an m dyad and conformity with Bernoullian statistics is proven if the experimental probabilities of triad or higher sequences then agree satisfactorily with those calculated using that value of P_m. More generally, Bernoullian statistics are obeyed if all sequence probabilities can be rationalised within experimental error by a single propagation probability according to the formulae in Table 1.8.

At the next level of complexity, the first-order Markov level (see chapter 2),

Table 1.8 Tacticity sequence probabilities expressed in terms of propagation probabilities for Bernoullian and first-order Markov statistics; the parameters P_m, u and v are defined in the text

Sequence	Bernoullian	First-order Markov
(m)	P_m	$u/(u+v)$
(r)	$(1-P_m)$	$v/(u+v)$
(mm)	$P_m{}^2$	$u(1-v)/(u+v)$
(mr)	$2P_m(1-P_m)$	$2uv/(u+v)$
(rr)	$(1-P_m)^2$	$v(1-u)/(u+v)$
(mmm)	$P_m{}^3$	$u(1-v)^2/(u+v)$
(mmr)	$2P_m{}^2(1-P_m)$	$2uv(1-v)/(u+v)$
(rmr)	$P_m(1-P_m)^2$	$uv^2/(u+v)$
(mrm)	$P_m{}^2(1-P_m)$	$u^2v/(u+v)$
(mrr)	$2P_m(1-P_m)^2$	$2uv(1-u)/(u+v)$
(rrr)	$(1-P_m)^3$	$v(1-u)^2/(u+v)$

the probability of forming m or r may depend on the structure of the preceding dyad, i.e. the statistics are described by four conditional probabilities, $P_{(m|m)}$, $P_{(r|m)}$, $P_{(m|r)}$ and $P_{(r|r)}$, where $P_{(x|y)}$ is the probability of adding an x dyad to a y dyad. Necessarily

$$P_{(m|m)} + P_{(r|m)} = 1 \tag{1.16}$$

$$P_{(r|r)} + P_{(m|r)} = 1 \tag{1.17}$$

Hence, only two of the probabilities are independent. For simplicity, the independent probabilities are defined as $u = P_{(m|r)}$ and $v = P_{(r|m)}$; the sequence probabilities are then given in terms of u and v by the expressions in Table 1.8. The probabilities of higher sequences are readily derived, e.g. the probability (Xm) is simply u(X) if Xm is symmetrical, or $2u$(X) if Xm is unsymmetrical. For this model, tetrad probabilities (at least) are necessary to confirm that it satisfactorily describes the sequence statistics.

Higher-order Markov models can also be applied [65]. In second-order Markov statistics, the probability of forming m or r depends on the structure of the previous two dyads. There is a total of eight conditional probabilities, of which four are independent. In order to confirm that this model is correct, it is necessary to have accurate pentad probabilities or longer.

Other propagation models are also possible. An example is an adaptation [48, 55, 65] of the Coleman–Fox [66] mechanism in which the polymerisation is considered to take place at two active sites, each with a different Bernoullian propagation probability (see chapter 2).

As an example of the statistical analysis of a tacticity sequence distribution, Table 1.9 gives the experimental distribution for poly(vinyl chloride) derived from Figure 1.19, together with the distribution calculated for Bernoullian statistics with $P_m = 0.456$. The agreement is within experimental error up to tetrads. A Bernoullian sequence distribution and a preference for syndio-

Table 1.9 Statistical analysis of the sequence distribution of free-radical poly(vinyl chloride); the calculated probabilities are for Bernoullian statistics with $P_m = 0.456$

Sequence	Probability	
	Experimental	Calculated
mm	0.21	0.21
mr	0.49	0.50
rr	0.30	0.30
mmm	0.08	0.10
mmr + *mrm*	0.32	0.34
rmr	0.15	0.14
mrr	0.26	0.27
rrr	0.19	0.16

tactic placement is characteristic of free-radical vinyl polymers. For singly substituted monomers $CH_2{=}CHX$, P_m is typically 0.4–0.5, but for poly(methyl methacrylate), P_m falls to about 0.25. Note that P_m may be temperature-dependent; an increase in temperature increases the probability of the less favoured placement [67]. For vinyl polymers prepared with ionic catalysts, the tacticity statistics frequently deviate from the Bernoullian model. For example, the sequence distribution for predominantly isotactic poly(methyl methacrylate) prepared using phenyl magnesium bromide was found to be consistent with either a second-order Markov model or with the Coleman–Fox model [65]. Presumably the dependence of the propagation probability on the tacticity of previous dyads is related to configuration-dependent complex formation between several monomer units and the metal ion.

1.8.4 Stereochemistry of poly(epoxides)

Poly(epoxides) are readily prepared by a ring-opening addition polymerisation catalysed by anionic, cationic or coordination catalysts. Consider the polymerisation of an epoxide bearing a single substituent:

$$n\left[\begin{array}{c}\text{R}\\ |\\ \text{CH}_2-\text{CH}\\ \diagdown\ \diagup\\ \text{O}\end{array}\right] \longrightarrow \left[\begin{array}{c}\text{R}\\ |\\ -\text{CH}_2-\text{CH}-\text{O}-\end{array}\right]_n$$

This chain contains a truly chiral centre and a chain with only a single configuration would be optically active. As far as NMR is concerned, it is the relative configuration that is important, and as for vinyl polymers, the stereochemical sequence structure may be described in terms of *m* (same configuration) and *r* (opposite configuration) dyad building blocks repre-

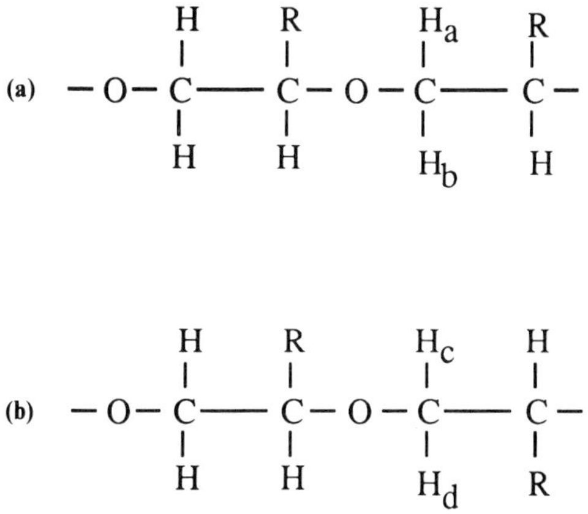

```
         H      R          Ha     R
         |      |          |      |
(a)  —O—C——————C—O—C——————C—
         |      |          |      |
         H      H          Hb     H


         H      R          Hc     H
         |      |          |      |
(b)  —O—C——————C—O—C——————C—
         |      |          |      |
         H      H          Hd     R
```

Figure 1.26 Dyad configurations of a poly(epoxide). (a) Isotactic; (b) syndiotactic.

sented as projection diagrams in Figure 1.26. Isotactic, syndiotactic and atactic chains would consist, respectively, of all *m* dyads, all *r* dyads, and an irregular sequence of *m* and *r* dyads. However, because of the additional oxygen atom, the tacticity of poly(epoxides) is manifest in NMR spectra in a number of different ways compared to vinyl polymers.

First, because the CHR carbon is truly chiral, the two protons in the CH_2 group are non-equivalent in both *m* and *r* dyads. It is therefore not possible to derive an absolute identification of tacticity from ^1H spectra as may be done for vinyl polymers. Second, because of the asymmetry in terms of chemical structure introduced by the additional backbone atom, no stereosequences of a particular length are identical. For example, the *mr* and *rm* configurations of a triad are no longer equivalent, so if a CH carbon responds to triad tacticity, four peaks would be expected. Third, again because of the extra backbone atom, the chemical shifts of both CH_2 and CHR groups may be sensitive to dyads, triads, tetrads, etc., depending on the structure of the side-group.

An example of tacticity effects in poly(epoxides) is shown in the backbone ^{13}C spectrum of anionically polymerised poly(butylene oxide) in Figure 1.27 [35]. The CH resonance is narrowly split into eight peaks of essentially equal intensity, consistent with chemical shift sensitivity at the tetrad level. Evidently the CH carbon senses the tacticity of its nearest neighbours in both directions in the chain, but senses the tacticity of its next-nearest neighbour in only one direction. The CH_2 carbon is split into four equal peaks, consistent with sensitivity to tacticity at the triad level. The fact that the triad and tetrad peaks are essentially equal in intensity indicates a perfectly random tacticity distribution, i.e., $P_m = 0.5$.

Unlike a vinyl monomer, the epoxide monomer is itself chiral. If the polymerisation proceeds by ring-opening at the CH_2–O bond, the chirality is

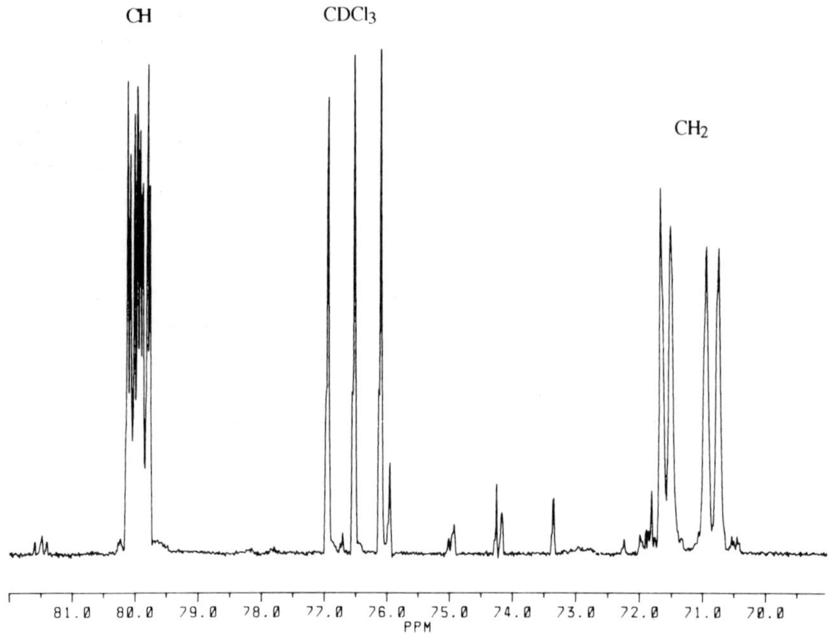

Figure 1.27 75.5 MHz ^{13}C spectrum of the backbone carbons of poly(butylene oxide) in $CDCl_3$. The minor peaks arise fron end-groups. Moderate resolution enhancement was applied.

preserved and an isotactic polymer will be produced from any catalyst if an enantiomerically pure monomer is used. This principle has been used to assist in the assignment of the ^{13}C spectrum of poly(propylene oxide) [32]. For an *R*/*S* monomer mixture, an isotactic polymer is produced if the polymerisation mechanism is enantiomer-selective. If the ring-opening proceeds by S_N2 substitution at the tertiary carbon, the configuration will be reversed. If the mechanism involves S_N1 substitution at the tertiary carbon, the configuration will be randomised.

1.8.5 Geometrical isomerism in polydienes

Cis/*trans* isomerism of backbone double bonds (more correctly *Z*/*E* isomerism in modern nomenclature) is a structural feature of polymers that may affect NMR spectra in a manner similar to tacticity in vinyl polymers. We will take the prototype chain, poly(1,4-butadiene) $(-CH_2CH{=}CHCH_2-)_n$ as an example. Denoting *cis* and *trans* isomers by *c* and *t*, the possible dyad, triad, etc. isomeric sequences can be constructed from *c* and *t* in the same way as *m* and *r* were used for vinyl polymer stereosequences. The possible dyads are *cc*, *ct*, *tc* and *tt*. In the *cc* and *tt* dyads, the two CH_2 carbons are equivalent whereas in the *ct* and *tc* dyads, the two CH_2 carbons are not equivalent.

Thus if the CH_2 ^{13}C chemical shifts depend on the structure of the two neighbouring double bonds, a total of four peaks would be expected, one assigned to each of *cc* and *tt*, and two assigned to *ct*/*tc*. Similarly, at the triad level, there are six distinct triads, four symmetrical (*ccc*, *tct*, *ctc* and *ttt*) and two unsymmetrical (*tcc*/*cct* and *ctt*/*ttc*). If the chemical shifts of the olefinic carbons of the central unit depend on triad structure, a total of eight peaks would be expected. This example shows that when relating NMR spectra to structure, due regard must be paid to the total number of distinct nuclear environments in a particular structure.

Experimentally, the chemical shifts of both ^1H [68] and ^{13}C [69, 70] nuclei in poly(1,4-butadiene) are found to depend almost entirely only on the structure at the monomer level. In the CH_2 region of the ^{13}C spectrum, *cis* units appear at 27.4 ppm and *trans* units at 32.8 ppm [69]. In the ^{13}C olefinic region, there is a slight dependence on sequence structure [70]; peaks from *cis* units at 128.8 ppm and *trans* units at 129.35 ppm are each split into a doublet of about 0.1 ppm due to a slight sensitivity to dyad structure. It should be noted that analysis of the spectra of poly(dienes) in terms of *cis*/*trans* isomerism is severely complicated by the occurrence of 1,2- as well as 1,4-addition.

1.9 Summary

In essence, the analytical power of NMR lies in the dependence of the nuclear resonance frequency on the chemical environment of the nucleus, thus allowing the characterisation of functional groups and sub-structures. In polymers, the features that may be identified include the chemical structure of the repeat unit, the structure of end-groups, the occurrence of inverted units (i.e. regio-regularity), chain branches, chain stereochemistry and the sequence structure of copolymers. Both ^1H and ^{13}C NMR are in widespread use, the former being much more sensitive than the latter, but suffering from a much lower dependence of resonance frequency on environment. NMR spectra frequently show fine structure, termed spin–spin coupling, arising from interactions between proximate nuclei, the form of the splitting giving information on the number of nuclei in neighbouring groups. Further detailed information on the structural relationship between different types of nuclei can be obtained by a variety of powerful pulse Fourier transform techniques. Subject to a few qualifications, especially in ^{13}C NMR, reliable quantitative information is readily obtainable from the resonance peak intensities.

In studying polymer stereochemistry, the form of the ^1H NMR spectrum gives a rapid, unequivocal identification of the structure of stereoregular vinyl polymers. For stereo-irregular polymers, the resonance peaks are frequently split into several lines from sequences of a few monomer residues of different stereochemistry. The relative peak intensities give statistical information on the stereochemical sequence distribution, leading to a detailed insight into the average chain structure that is not available from any other technique.

References

1. R.J. Abraham, J. Fisher and P. Loftus, *Introduction to NMR Spectroscopy*, Wiley, Chichester (1988).
2. J.W. Akitt, *NMR and Chemistry. An Introduction to Modern NMR Spectroscopy*, 3rd edition, Chapman & Hall, London (1992).
3. F.A. Bovey, L.W. Jelinski and P.A. Mirau, *Nuclear Magnetic Resonance Spectroscopy*, 2nd edition, Academic Press, San Diego (1988).
4. R.K. Harris, *Nuclear Magnetic Resonance Spectroscopy. A Physicochemical View*, Pitman, London (1983).
5. M.L. Martin, J.-J. Delpuech and G.J. Martin, *Practical NMR Spectroscopy*, Heyden, London (1980).
6. A.E. Derome, *Modern NMR Techniques for Chemistry Research*, Pergamon, Oxford (1987).
7. J.K.M. Sanders and B.K. Hunter, *Modern NMR Specctroscopy: A Guide for Chemists*, 2nd edition, Oxford University Press, Oxford (1993).
8. R.R. Ernst, G. Bodenhausen and A. Wokaun, *Principles of Nuclear Magnetic Resonance in One and Two Dimensions*, Oxford University Press, Oxford (1990).
9. W.S. Brey (Editor), *Pulse Methods in 1D and 2D Liquid-Phase NMR*, Academic Press, New York (1988).
10. H. Kessler, M. Gehrke and C. Griesinger, *Angew. Chem., Int. Ed. Engl.* **27** (1988) 490.
11. F.W. Wehrli, A.P. Marchand and S. Wehrli, *Interpretation of ^{13}C NMR Spectra*, Wiley, New York (1988).
12. E. Breitmeier and W. Voelter, *Carbon-13 NMR Spectroscopy*, 3rd edition, VCH Publishers, Weinheim, Germany (1987).
13. H.-O. Kalinowski, S. Berger and S. Braun, *Carbon-13 NMR Spectroscopy*, Wiley, Chichester (1988).
14. A.E. Tonelli, *NMR Spectroscopy and Polymer Microstructure. The Conformational Connection*, VCH Publishers, New York and Weinheim, Germany (1989).
15. F.A. Bovey and L.W. Jelinski, *Chain Structure and Conformation of Macromolecules*, Academic Press, New York (1983).
16. J.C. Randall, *Polymer Sequence Determination. Carbon-13 NMR Method*, Academic Press, New York (1977).
17. F.A. Bovey, Structure of chains by solution NMR spectroscopy, in *Comprehensive Polymer Science*, Vol. 1, eds. C. Booth and C. Price, Pergamon Press, Oxford (1989), Chapter 17.
18. F.A. Bovey and P.A. Mirau, *Acc. Chem. Res.* **21** (1988) 37.
19. L.F. Johnson and W.C. Jankowski, *Carbon-13 NMR Spectra*, Wiley, New York (1972).
20. W. Bremser, L. Ernst, W. Fachinger, R. Gerhards, A. Hardt and P.M.E. Lewis, *^{13}C NMR Spectral Data*, 4th edition, VCH Publishers, New York (1987) (microfiche).
21. Q.T. Pham, R. Petiaud and H. Wotan, *^{1}H and ^{13}C NMR Spectra of Polymers*, Vols. 1 and 2, Wiley, Chichester (1983).
22. E. Pretsch, T. Clerc, J. Seibl and W. Simon, *Tables of Spectral Data for Structure Determination of Organic Compounds*, Springer-Verlag, Berlin (1983).
23. C.J. Pouchert and J. Behnke, *The Aldrich Library of ^{13}C and ^{1}H FT-NMR Spectra*, Aldrich Chemical Company, Milwaukee (1992).
24. P.L. Corio, *Structure of High-Resolution NMR Spectra*, Academic Press, New York (1966).
25. A.D. English and O.T. Garza, *Macromolecules* **12** (1979) 352.
26. D.L. Rabenstein and T.T. Nakashima, *Anal. Chem.* **51** (1979) 1465A.
27. C. Lecocq and J.-Y. Lallemand, *J. Chem. Soc., Chem. Commun.* (1981) 150.
28. S.L. Patt and J.N. Shoolery, *J. Magn. Reson.* **46** (1982) 535.
29. D.M. Doddrell, D.T. Pegg and M.R. Bendall, *J. Chem. Phys.* **77** (1982) 2745.
30. O.W. Sørensen, S. Donstrup, H. Bildsøe and H.J. Jakobsen, *J. Magn. Reson.* **55** (1983) 347.
31. D.T. Pegg and M.R. Bendall, *J. Magn. Reson.* **60** (1984) 347.
32. F.C. Schilling and A.E. Tonelli, *Macromolecules* **19** (1986) 1337.

33. F. Heatley, Y.-Z. Luo, J.-F. Ding, R.H. Mobbs and C. Booth, *Macromolecules* **21** (1988) 2713.
34. R.A. Newmark, *Appl. Spectrosc.* **39** (1985) 507.
35. F. Heatley, G.-E. Yu, W.-B. Sun, E.J. Pywell, R.H. Mobbs and C. Booth, *Eur. Polym. J.* **26** (1990) 583.
36. F. Heatley, G.-E. Yu, M.D. Draper and C. Booth, *Eur. Polym. J.* **27** (1991) 471.
37. J.E. Meier and A.G. Marshall, *Biol. Magn. Reson.* **9** (1990) 199.
38. A.G. Redfield, S.D. Kunz and E.K. Ralph, *J. Magn. Reson.* **19** (1975) 114.
39. P. Plateau and M. Gueron, *J. Am. Chem. Soc.* **104** (1982) 7310.
40. P.J. Hore, *J. Magn. Reson.* **55** (1983) 283.
41. A.G. Ferrige and J.C. Lindon, *J. Magn. Reson.* **31** (1978) 337.
42. A. De Marco and K. Wüthrich, *J. Magn. Reson.* **24** (1976) 201.
43. M. Guéron, *J. Magn. Reson.* **30** (1978) 515.
44. I. Ando, T. Yamanobe and T. Asakura, *Prog. NMR Spectrosc.* **22** (1990) 349.
45. P.A. Mirau, F.A. Bovey, A.E. Tonelli and S.A. Heffner, *Macromolecules* **20** (1987) 1701.
46. F.A. Bovey and G.V.D. Tiers, *J. Polym. Sci.* **44** (1960) 173.
47. M. Farina, *Topics Stereochem.* **17** (1987) 1.
48. H.L. Frisch, C.L. Mallows and F.A. Bovey, *J. Chem. Phys.* **45** (1966) 1565.
49. R.C. Ferguson, *Am. Chem. Soc., Div. Polym. Chem., Polym. Preprints* **8**(2) (1967) 1026.
50. F. Heatley and F.A. Bovey, *Macromolecules* **1** (1968) 303.
51. F.A. Bovey, F.P. Hood, E.W. Anderson and L.C. Snyder, *J. Chem. Phys.* **42** (1965) 3900.
52. F. Heatley and F.A. Bovey, *Macromolecules* **1** (1968) 301.
53. A.E. Tonelli and F.C. Schilling, *Acc. Chem. Res.* **14** (1981) 233.
54. C.J. Carman, *Macromolecules* **6** (1973) 725.
55. I.R. Peat and W.F. Reynolds, *Tetrahedron Lett.* **14** (1972) 1359.
56. G. Di Silvestro, P. Sozzani, G. Savare and M. Farina, *Macromolecules* **18** (1985) 928.
57. D.W. Ovenall, *Macromolecules* **17** (1984) 1458.
58. F.C. Schilling, F.A. Bovey, M.D. Bruch and S.A. Kozlostetski, *Macromolecules* **18** (1985) 1418.
59. F. Heatley and F.A. Bovey, *Macromolecules* **2** (1969) 241.
60. J.C. Randall, *J. Polym. Sci., Polym. Phys. Ed.* **13** (1975) 889.
61. K. Matsuzaki, T. Uryu, K. Osada and T. Kawamura, *Macromolecules* **5** (1972) 816.
62. G. Moad, E. Rizzardo, D.H. Solomon, S.R. Johns and R.I. Willing, *Macromolecules* **19** (1986) 2494.
63. M.W. Crowther, N.M. Szeverenyi and G.C. Levy *Macromolecules* **19** (1986) 1333.
64. K. Hikichi and M. Yasuda, *Polym. J.* **19** (1987) 1003.
65. H.L. Frisch, C.L. Mallows, F. Heatley and F.A. Bovey, *Macromolecules* **1** (1968) 533.
66. B.D. Coleman and T.G. Fox, *J. Chem. Phys.* **38** (1963) 1065; *J. Am. Chem. Soc.* **85** (1963) 1241.
67. F.A. Bovey, F.P. Hood, E.W. Anderson and R.L. Kornegay, *J. Phys. Chem.* **71** (1967) 312.
68. H. Sato and Y. Tanaka, *J. Polym. Sci., Polym. Chem. Ed.* **17** (1979) 3551.
69. H. Sato, K. Takebayashi and Y. Tanaka, *Macromolecules* **20** (1987) 2418.
70. K.F. Elgert, G. Quack and B. Stützel, *Polymer* **16** (1975) 154.

2 Statistical analysis of copolymer sequence distribution

I.R. HERBERT

2.1 Introduction

The properties of copolymers depend not only on the comonomer composition but also on the sequence distribution of the constituent comonomers, which can range from alternating through random to blocky. Techniques that are able to determine comonomer sequence distribution are therefore of special interest to the polymer chemist. It has already been revealed in chapter 1 that NMR spectroscopy is particularly sensitive to stereochemical sequence distribution, and it is clear that by analogous procedures, it can also be used to count the number of different types of monomer sequences present in a copolymer (e.g. AAA, AAB, BAB,...). A knowledge of the relative abundances of monomer sequences provides a means of characterising the average microstructure of a copolymer. Thus, for example, from the relative abundances of the three dyad sequences in an AB copolymer (namely, AA, AB and BB) the number-average sequence lengths for the A and B comonomers can be determined immediately. This sequence information describes the degree to which the comonomers are incorporated in the polymer in random, alternating or blocky fashion.

As an extension to this type of sequence information, it has become common practice to check for conformity of the observed sequence distribution to an appropriate statistical model. The rationale for this is that, as well as providing a more complete description of copolymer microstructure, conformity to a particular model can impart important information concerning the mechanism of polymer propagation. In addition, in favourable cases, the relative reactivities of the comonomers can also be established. The more traditional methods for assessing comonomer reactivities are significantly more laborious than NMR methods and often relatively crude.

Some polymer systems of both academic and commercial interest are derived from chemical modification of preformed polymers. For example, poly(vinyl alcohol) is prepared by hydrolysis of poly(vinyl acetate) because vinyl alcohol monomer does not exist. NMR spectroscopy again plays an important role in characterising such materials. In cases where chemical modification is incomplete, these systems fall into the category of copolymers comprising modified and unmodified monomer repeat units. Statistical analyses of the

sequence distribution of such copolymers can lead to important conclusions concerning the mechanisms and kinetics of chemical modification.

The purpose of this chapter, therefore, is to describe the basic concepts of the statistical analysis of copolymer sequence distribution. The necessary relationships between various comonomer sequence abundances are introduced, along with simple statistical models based on monomer addition probabilities. The relationships between the statistical models and propagation models based on reactivity ratios are discussed. The use of these models is then illustrated by means of selected examples. Techniques for extracting sequence information from *in situ* NMR measurements are also described. Finally, the statistical analysis of chemically modified polymers is introduced with examples.

2.2 Copolymerisation statistics and models

The development of statistical analyses of polymers was pioneered by Bovey [1, 2] and Price [3] and has also been discussed in detail by Randall [4]. Here, the intention is to discuss only the principal elements of copolymer statistics required for the analysis of the various sequence abundances observed using NMR spectroscopy. For the full derivation of copolymer statistics, the interested reader is referred to the original work of these authors. It should also be noted that most of what follows is limited to A/B copolymer systems. Whilst it is relatively straightforward to extend the analysis to copolymers containing more than two comonomers, the number of equations required to define the system increases dramatically. Furthermore, for such systems, it proves increasingly difficult to assign their NMR spectra in the detail required for complete analysis.

Before proceeding, it is worth while outlining once more the possible types of copolymer chain. Comonomer units in an A/B copolymer may be incorporated in random fashion, regular alternating fashion or in block structures, as indicated in Figure 2.1. Of course, these copolymer types represent 'extremes' of structure. In practice, many so-called random or statistical copolymers actually show some tendency for their individual comonomers to alternate or block together. In this chapter, we are primarily concerned with these less

Random —B—A—B—B—A—B—A—A—A—B—B—A—A—B—

Alternating —A—B—A—B—A—B—A—B—A—B—A—B—A—B—

Block —A—A—A—A—A—A—A—B—B—B—B—B—B—B—

Figure 2.1 Types of A/B copolymer.

than idealised types of copolymer. Copolymers often give rise to NMR spectra that are markedly more complex than those expected from a summation of the corresponding homopolymer spectra. It is this additional complexity which provides a basis for the measurement of the average of sequence distribution of the mixed polymer chains. In addition to comonomer sequence, it should be noted that copolymer spectra can also exhibit effects due to tacticity, geometrical isomerism and regiosequence. These complications to copolymer statistical analysis are illustrated by examples where appropriate. Whilst technologically important, block copolymers are not discussed further since they do not possess a distribution of sequence types.

2.2.1 Copolymer number-average sequence lengths and necessary relationships

As shown in Figure 2.1, any copolymer can be described by a sequential arrangement of A and B units. Such an arrangement can be divided up into combinations of two, three, or more units, corresponding to dyad, triad and higher sequences. For an A/B copolymer, there are three types of dyad sequence, AA, BB, and AB. (Note that the heterogeneous AB dyad is equivalent to the BA dyad, given that the polymer chain direction normally cannot be specified. Any reference to the AB dyad will be taken to mean AB + BA dyad.) Similarly, there are six possible triad sequences, AAA, AAB, BAB, ABA, ABB, and BBB. The ability of NMR spectroscopy to measure the concentrations of either these dyad or triad (or sometimes higher *n*-ad) sequences enables characterisation of the chain microstructure in terms of the number-average sequence lengths of monomers A and B (i.e. the average length of sequences of adjacent A or B units).

Consider a model chain segment comprising equal amounts of monomers A and B:

–A–B–A–A–B–A–B–A–A–A–B–A–B–B–A–B–B–B–A–B–

Inspection reveals that there are 12 AB dyads, 3 AA dyads and 3 BB dyads (remembering that AB is equivalent to BA). It can be shown that the number-average sequence lengths, $\tilde{n}_A$ and $\tilde{n}_B$ (simply the average length of sequences comprising only A and only B units, respectively), for any copolymer are given by

$$\tilde{n}_A = \frac{2N_{AA} + N_{AB}}{N_{AB}} \tag{2.1}$$

$$\tilde{n}_B = \frac{2N_{BB} + N_{AB}}{N_{AB}} \tag{2.2}$$

where N is the abundance of the sequence denoted by the subscript. Thus, for the model chain segment, $\tilde{n}_A = 18/12 = 1.5$. Similarly, $\tilde{n}_B$ is also 1.5. It can also be shown that for a truly random chain

$$\tilde{n}_A = \frac{N_A + N_B}{N_B} \tag{2.3}$$

$$\tilde{n}_B = \frac{N_A + N_B}{N_A} \tag{2.4}$$

where the terms N refer to the abundances of monomer units A and B [4]. Thus, the number-average sequence lengths, $\tilde{n}_A$ and $\tilde{n}_B$, for a 50/50 purely random copolymer are both equal to 2.0. Comparison of actual number-average sequence lengths with those for the truly random case therefore provides a measure of the tendency of a copolymer to deviate from randomness. For the model chain segment, number-average sequence lengths of less than 2.0 indicate that there is a tendency towards alternation of the two comonomers since in the limiting case of perfectly alternating 50/50 copolymer, the two number-average sequence lengths would both be unity. In contrast, number-average sequence lengths greater than 2.0 for a 50/50 copolymer would indicate a tendency for the comonomers to form blocks. Clearly, therefore, calculation of number-average sequence lengths from NMR data serves as a simple measure of copolymer microstructure, without recourse to any form of statistical model to describe polymerisation.

If comonomer and sequence abundances are expressed as mole fractions, equations (2.1) and (2.2) reduce to the following:

$$\tilde{n}_A = \frac{2(A)}{(AB)} \tag{2.5}$$

$$\tilde{n}_B = \frac{2(B)}{(AB)} \tag{2.6}$$

where (AB) is the AB dyad abundance expressed as a fraction of total dyad abundance. (For the remainder of the chapter, any sequence shown in parentheses is taken as the fractional abundance of that sequence.)

A similar set of relationships between triad abundances and number-average sequence lengths has also been derived [1–4]. The derivation of number-average sequence lengths in terms of dyad and triad distributions is intimately related to a range of necessary relationships between the various monomer and sequence abundances. Some of these are given in Table 2.1, it being noted that they are analogous to the stereochemical *n*-ad relationships in chapter 1. These relationships provide a means for checking the validity of sequence abundances as determined by NMR spectroscopy, and can prove particularly useful in cases where there are uncertainties over NMR peak assignments.

Table 2.1 Some necessary relationships for copolymers [4]

$$(\mathrm{A}) = (\mathrm{AA}) + \frac{(\mathrm{AB})}{2} = (\mathrm{AAA}) + (\mathrm{AAB}) + (\mathrm{BAB})$$

$$(\mathrm{B}) = (\mathrm{BB}) + \frac{(\mathrm{AB})}{2} = (\mathrm{BBB}) + (\mathrm{ABB}) + (\mathrm{ABA})$$

$$(\mathrm{AA}) = (\mathrm{AAA}) + \frac{(\mathrm{AAB})}{2}$$

$$(\mathrm{BB}) = (\mathrm{BBB}) + \frac{(\mathrm{ABB})}{2}$$

$$(\mathrm{AB}) = 2(\mathrm{ABA}) + (\mathrm{ABB}) = 2(\mathrm{BAB}) + (\mathrm{AAB})$$

2.2.2 Statistical models

In order to derive further information from NMR sequence data, it is necessary to relate the sequence distribution to some form of statistical model describing copolymerisation. The basis of these models is the concept of monomer addition probabilities. It will be apparent that some of these models are closely related to those used to describe configurational effects in homopolymers (e.g. see chapter 1).

The Bernoullian model. The Bernoullian model is the simplest of the statistical models used to describe copolymerisation, whereby the addition of each comonomer to a growing polymer chain is regarded as a random process. Thus, the framework for the model is that the probability of addition of a given monomer unit to a growing polymer chain is only dependent upon the mole fraction of the monomer in the feed.

$$-\bullet + \mathrm{A} \xrightarrow{P_\mathrm{A}} -\mathrm{A}\bullet$$

$$-\bullet + \mathrm{B} \xrightarrow{P_\mathrm{B}} -\mathrm{B}\bullet$$

The *Bernoullian process* is therefore defined for a copolymer by two transition probabilities, P_A and P_B, which reflect the mole fractions of monomers A and B within the resulting copolymer. Given the two addition probabilities, the mole fraction of any given sequence can be calculated straightforwardly. Thus, for example, the abundance of an AA dyad is given by P_A^2, whilst that of an AB dyad is equal to $2P_\mathrm{A}P_\mathrm{B}$ (the factor of two arises because the AB dyad represents both AB and BA sequences). Table 2.2 shows the set of Bernoullian expressions for the three dyad and six triad sequences in an A/B copolymer.

Given a set of sequence distribution data from NMR measurements on a copolymer, it is normal practice to test how well that set matches the data

Table 2.2 Bernoullian model dyad and triad sequence abundance [4]

$P_A + P_B = 1$

Dyads
$(AA) = P_A^2$
$(AB) = 2P_A(1 - P_A)$
$(BB) = (1 - P_A)^2$

Triads
$(AAA) = P_A^3$
$(AAB) = 2P_A^2(1 - P_A)$
$(BAB) = P_A(1 - P_A)^2$
$(ABA) = P_A^2(1 - P_A)$
$(ABB) = 2P_A(1 - P_A)^2$
$(BBB) = (1 - P_A)^3$

calculated from monomer mole fractions using Bernoullian expressions, such as those given in Table 2.2. (These are analogous to the expressions for Bernoullian stereochemical sequence distributions in chapter 1.)

Only one independent variable is necessary to describe the Bernoullian process in full since the sum of P_A and P_B is equal to unity. Dyad distribution data contain two independent observations which is sufficient to check for conformity to *Bernoullian statistics*. (Although there are three dyad types, the mole fraction of one of them is always defined by the mole fractions of the other two since the mole fraction of the total dyad is, of course, equal to one, i.e. there are only two independent observations.) However, a better test for conformity is found in a triad distribution since this contains five independent observations (there are six triads).

It is important to note here that random copolymers which show conformity to Bernoullian statistics can result from *non-Bernoullian processes* under some sets of circumstances. In order to distinguish between random copolymers resulting from Bernoullian and non-Bernoullian processes, it is necessary to have access to additional information, e.g. comonomer feed information, copolymer composition versus conversion data, and so on. Examples of random copolymers resulting from non-Bernoullian processes are given in section 2.3.

When Bernoullian statistics apply, the expressions for number-average sequence lengths, equations (2.5) and (2.6), reduce simply to

$$\tilde{n}_A = \frac{1}{P_B} \tag{2.7}$$

$$\tilde{n}_B = \frac{1}{P_A} \tag{2.8}$$

The first-order Markov model. In the first-over Markov model, the probability of monomer addition depends upon the identity of the preceding monomer unit. Hence, for an A/B copolymer, there are four possible propagation steps defined by four monomer addition probabilities:

$$—\mathrm{A}\bullet + \mathrm{A} \xrightarrow{P_{AA}} —\mathrm{AA}\bullet$$

$$—\mathrm{A}\bullet + \mathrm{B} \xrightarrow{P_{AB}} —\mathrm{AB}\bullet$$

$$—\mathrm{B}\bullet + \mathrm{A} \xrightarrow{P_{BA}} —\mathrm{BA}\bullet$$

$$—\mathrm{B}\bullet + \mathrm{B} \xrightarrow{P_{BB}} —\mathrm{BB}\bullet$$

Here, P_{AA} represents the probability of an A unit adding to an A chain end, P_{AB} is the probability of a B monomer adding to an A chain end, and so on. Since an A chain end can only add an A or B monomer, then the following normalisation condition applies:

$$P_{AA} + P_{AB} = 1 \tag{2.9}$$

Similarly,

$$P_{BB} + P_{BA} = 1 \tag{2.10}$$

First-order Markov processes are therefore defined by two independent addition probabilities. Although the propagation steps shown above depict free radical polymerisation, the statistical models are equally applicable to other types of chain growth as found, for example, in ionic and Ziegler–Natta polymers (see section 2.3.4).

It can be shown [4] that the probabilities of finding either A or B monomer units (i.e. the mole fractions of A and B) in a first-order Markov chain are given by

$$(\mathrm{A}) = \frac{P_{BA}}{P_{AB} + P_{BA}} \tag{2.11}$$

and

$$(\mathrm{B}) = \frac{P_{AB}}{P_{AB} + P_{BA}} \tag{2.12}$$

The mole fraction of any given sequence can therefore now be determined. Take an AA dyad, for example. This arises from addition of an A unit to an A chain end. The probability of an A chain end is given by equation (2.11) and the probability of an A unit adding to an A unit is P_{AA} which is equivalent to $(1 - P_{AB})$. Hence, the mole fraction of the AA dyad is simply the product of these two probabilities:

$$(\mathrm{AA}) = \frac{P_{BA}(1 - P_{AB})}{P_{AB} + P_{BA}} \tag{2.13}$$

Table 2.3 First-order Markov model dyad and triad sequence abundances [4]

$P_{AA} + P_{AB} = 1, \quad P_{BB} + P_{BA} = 1$

Dyads

$$(AA) = \frac{P_{BA}(1 - P_{AB})}{(P_{AB} + P_{BA})}$$

$$(AB) = \frac{2P_{BA}P_{AB}}{(P_{AB} + P_{BA})}$$

$$(BB) = \frac{P_{AB}(1 - P_{BA})}{(P_{AB} + P_{BA})}$$

Triads

$$(AAA) = \frac{P_{BA}(1 - P_{AB})^2}{(P_{AB} + P_{BA})} \qquad (ABA) = \frac{P_{AB}P_{BA}^2}{(P_{AB} + P_{BA})}$$

$$(AAB) = \frac{2P_{AB}P_{BA}(1 - P_{AB})}{(P_{AB} + P_{BA})} \qquad (ABB) = \frac{2P_{AB}P_{BA}(1 - P_{BA})}{(P_{AB} + P_{BA})}$$

$$(BAB) = \frac{P_{AB}^2 P_{BA}}{(P_{AB} + P_{BA})} \qquad (BBB) = \frac{P_{AB}(1 - P_{BA})^2}{(P_{AB} + P_{BA})}$$

Expressions for the mole fractions of longer sequences can be built up in a similar fashion. Expressions for first-order Markov dyad and triad distributions are given in Table 2.3, and can be compared with the analogous expressions for stereochemical sequence distributions in chapter 1.

As with the Bernoullian model, comparison between an observed and calculated sequence distribution is required to check for conformity to first-order Markov statistics. Obviously, with only two independent observations, a dyad distribution is insufficient for determining the two independent probabilities of the model. In contrast, a triad distribution provides five independent observations, so this can be used to check conformity to first-order Markov statistics. Trial values of the monomer addition probabilities can be obtained by taking appropriate combinations of the expressions shown in Table 2.3. For example, P_{AB} is given by

$$P_{AB} = \frac{(AAB)}{2(AAA) + (AAB)} \tag{2.14}$$

These trial values are then used to calculate the full triad (or higher n-ad) distribution for comparison with the observed distribution.

Number-average sequence lengths for a first-order Markov copolymer are as follows:

$$\bar{n}_A = \frac{1}{P_{AB}} \tag{2.15}$$

and

$$\tilde{n}_B = \frac{1}{P_{BA}} \tag{2.16}$$

The four comonomer addition probabilities prove to be useful characteristics since they define the tendency of a copolymer towards either block formation or alternation. For example, in the case of $P_{AB} > P_{BB}$ and $P_{BA} > P_{AA}$, the monomers tend to add to the growing chain in an alternating fashion. If the above inequalities are reversed, like sequences of additions are preferred such that the comonomers tend to form blocks. Finally, there is a third case: if $P_{AB} = P_{BB}$ and $P_{BA} = P_{AA}$, the system reduces to Bernoullian statistics.

Other statistical models. The second-order Markov model is sometimes also applied to copolymers. Here, the probability of addition of a given monomer depends not only on the identity of the chain end monomer, but also on the nature of the preceding or penultimate monomer unit. As there are then four possible types of chain end to consider (namely, –AA, –AB, –BA, and –BB), there are eight addition probabilities which describe addition of the A and B monomers (e.g. P_{BAB} represents the probability of B adding to a –BA chain end). As with the first-order Markov case, only half of these are independent because $(P_{AAA} + P_{AAB}) = 1$, $(P_{ABA} + P_{ABB}) = 1$, and so on. Equations for determining the various sequence mole fractions are not given here; for further details see [5]. An example of a polymer conforming to this model is discussed in section 2.3. Higher-order Markov models can also be developed.

Other models sometimes invoked include two-component models based on combinations of the models discussed above [5] and the complex participation model [6]. Examples of the use of these are given in section 2.3. The complex participation model is a modification of the first-order Markov model to take account of the formation of A–B comonomer complexes which compete with monomer during polymerisation. Thus, four propagation steps in addition to those shown for the first-order Markov model are required to describe addition of A–B and B–A complexes (i.e. it can add either way round) to the two types of growing chain and (either A or B). The monomer and comonomer complex addition probabilities are then related to the equilibrium constant for complex formation. As might be expected, this model has been applied particularly to systems that show a marked tendency towards alternation of their comonomers [7]. A probabilistic description of the complex participation model has been given by Cais *et al.* [6].

2.2.3 Statistical models and polymer propagation

The statistical models discussed above were developed essentially as a means of describing polymer microstructure as observed by NMR spectroscopy. However, it is certainly no coincidence that these models are intimately related to more traditional 'kinetic' models describing polymer chain propagation. Thus, the first-order Markov model is simply a re-statement of the terminal model for copolymerisation proposed independently in 1944 by Mayo and Lewis [8], and by Alfrey and Goldfinger [9]. Similarly, the second-order Markov model is equivalent to the penultimate model for copolymer propagation. The Bernoullian statistical model merely corresponds to a special case of the terminal model.

It is worth while examining here the links between these two formalisms for polymer propagation (i.e. probabilistic and kinetic) because, under appropriate conditions, NMR-derived sequence distributions can be used to determine the relative reactivities of pairs of comonomers. Whilst more traditional techniques are available to do this, they prove to be very laborious and rather insensitive. In addition, of course, the NMR method can also provide insight into the mechanism of copolymer propagation often not available using traditional methods.

Since the terminal model is the one most frequently used to describe copolymerisation, its relationship with the corresponding first-order Markov model is examined in some detail.

In general, during copolymerisation, the two reacting monomers are not incorporated into the polymer chain in the same proportion as in the initial comonomer mixture. The reason for this is that the two monomers generally have different reactivities towards the growing chain ends. In the terminal model, it is assumed that the nature of the monomer at the growing chain end determines relative comonomer reactivity. In other words, there are four types of propagating step, each defined by a different rate constant:

$$\text{—A}\bullet + \text{A} \xrightarrow{k_{AA}} \text{—AA}\bullet$$

$$\text{—A}\bullet + \text{B} \xrightarrow{k_{AB}} \text{—AB}\bullet$$

$$\text{—B}\bullet + \text{A} \xrightarrow{k_{BA}} \text{—BA}\bullet$$

$$\text{—B}\bullet + \text{B} \xrightarrow{k_{BB}} \text{—BB}\bullet$$

These propagation steps are entirely analogous to those given for the first-order Markov model, except that addition probabilities have been replaced by rate constants. Mayo and Lewis derived the following differential equation to describe terminal model copolymerisation [8]:

$$\frac{\mathrm{d}A}{\mathrm{d}B} = \frac{A\,(r_A A + B)}{B\,(r_B B + A)} \tag{2.17}$$

where A and B represent the number of moles of the two monomers in the reaction mixture, and r_A and r_B are comonomer reactivity ratios defined as

$$r_A = \frac{k_{AA}}{k_{AB}} \tag{2.18}$$

$$r_B = \frac{k_{BB}}{k_{BA}} \tag{2.19}$$

The differential term, dA/dB, in equation (2.17) gives the ratio of the rates at which the two monomers enter the polymer. In other words, it represents the comonomer composition of polymer chains forming at any given instant.

The comonomer reactivity ratios are especially useful characteristics for a pair of monomers since a knowledge of their values allows polymer composition and microstructure to be predicted over the full range of monomer feeds. It can be shown that the monomer addition probabilities of the first-order Markov model are related to the reactivity ratios by the following expressions:

$$P_{AB} = \frac{1}{1 + r_A x} \tag{2.20}$$

$$P_{BA} = \frac{1}{1 + \dfrac{r_B}{x}} \tag{2.21}$$

where x represents the mole ratio of the two monomers, A/B, in the reaction mixture. In conclusion, then, if the sequence distribution of a copolymer conforms to first order Markov statistics, its comonomer reactivity ratios can be established via equation (2.20) and (2.21). However, there is one extremely important condition which has to apply for this to be true. The monomer addition probabilities, as defined above, depend upon the mole ratio, x, of the two comonomers. Since two comonomers will generally exhibit differing reactivities, the reaction mixture will become depleted in the more reactive monomer, i.e. the ratio x will change during the course of polymerisation. This drift in monomer feed composition is well known. Hence, the monomer addition probabilities will also change throughout the polymerisation. For this reason, reactivity ratios determined from sequence distribution data are only valid for polymers where monomer conversion has been restricted to less than approximately 5% (i.e. where drift in monomer feed composition is not significant). At higher conversions, the copolymer composition and microstructure become increasingly more heterogeneous. In these cases, the sequence distribution determined by NMR spectroscopy represents only some average view of the microstructure, taken over an ensemble of differing polymer chains. The observed sequence distribution therefore belies the true nature of the microstructure of the polymer. These points are sometimes overlooked in NMR studies of sequence distribution. The ability to fit an observed sequence distribution to, say, the first-order Markov model does not imply that the polymer propagation conforms to the terminal model; that is only valid if the polymer has been prepared under the constraint of low conversion or special steps have been taken to maintain a constant monomer ratio during polymerisation.

Equations (2.20) and (2.21) can also be used, of course, to determine comonomer addition probabilities from reactivity ratios derived by some other means. From these, copolymer sequence distribution can be predicted using expressions such as those in Table 2.3.

For the terminal model, some of the limiting cases for reactivity ratios and addition probabilities, corresponding to the three 'extremes' of copolymer type, are shown below:

$$\begin{array}{lll} \text{Alternating:} & r_A = r_B = 0, & P_{AB} = P_{BA} = 1 \\ \text{Random:} & r_A r_B = 1, & P_{AB} + P_{BA} = 1 \\ \text{Block:} & r_A r_B > 1, & P_{AB} = P_{BA} = 0 \end{array}$$

Note, in addition, that the special case $r_A = r_B = 1$ corresponds to a random copolymer formed by a Bernoullian process.

Relationships between second-order Markov addition probabilities and penultimate model reactivity ratios have also been derived and can be found in [5]. Similarly, for the complex participation model, Cais *et al.* have derived expressions which enable calculation of any given sequence from a knowledge of reactivity ratios and the equilibrium constant for complex formation determined by other methods [6].

In the next section, examples are given of how these statistical models have been used to examine NMR-derived sequence information so revealing the details of copolymer microstructure and propagation.

2.3 Examples of the use of copolymer statistics

This section is not intended as an exhaustive review of the literature concerned with NMR studies of copolymers. Instead, examples have been chosen to reflect and reinforce the principles described in the preceding section. It seems logical here to treat these examples approximately in the order in which the statistical models were introduced.

2.3.1 Copolymers with Bernoullian sequence distributions

As stated earlier, copolymers whose sequence distributions can be described by Bernoullian statistics (i.e. so-called random or statistical copolymers) can result either from Bernoullian or non-Bernoullian processes. Those formed by genuine Bernoullian mechanisms are relatively rare, reflecting the likelihood of a pair of comonomers possessing similar reactivities towards like and unlike chain ends. A particularly interesting example is a series of poly-(acrylamide-co-sodium acrylates) prepared in inverse micro-emulsions [10]. Here, polymer composition and triad sequence distribution were examined as a function of conversion for a range of starting comonomer ratios, using carbon-13 NMR spectroscopy. Average copolymer composition was shown to be independent of the degree of conversion in contrast to expectations based on reactivity ratios derived from separate inverse emulsion polymerisation studies. Moreover, the observed triad distributions conformed to the same Bernoullian statistics at both low and high conversions, as shown in Figure 2.2. The comonomer reactivity ratios are therefore close to unity. The implication of these observations is that the Bernoullian behaviour is the result of a true Bernoullian mechanism for polymerisation. Candau *et al.* [10] were able to conclude from this work that, in these inverse micro-emulsions, nucleated polymerising particles grow by collision with non-nucleated monomer micelles rather than by diffusion of the monomers through the organic phase.

Several examples of NMR studies of copolymers that exhibit Bernoullian sequence distributions but arise from non-Bernoullian mechanisms have been reported. Komoroski and Schockcor [11], for example, have characterised a range of commercial vinyl chloride (VC)/vinylidene chloride (VDC) copolymers using carbon-13 NMR spectroscopy. Although these polymers were prepared to high conversion, the monomer feed was continuously adjusted to maintain a constant comonomer composition. Full triad sequence distributions were determined for each sample. These were then compared with distributions calculated using Bernoullian and first-order Markov statistics; the better match was observed with the former. Independent studies on the variation of copolymer composition with feed composition have indicated that the VDC/VC system exhibits terminal model behaviour, with reactivity ratios $r_{VDC} = 3.2$ and $r_{VC} = 0.3$ [12]. As the product of these reactivity ratios is close to unity, sequence distributions that are approximately Bernoullian are expected.

Copolymers of acrylamide with acrylic acid and with a quaternary ammonium acrylate, both types prepared in solution to high conversion, have also been investigated using carbon-13 NMR spectroscopy [13, 14]. The two systems give rise to Bernoullian-like triad sequence distributions, contrary to expectations based on known reactivity ratios. In the case of the acrylamide/quaternary ammonium acrylate system, correlations of copolymer composition versus feed composition indicate that the terminal model applies and that both of the reactivity ratios are significantly less than one [15]. Similar results have also been found for the acrylamide/acrylic acid system [13]. In both of the NMR studies, the observed Bernoullian-like sequence distributions were attributed to compositional heterogeneity resulting from the high degree of monomer to polymer conversion. As expected, therefore, the average NMR

(a)

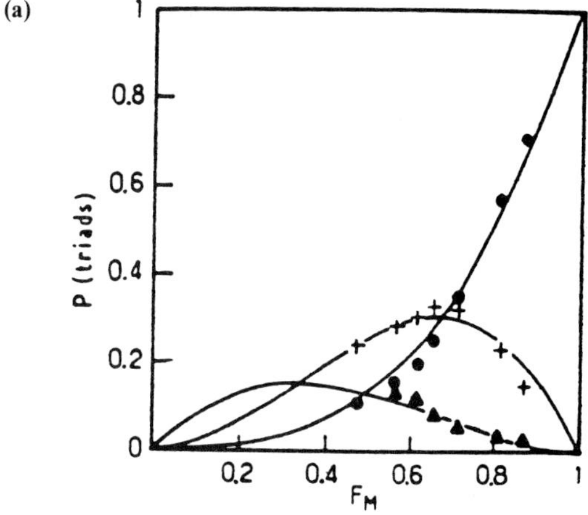

(b)

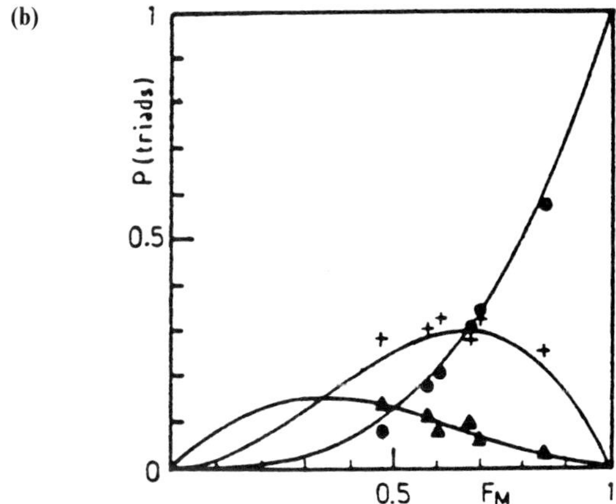

Figure 2.2 Proportions of acrylamide-centred triad sequences for a series of copolymers of sodium acrylate (A) and acrylamide (M), prepared in inverse microemulsions. (a) High conversion polymers; (b) low conversion polymers. The curves represent triad proportions calculated using Bernoullian statistics; symbols represent experimental triad data: (▲) AMA; (+) AMM; (●) MMM; F_M is the mole fraction of M in the copolymer. Reprinted with permission from [10]. © (1986) American Chemical Society.

view of a copolymer prepared to high conversion bears little resemblance to the true, often highly heterogeneous, nature of its microstructure.

2.3.2 Copolymers with first-order Markov sequence distributions

The vast majority of copolymers described in the literature conform to the terminal model for copolymerisation and therefore exhibit sequence distributions which will in principle conform to first-order Markov statistics. Of the copolymers that have had their microstructure characterised using NMR techniques, many are based on acrylate or methacrylate comonomers [16–19]. The microstructure of such systems is then determined, not only by sequence, but also by tacticity effects (see chapter 1). Whilst this adds an extra level of complexity to analysis of sequence distribution, the additional effect of tacticity can usually be dealt with relatively simply. (It is noteworthy that the main limitation to this type of study is often the overlap of sequence and tacticity effects in the NMR spectra of such systems which severely hinders peak assignment.)

Román and Valero [17] have examined the microstructure of ethyl acrylate/methyl methacrylate (A/M) copolymers using carbon-13 NMR spectroscopy. Here, there are ten distinct M-centred triads, as shown in Figure 2.3, because of the effects of tacticity [2]. Likewise, ten A-centred sequences also need to be considered. Note that the normal degeneracy of asymmetrical sequences (e.g. AMM/MMA) is lifted in the case of *mr* tacticity. A series of polymers, covering a range of monomer feed ratios and all prepared to low conversion, was studied. Whilst it is not possible to observe distinct signals for all 20 triads, a sufficient number of them (most notably via the A and M carbonyl resonances and the M α-methyl resonances) can be resolved and assigned, thereby allowing a detailed examination of the polymerisation statistics. The

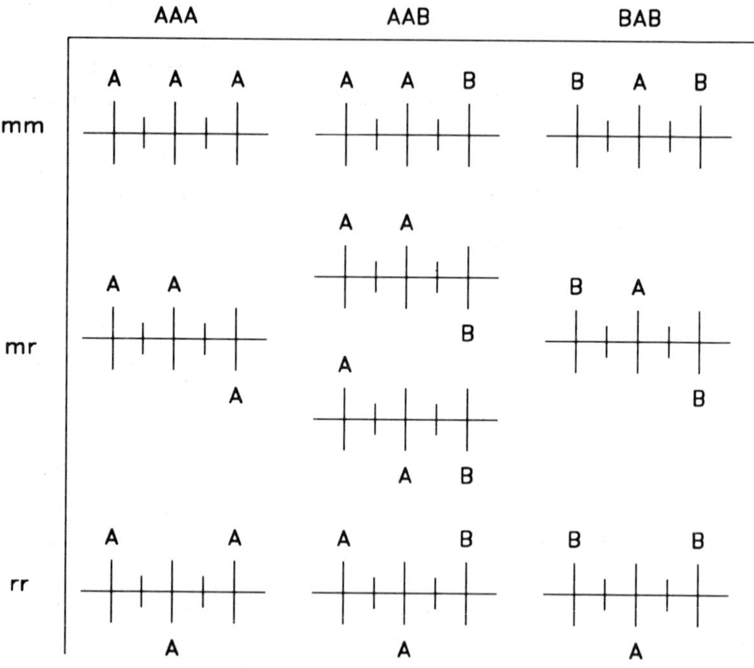

Figure 2.3 Configurations of copolymer triad sequences.

sequence distribution can be treated essentially independently from tacticity and it was shown that the polymer conforms fairly well to first-order Markov statistics or terminal model polymerisation. The stereoregularity of the polymers was treated using Bernoullian statistics based on isotacticity and co-isotacticity parameters, σ_{MM}, σ_{MA}, σ_{AM}, and σ_{AA}, as originally defined by Bovey and Tiers [20]. These σ_{xy} parameters represent the (Bernoullian) probability of generating a *meso* dyad between a growing chain ending in x and incoming monomer y. Thus, for example, the probability of finding an isotactic MMM triad is simply given by first-order Markov probability of finding any MMM triad multiplied by σ_{MM}^2:

$$(M_m M_m M) = \sigma_{MM}^2 \frac{P_{AM}(1 - P_{MA})^2}{P_{AM} + P_{MA}} \qquad (2.22)$$

Assuming that $\sigma_{AM} = \sigma_{MA}$, Román and Valero were able to determine the values of the tacticity parameters for these A/M copolymers from the NMR data ($\sigma_{MM} = 0.24$, $\sigma_{AA} = 0.31$, $\sigma_{AM} = \sigma_{MA} = 0.41$).

Styrene/methyl methacrylate (S/M) copolymers have been examined in detail using proton NMR spectroscopy [18, 19]. The methoxy protons give rise to a set of resonances which span a chemical shift range of some 1.5 ppm as a result of chemical shift sensitivity to a combination of sequence effects and tacticity effects. Uebel and co-workers [18, 19] have assigned this complicated region of the 270 MHz proton NMR spectrum by combining statistical and kinetic models for polymerisation. A series of S/M polymers were prepared, each to low conversion, covering a range of monomer feed ratios. Compositional data obtained for the polymers were used to determine the terminal model reactivity ratios ($r_M = 0.45$, $r_S = 0.50$) via the Kelen–Tüdõs method [21]. From these reactivity ratios, first-order Markov monomer addition probabilities were determined (see equations (2.20) and (2.21)) for each polymer. The triad sequence distributions of all the polymers were then calculated using first-order Markov relationships. Starting with a set of literature assignments, Uebel and co-workers [18, 19] compared these calculated sequence distributions with those that were observed. Examination of the large discrepancies between the two sets of data led them to re-assign the methoxy peaks in the proton spectrum in order to obtain a good fit between observed and calculated distributions.

Fukuda *et al.* [22] have carried out an extensive study of the kinetics of the S/M copolymer system. They showed that, although copolymer compositions conform to the terminal model, rate constants for propagation are best represented by the penultimate model. There must therefore be some doubt over the assignments of Uebel and Dinan [18] since these were made on the premise of terminal model conformity.

To summarise, NMR studies of copolymers can be used to gain information on the propagation mechanisms relating to both comonomer sequence and configuration. However, such studies require NMR spectra in which a sufficient number of the various sequence types are resolved and correctly assigned.

2.3.3 Penultimate model polymers and complex participation

There are several cases where NMR spectroscopy has been used to investigate copolymers which deviate from the terminal model for copolymerisation (see also chapter 3). For example, Hill and co-workers [23, 24] have examined sequence distributions in a number of low conversion styrene/acrylonitrile (S/A) copolymers using carbon-13 NMR spectroscopy. Previous studies on this copolymer system, based on examination of the variation of copolymer composition with monomer feed ratio, indicated significant deviation from the terminal model. In order to explain this deviation, propagation conforming to the penultimate (second-order Markov) and antepenultimate (third-order Markov) models had been proposed [25–27]. Others had invoked the complex participation model as the cause of deviation [28]. From their own copolymer/comonomer composition data, Hill *et al.* [23] obtained best-fit reactivity ratios for the terminal, penultimate, and the complex participation models using non-linear methods. After application of the statistical F-test, they rejected the terminal model as an inadequate description of the data in comparison to the other two models. However, they were unable to discriminate between the penultimate and complex participation models. Attention was therefore turned to the sequence distribution of the polymer.

In the carbon-13 NMR spectrum of this system, each of the triads gives rise to a well-resolved signal as shown in Figure 2.4 (A-centred sequence abundances were determined from the nitrile resonances; S-centred sequence abundances were obtained from the styrene quaternary aromatic carbon resonances). From the observed triad distributions for each sample, various number-fractions of styrene and acrylonitrile sequences were calculated. For example, N_{SAAS}, which represents the number-fraction of acrylonitrile sequences of length two, is given by

$$N_{\mathrm{SAAS}} = \frac{(\mathrm{SAS})^2}{[2(\mathrm{SAS}) + (\mathrm{AAS})][2(\mathrm{AAA}) + (\mathrm{AAS})]} \tag{2.23}$$

A plot of N_{SAAS} as a function of the mole fraction of acrylonitrile in each of the polymers studied is shown in Figure 2.5. The curves shown are those predicted from the reactivity ratios derived from copolymer/comonomer composition data for the three types of propagation model. Note that two types of predicted curve are shown for the complex participation model. One was obtained using a literature value of the equilibrium constant, K_x, for complex formation; the other was derived using the equilibrium constant obtained from an unrestricted non-linear fit to the copolymer/comonomer composition data. Clearly, by far the better fit between predicted and

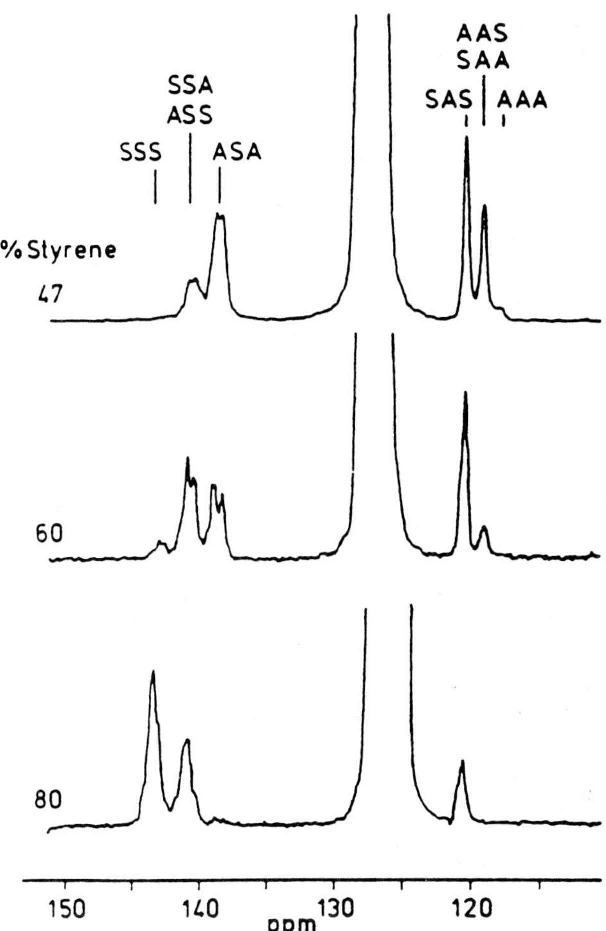

Figure 2.4 Carbon-13 NMR spectra of the aromatic and nitrile carbon region of styrene (S) and acrylonitrile (A) copolymers, showing full triad assignments. Reprinted with permission from [23]. © (1982) American Chemical Society.

experimental data is given by the penultimate model. Similar, good agreement with this model was also observed for all other number-fractions of sequence examined. The observation of a fit to the penultimate model is consistent with a mechanism proposed by others [29]. In this, repulsive electrostatic forces between highly polar nitrile groups are thought to affect the addition of nitrile monomer to polymer chains containing nitrile units near their growing ends. It was therefore concluded that, whilst comonomer complexes may be present, they do not participate to any significant extent in the polymerisation. It is appropriate to note here that Ebdon, who has recently reviewed the evidence for involvement of comonomer complexes in highly

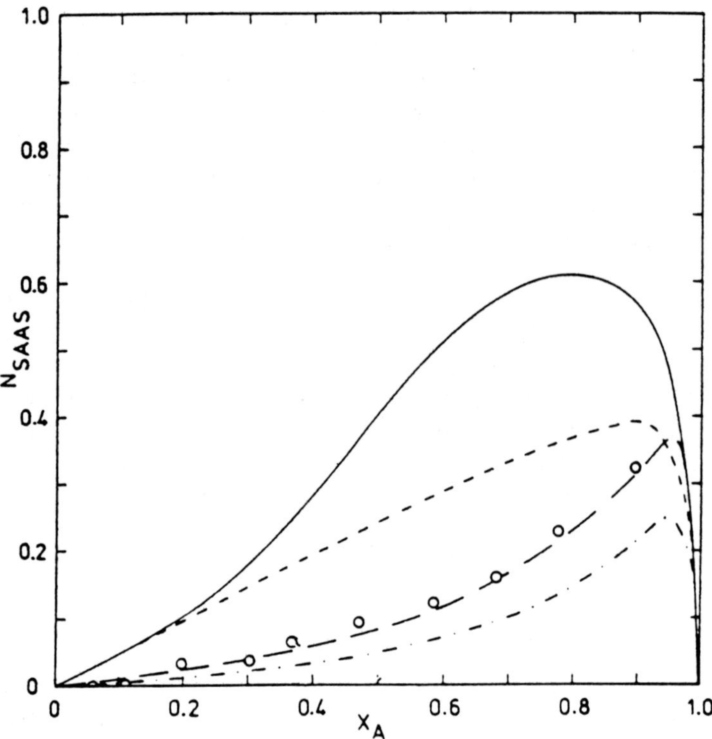

Figure 2.5 Number fraction of acrylonitrile sequences of length two versus acrylonitrile mole fraction, X_A, in acrylonitrile/styrene copolymers: (○), experimental data; (———) complex participation model, K_x of 9.98 from unrestricted fit; (----) complex participation model, K_x fixed at 0.52; (—–), penultimate model; (–·–·–) terminal model. Reprinted with permission from [23]. © (1982) American Chemical Society.

alternating systems, concluded that cross-propagation is the predominant mechanism by which alternation occurs rather than involvement of complexes [7, 30].

2.3.4 Other copolymers

All of the examples discussed so far have been free-radically polymerised systems. To conclude this section, some copolymers prepared by other routes are described briefly.

Significant interest has been shown in the use of NMR spectroscopy to study copolymers of ethylene with, for example, propylene, butene, or hexene [5, 31–35]. This no doubt reflects the technological importance of these polymers, which are prepared using Ziegler–Natta catalysts. Whist these polymers can give rise to relatively complex carbon-13 NMR spectra, they show a marked chemical shift sensitivity to sequence; indeed, signals due to heptads can sometimes be observed. In addition, detailed chemical shift assignments can be made using either empirical rules for shift behaviour or the *gamma-gauche* method developed principally by Tonelli [36] (see also chapter 4). The large number of sequence observations for these polymers means that some fairly stringent tests of propagation mechanism can be made. These tests tend to involve non-linear least squares optimisation of monomer addition probabilities by comparison of computer-calculated and observed signal intensities. Often, with these copolymers it is necessary to invoke 'two-site' models to explain the observed sequence distribution. Inoue *et al.* [31], for example, have examined the sequence distribution of an ethylene-propylene copolymer, prepared using a $TiCl_3/Et_2AlCl$ catalyst system. They found that the observed distribution was best explained by a combination of two distinct sets of first-order Markov statistics, in keeping with two different catalytic sites for polymerisation. The best-fit monomer addition probabilities for the two sites are as follows:

Site 1: $P_{EE} = 0.04, \quad P_{PP} = 0.94$

Site 2: $P_{EE} = 0.45, \quad P_{PP} = 0.42$

(E = ethylene; P = propylene)

Clearly, propylene is preferentially polymerised at site 1, while at site 2 ethylene and propylene copolymerise to give an approximately random copolymer.

In a similar study by Cheng [32] on ethylene/1-butene copolymers, a two-site model was also invoked. In contrast, he found that the microstructure of ethylene/1-hexene copolymers could be described adequately using a one-site first-order Markov model [33]. Regio-irregularity in polypropylene has also been treated by Cheng using Markov statistics (see [5] and references therein).

Using NMR data, statistical methods have also been used to characterise the microstructure of polymers prepared by step-growth polymerisation. Sequences in copolyamides, for example, have been examined using carbon-13 and nitrogen-15 NMR spectroscopy; cases of alternating, random and blocky sequence distributions can be readily distinguished [37, 38]. Bunn [39] has determined the sequence distributions of a number of aryl ether sulphone copolymers. Carbon-13 NMR measurements, in combination with statistical methods, revealed that transetherification occurs during polymerisation to yield essentially random copolymers. In the case of some related copolymers, based on various aryl ether sulphone and aryl ether ketone comonomers, sequence distributions were used to determine the degree of transetherification. This was found to correlate with polymer solubility. Another structure–property relationship has been examined by Staubli *et al.* [40]. They were able to relate the glass transition temperatures of a series of poly(anhydride-co-imide) systems containing asymmetric monomers to the sequence distribution measured by NMR spectroscopy.

Havens and Reimer have used carbon-13 NMR measurements to study the sequence distribution in a series of aryl ether ketone copolymers [41]. These copolymers were based on terephthaloyl chloride (T), 1,4-diphenoxybenzene (B), and diphenyl ether (E) and were prepared by a Friedel–Crafts reaction using either HF/BF_3 or buffered $AlCl_3$ as the catalyst. The copolymers formed are comprised of T units alternating with either E or B units.

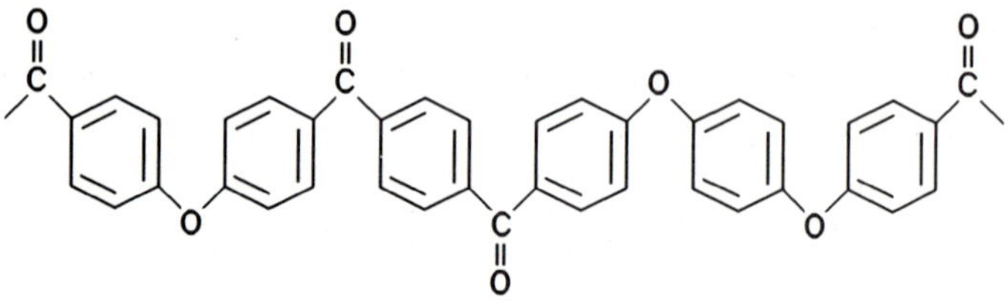

(E) (T) (B)

In the carbon-13 NMR spectra of these samples, the terephthaloyl quaternary carbons give rise to signals due to –E–T–E–, –E–T–B–, and –B–T–B– sequences. These are essentially equivalent to *dyad* sequences since the T units can be ignored because, of necessity, they link every adjacent pair of ether units. (Note that the approach shown here is somewhat different from that of Havens and Riemer; however, the results are the same.) Number-average sequence lengths for the B and E units can thus be calculated using equations (2.5) and (2.6). Havens and Riemer found that polymers prepared using HF/BF_3 are essentially random. In contrast, the number-average sequence lengths for polymers prepared using buffered $AlCl_3$ as catalyst are appreciably higher than those predicted on the basis of random statistics, indicating a definite blocky character in the copolymers.

To conclude this section, an example is given of the application of statistical models to examine double bond sequences in polymers with unsaturated backbones. The presence of backbone unsaturation leads to the occurrence of *cis/trans* isomerism and this can be treated entirely analogously to comonomer sequence. Ivin [42] and others have used NMR spectroscopy extensively to study the microstructure of polymers prepared by transition metal catalysed ring-opening metathesis polymerisation (ROMP) such as polynorbornene (PNB).

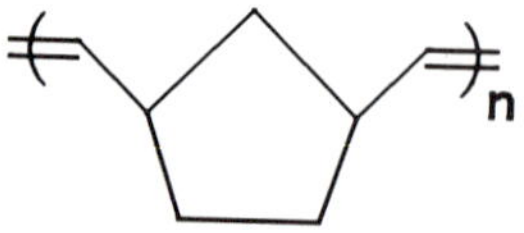

(PNB)

Polymers of this type generally give rise to carbon-13 NMR spectra which show sensitivity to at least dyad *cc*, *tc*, and *tt* sequences (where $t = trans$, $c = cis$), therefore allowing the distribution of *c* and *t* bonds to be determined. From an examination of his own results, and those of others, Ivin found that ROMP polymers with less than 35% *cis* double bonds had a random distribution of *cis*/*trans* bonds. In contrast, polymers with a *cis* double bond fraction greater than 50% show a marked tendency towards a blocky distribution. Ivin concluded that for polymers with high *cis* content, there are at least two kinetically distinct propagating species, whereas for low *cis* polymers there is only one. Explanations for this behaviour have been proposed which are based on steric constraints about the transition metal to which the growing chain end is coordinated.

2.4 *In situ* methods and simulation techniques

In a number of the examples discussed in the preceding section, comonomer reactivity ratios were used to predict sequence distributions. A number of procedures exist for deriving reactivity ratios based on copolymer/comonomer composition data. Recently, a new method for determining reactivity ratios, based on *in situ* NMR measurements has been derived. This method is described. In addition, some of the mathematical techniques available to calculate sequence distributions using reactivity ratios are mentioned briefly, since their use can impinge on a number of the NMR studies of sequence distributions.

The Mayo–Lewis equation [8] describing terminal model binary copolymerisation was given in section 2.2.3 and is also given below:

$$\frac{\mathrm{d}A}{\mathrm{d}B} = \frac{A\,(r_{\mathrm{A}}A + B)}{B\,(r_{\mathrm{B}}B + A)} \tag{2.24}$$

The terms A and B represent the number of moles of the two comonomers in the feed at any given instant, and r_{A} and r_{B} are the comonomer reactivity ratios defined earlier. The differential $\mathrm{d}A/\mathrm{d}B$ is sometimes termed the 'instantaneous' copolymer composition since it represents the composition of polymer chains forming at any instant. A variety of methods have been developed to determine the reactivity ratios. Most of these methods are based on the assumption that for conversions up to approximately 5%, the ratio of the two monomers in the feed does not change appreciably. Thus, equation (2.24) can be rewritten as

$$\frac{F_{\mathrm{A}}}{F_{\mathrm{B}}} = \frac{A\,(r_{\mathrm{A}}A + B)}{B\,(r_{\mathrm{B}}B + A)} \tag{2.25}$$

where F_{A} and F_{B} represent the number of moles of monomers A and B in the polymer. A series of polymers, prepared under conditions of low ($< 5\%$)

conversion using a range of comonomer feed ratios, is isolated. The composition of each copolymer is then determined. (Proton NMR spectroscopy is usually the technique of choice for this.) The composition versus monomer feed data are then fitted to equation (2.25), or some rearrangement of it, using linear (e.g. Fineman–Ross [43], Kelen–Tüdős [21]) or non-linear (e.g. Tidwell–Mortimer [44]) methods in order to determine best-fit reactivity ratios. These methods prove to be somewhat laborious.

A new method has recently been devised for determining comonomer reactivity ratios [45], following a suggestion by Maitland that proton NMR spectroscopy can be used as a convenient monitor of free-radical copolymerisation reactions performed *in situ* in an NMR tube [46]. From the

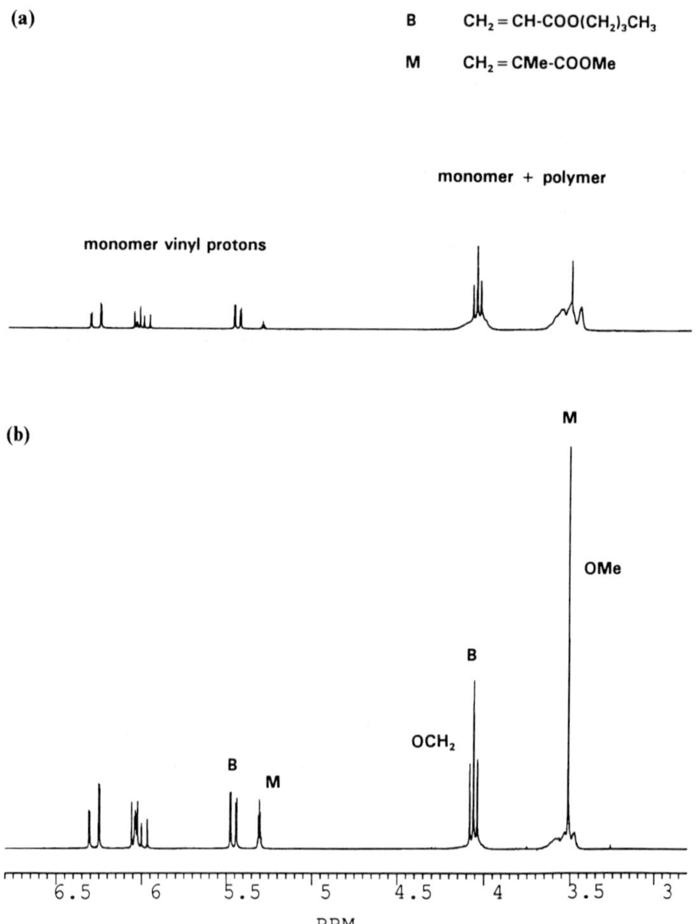

Figure 2.6 Proton NMR spectra (2.8–6.8 ppm) of a polymerising mixture of butyl acrylate (B) and methyl methacrylate (M) taken at (a) 90% and (b) 45% conversion.

relative intensities of appropriate proton resonances (see Figure 2.6), it is a straightforward matter to measure comonomer mole fractions, A and B, at stages during a polymerisation. A plot of A versus B is then a curve whose differential is given by the Mayo–Lewis equation (2.24). In other words, this curve is a representation of the integrated form of the Mayo–Lewis equation. The problem then is to fit the A–B dataset to an integrated form of this equation in order to obtain estimates of the reactivity ratios. This can be achieved using numerical integration of equation (2.24) in conjunction with non-linear least-squares fitting. (Although the Mayo–Lewis equation has been integrated analytically, the resulting expressions are somewhat unwieldy.)

For convenience, a modified form of the differential equation derived by Skeist [47] was used in this method:

$$\frac{\mathrm{d}A}{\mathrm{d}Z} = \frac{(r_A A + Z - A)}{\left[r_A A + 2Z - 2A + r_B \frac{(Z-A)^2}{A}\right]} \tag{2.26}$$

where Z is the total mole fraction of unreacted monomer (i.e. $Z = A + B$; fractional conversion $= (1 - Z)$). Numerical integration of (2.26) was achieved using the Runge–Kutta–Nystroem algorithm; non-linear least squares fitting was performed using the simplex algorithm.

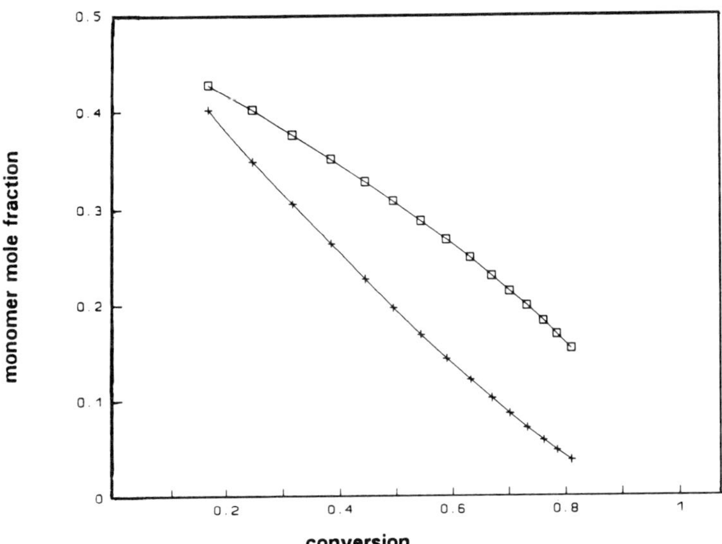

Figure 2.7 Plots of the mole fractions of unreacted monomer in a polymerising mixture of butyl acrylate (B) and methyl methacrylate (M) as a function of conversion. The data were generated from *in situ* proton NMR measurements. The starting monomer mole fractions were: M, 0.49; B, 0.51.

Figure 2.7 shows a plot of the mole fractions of unreacted monomer as a function of conversion for a 51/49 butyl acrylate (B)/methyl methacrylate (M) solution-state copolymerisation performed *in situ* in an NMR tube. The solid lines represent 'best-fit' curves based on the following reactivity ratios: $r_M = 1.38, r_B = 0.23$. This procedure offers significant advantages over classical methods for determining reactivity ratios in terms of its speed, ease of experimentation, and the fact that it utilises data obtained over the full range of conversion. It is anticipated that the technique will also be applicable to systems which deviate from the terminal model for polymerisation.

For completeness, it is worth noting that there is a caveat to the above procedure. Consider the case of a pair of comonomers where the starting monomer mole fractions are both 0.5. If the two reactivity ratios are both unity (i.e. the polymerisation is a Bernoullian process), a plot of A against B (or A against Z) will be linear. However, this plot will be indistinguishable from one derived from a pair of comonomers whose reactivity ratios are both zero (i.e. they form an alternating copolymer), or indeed any pair with equal reactivity ratios. To differentiate these cases, it is necessary to make a judicious choice of starting monomer mole fractions: the cases above will be readily distinguishable using starting moles fractions of, say, 0.8 and 0.2.

Methods for simulation of sequence distribution. It often proves beneficial to be able to predict sequence distribution from reactivity ratio data. For example, it may be necessary to compare the sequence distribution determined by NMR spectroscopy with one determined from reactivity ratio data obtained by some other technique. Relationships which enable monomer addition probabilities to be calculated from reactivity ratios have already been given (equations (2.20), (2.21)). The monomer addition probabilities can in turn be used to calculate sequence distribution as discussed earlier. This approach is only valid, however, for low conversion polymers. As stated before, as conversion increases, the ratio of the two unreacted monomers drifts because one monomer is normally more reactive than the other. Thus, the monomer addition probabilities are continuously changing as the polymerisation proceeds.

Computer-based methods have therefore been developed to predict or calculate the microstructural features of copolymers over a range of conversions (e.g. see [48]). Generally, these methods fall into one of two categories, either stochastic or numerical. The stochastic methods are used to simulate growth of polymer chains by comparing the probability of the addition of a given monomer with computer-generated random numbers. The numerical methods essentially rely on numerical integration of differential equations describing copolymerisation.

Stochastic methods have been applied to compare observed and predicted sequence distribution in polymers prepared to high conversion. For example, the sequence distributions of a range of styrene/methyl methacrylate copolymers, determined using carbon-13 NMR spectral editing techniques, are similar to those predicted stochastically [49]. Indeed, in this particular case, the sequence distributions are comparable with those for low conversion polymers. In butyl acrylate/methyl methacrylate copolymers, the sequence distributions (both predicted and observed) for high conversion polymers are close to random, which is in marked contrast to those predicted for low conversion polymers. Here, significant drift occurs in the comonomer feed ratio during polymerisation resulting in a non-uniform product, both in terms of composition and sequence distribution. Of course, the NMR spectrum of the final sample only provides an average view of this sequence heterogeneity.

2.5 Polymer modification

It is often desirable to perform some kind of chemical modification to polymers once they have formed. Acetylation of cellulose, for example, represents the chemical modification of a naturally occurring polymer (see chapter 3) to give the technologically useful cellulose diacetate and triacetate polymers. Poly(vinyl alcohol) (PVA) is a well-known example of a polymer that can only be formed by chemical modification since vinyl alcohol monomer does not exist (except as its keto form, acetaldehyde). Instead, PVA is formed by hydrolysis of poly(vinyl acetate) (PVAc). Partly hydrolysed PVAc is, of course, simply a copolymer of VA and VAc. As another example, ethylene/vinyl chloride copolymers can be prepared by reductive elimination of chlorine from poly(vinyl chloride) (PVC). The driving force here is that, although ethylene and vinyl chloride can be copolymerised directly, the normal routes to these polymers give insufficient control over composition and sequence distribution.

The microstructures of copolymers prepared by chemical modification are obviously amenable to study using exactly the same techniques that are applied to copolymers prepared by more normal routes. Here, part of the value of such studies is that they can impart information concerning the mechanism of the chemical modification reactions. Some examples of polymer modification are discussed in this section.

Copolymers of ethylene/vinyl chloride (E/V), prepared by reductive dechlorination of PVC using tri-n-butyltin hydride, have been studied extensively and this work has recently been described in some depth (see [36, chapter 9] and references therein). Only a brief description of this system is given here. The carbon-13 NMR spectra of E/V copolymers have been assigned in detail using the *gamma-gauche* effect method (see chapter 4). Thus, it proves possible to determine the abundances of all six comonomer triads. Furthermore, the tacticity of VV dyad sequences (r and m) can also be measured. For lightly reduced polymers, an almost random distribution of comonomers was found. However, for higher degrees of reduction, examination of sequence distribution indicates that V-unit chlorine atoms adjacent to other V units are removed preferentially. This leads to an increase in the number of EVE and VEV triads in comparison to that expected for random dechlorination, i.e. the polymers show a tendency towards alternation of E and V units. In terms of tacticity, the ratio of *r* and *m* dyads in the starting polymer is 1.27. For the polymer in which 79% of chlorine atoms have been removed, this ratio is 4.2. The *m*-dyads are therefore preferentially reduced. The *r*/*m* ratios can also be used to examine the ratio of rate constants governing reduction of *m* and *r* dyads. It was found that this was independent of the degree of reduction. Thus, NMR analysis of microstructure can provide information on the kinetics of polymer modification. Similar studies have also been performed on copolymers prepared by reductive debromination of poly(vinyl bromide) [50].

Poly(vinyl alcohol-co-vinyl acetate) polymers are surface active species which can be used to stabilise latex and oil in water dispersions. In order to understand the properties of these materials, it is necessary that their sequence distributions are well characterised. A number of NMR studies on the microstructure of PVA/PVAc copolymers have been made [51–53] (see also chapter 3). Moritani and Fujiwara [51], for example, have used proton and carbon-13 NMR spectroscopy to extract dyad distributions for a range of copolymers with different degrees of deacetylation. Samples were prepared using one of three routes: direct saponification of PVAc; alcoholyis of PVAc using sodium methoxide; and reacetylation of PVA. From the polymer composition and the dyad distribution, the parameter η was calculated for each polymer as follows:

$$\eta = \frac{(\mathrm{AB})}{2(\mathrm{A})(\mathrm{B})} \tag{2.27}$$

where A is vinyl alcohol and B is vinyl acetate. This parameter η is a measure of the blocky character of each polymer since it is simply the ratio of observed AB dyad abundance against that predicted by Bernoullian statistics. A value of η equal to one therefore indicates a random distribution of comonomers. Values less than one indicate blocky copolymers, whilst values greater than one indicate a tendency towards comonomer alternation. Significant differences were observed between the different types of polymer studied. Thus, reacetylated copolymers gave η values close to one, indicating approximately random microstructures. In contrast, the samples prepared by direct saponification ($\eta = 0.46 \pm 0.06$) and alcoholysis ($\eta = 0.56 \pm 0.08$) had blocky microstructures. The results of Moritani and Fujiwara [51] are presented graphically in Figure 2.8. Here, the A–Ac dyad abundance is plotted against the mole fraction of vinyl alcohol in the polymer. In addition to observing differences in microstructure in samples prepared by different routes, it was also noted that η decreased overall as the level of alcohol groups increased in the samples prepared both by direct saponification and by alcoholysis.

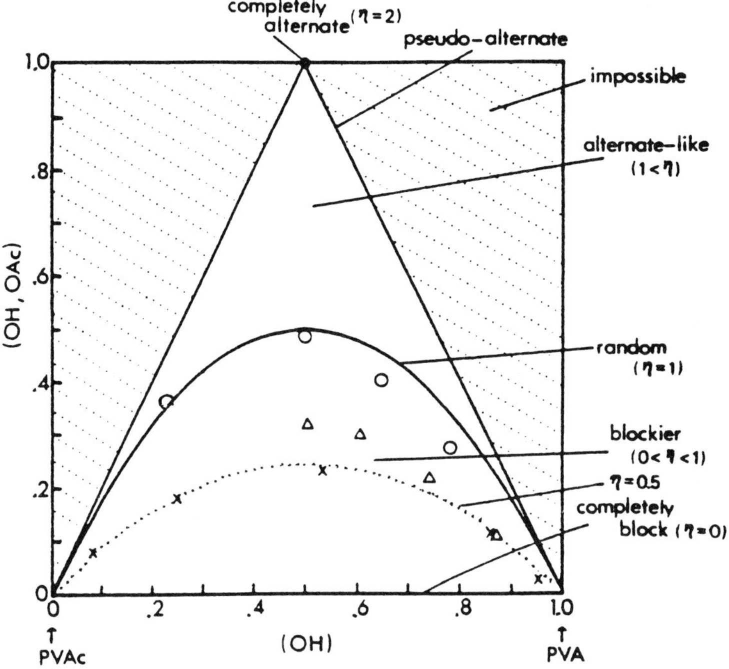

Figure 2.8 Proportions of (OH, OAc) dyad as a function of hydroxyl group content for sets of poly(vinyl alcohol-co-vinyl acetate) copolymers prepared by different routes. (×) Saponification; (△) alcoholysis; (○) reacetylation. The solid lines show calculated dyad proportions for ideal cases of alternating, random, and block copolymers. Reprinted with permission from [51]. © (1977) American Chemical Society.

These results are in keeping with predictions of microstructure based on a kinetic model describing PVAc hydrolysis. In this model for acid- or base-catalysed saponification, the catalyst interacts strongly with free hydroxyl groups and this induces successive hydrolysis of nearest-neighbour acetyl groups, so leading to a blocky microstructure.

2.6 Summary

The ability of NMR spectroscopy to distinguish between dyads, triads, and higher *n*-ad sequences in copolymers makes it an especially powerful tool for the polymer chemist interested in the fine details of molecular structure. In this chapter, we have seen that simple expressions derived from the relative abundances of various sequence types allow characterisation of the copolymer microstructure in terms of number-average sequence lengths, without any recourse to a statistical model. A more detailed examination of copolymer sequence distribution enables microstructure to be related to models based on comonomer addition probabilities. From a fit of sequence distribution to a given model, the abundance of any chosen sequence can be calculated. Moreover, conformity to a particular model then provides information on the mechanism of polymer propagation and the relative reactivities of the comonomers. Indeed, these statistical models are intimately related to the more traditional propagation models based on reactivity ratios. We have seen that these models and techniques are applicable to a wide range of copolymer types, including systems prepared by chemical modification of polymers. We have also seen that *in situ* NMR measurements can be used as a simple method for monitoring copolymerisation and studying comonomer reactivities.

For the future, microstructural studies of the type described here will necessarily rely heavily upon the quality of the NMR data that are available. Greater use of high-field instruments and the increasing application of two-dimensional NMR techniques to synthetic polymers will certainly help here. However, a well-resolved spectrum is of little use if it cannot be assigned. For this, we must look to an increase in the application of and the refinement of techniques used so successfully to predict chemical shifts in a number of homopolymers.

References

1. F.A. Bovey, *Acc. Chem. Res.* **1** (1968) 175.
2. F.A. Bovey, *High Resolution NMR of Macromolecules*, Academic Press, New York (1972).
3. F.P. Price, *J. Chem. Phys.* **36** (1962) 209.
4. J.C. Randall, *Polymer Sequence Determination—Carbon-13 NMR Method*, Academic Press, New York (1977).
5. H.N. Cheng, in *Modern Methods of Polymer Characterisation*, eds. H.G. Barth and J.W. Mays, Wiley–Interscience, New York (1991).
6. R.E. Cais, R.G. Farmer, D.J.T. Hill and J.H. O'Donnell, *Macromolecules* **12** (1979) 835.
7. J.R. Ebdon, *Macromol. Chem. Macromol. Symp.* **10/11** (1987) 441.
8. F.R. Mayo and F.M. Lewis, *J. Am. Chem. Soc.* **66** (1944) 1594.
9. T. Alfrey and G. Goldfinger, *J. Chem. Phys.* **12** (1944) 205.
10. F. Candau, Z. Zekhnini and F. Heatley, *Macromolecules* **19** (1986) 1895.
11. R.A. Komoroski and J.P. Schockcor, *Macromolecules* **16** (1983) 1539.
12. G. Talamini and E.G. Peggion, in *Kinetics and Mechanism of Polymerization*, Vol. 1, ed. G.E. Ham, Marcel Dekker, New York (1967).
13. N.D. Truong, J.C. Galin, J. François and Q.T. Pham, *Polymer* **27** (1986) 467.
14. F. Lafuma and G. Durand, *Polym. Bull.* **21** (1989) 315.
15. H. Tanaka, *J. Polym. Sci., Part A* **24** (1986) 29.
16. G.S. Kapur and A.S. Brar, *Polymer* **32** (1991) 1112.
17. J.S. Román and M. Valero, *Polymer* **31** (1990), 1216.
18. J.J. Uebel and F.J. Dinan, *J. Polym. Sci., Polym. Chem. Ed.* **21** (1983) 2427.
19. L.T. Kale, K.F. O'Driscoll, F.J. Dinan and J.J. Uebel, *J. Polym. Sci., Part A, Polym. Chem. Ed.* **24** (1986) 3145.
20. F.A. Bovey and G.V.D. Tiers, *J. Polym. Sci.* **44** (1960) 173.
21. T. Kelen and F. Tüdõs, *J. Macromol. Sci.-Chem.* **A9** (1975) 1.
22. T. Fukuda, Y. Ma, and H. Inagaki, *Macromolecules* **18** (1985) 17.

23. D.J.T. Hill, J.H. O'Donnell and P.W. O'Sullivan, *Macromolecules* **15** (1982) 960.
24. P.F. Barron, D.J.T. Hill, J.H. O'Donnell and P.W. O'Sullivan, *Macromolecules* **17** (1984) 1967.
25. G.E. Ham, *J. Polym. Sci.* **14** (1954) 87.
26. A. Guyot and J. Guillot, *J. Macromol. Sci. Chem.* **A1** (1967) 793.
27. A. Guyot and J. Guillot, *J. Macromol. Sci. Chem.* **A2** (1968) 889.
28. B. Sadner, S. Kraemer and B.T. An, *Acta Polym.* **30** (1979) 265.
29. L.J. Young, in *Copolymerization*, ed. G.E. Ham, Interscience, New York (1964).
30. J.R. Ebdon and C.R. Towns, *J. Macromol. Sci., Rev. Macromol. Chem. Phys.* **C26** (1986) 523.
31. Y. Inoue, T. Hayashi, R. Chûjô and T. Asakura, *Polymer* **29** (1988) 1848.
32. H.N. Cheng, *Macromolecules* **24** (1991) 4813.
33. H.N. Cheng, *Polym. Bull.* **26** (1991) 325.
34. V.B.F. Mathot and Ch. C.M. Fabrie, *J. Polym. Sci. B, Polym. Phys.* **28** (1990) 2487.
35. V.B.F. Mathot, Ch. C.M. Fabrie, G.P.J.M. Tiemersmathoone and G.P.M. Van der Velden, *J. Polym. Sci., B, Polym. Phys.* **28** (1990) 2509.
36. A.E. Tonelli, *NMR Spectroscopy and Polymer Microstructure*, VCH, New York (1989).
37. H.R. Kricheldorf and W.E. Hull, *J. Polym. Sci., Polym. Chem.* **16** (1978) 2253.
38. H.R. Kricheldorf, S.V. Joshi and W.E. Hull, *J. Polym. Sci., Polym. Chem.* **20** (1982) 2791.
39. A. Bunn, *Br. Polym. J.* **20** (1988) 307.
40. A. Staubli, E. Mathiowitz and R. Langer, *Macromolecules* **24** (1991) 2291.
41. J.R. Havens and K.B. Reimer, *J. Polym. Sci. A, Polym. Chem.* **27** (1989) 565.
42. K.J. Ivin, *Olefin Metathesis*, Academic Press, New York (1983).
43. M. Fineman and S.D. Ross, *J. Polym. Sci.* **5** (1950) 259.
44. P.W. Tidwell, and G.A. Mortimer, *J. Polym. Sci. A* **3** (1965) 369.
45. I.R. Herbert, in preparation.
46. D.J. Maitland, personal communication.
47. I. Skeist, *J. Am. Chem. Soc.* **68** (1946) 1781.
48. H.J. Harwood, in *Data Processing in Chemistry*, ed. Z. Hippe, Elsevier, New York (1980).
49. I.R. Herbert, unpublished results.
50. R.E. Cais and J.M. Kometani, *Macromolecules* **15** (1982) 954.
51. T. Moritani and Y. Fujiwara, *Macromolecules* **10** (1977) 532.
52. G. Van der Velden and J. Beulen, *Macromolecules* **15** (1982) 1071.
53. S. Toppet, P.J. Lemstra and G. Van der Velden, *Polymer* **24** (1983) 507.

3 Solution-state NMR determination of polymer end-groups, substituents and minor structures

J.C. BEVINGTON, J.R. EBDON and T.N. HUCKERBY

3.1 Introduction

3.1.1 General remarks

The physical and chemical behaviour of polymers can be greatly affected by structural irregularities present in small numbers in the macromolecules; these irregularities include end-groups, branch-points and unusual units. In most cases, these features arise during the polymerization and their numbers depend upon the precise conditions under which the polymer is made. It is important to have information about structural irregularities not only to be able to correlate the properties of polymeric materials with their structures and the methods of preparation, but also to understand all the reactions involved in a polymerization. With sufficient knowledge, it may be possible to select conditions for polymerization so that undesirable groupings can be reduced in number or even eliminated.

Many physical and chemical techniques have been developed to detect and to identify abnormal groups in polymers. NMR examination of polymers in solution has contributed very significantly and in many cases has led to information otherwise inaccessible; in this chapter, some examples are presented. Most attention is paid to end-groups formed during initiation of polymerization since general principles can be illustrated very well. Consideration is given also to other sources of end-groups, such as reactions of growing chains with retarders, and also scission of main chains. Other topics include regioselectivity during the growth of macromolecules, the development of branches and the formation of 'foreign' units during polymerization or subsequently by reactions of the polymers. No attempt has been made to mention every report on the application of NMR methods for study of abnormal groups in polymers; the subjects selected for discussion and the references chosen for citation are thought to be among the most important and to be illustrative.

Identification of the abnormal groups in a polymer may not be sufficient; it is often necessary also to determine the numbers of such groups for, say, 1000 'normal' units. NMR techniques can give the required information but precautions are needed if reliable results are to be obtained. The problems can be illustrated by considering the use of ^{13}C NMR to compare the number of end-groups derived from a carbon-containing initiator with the number of monomeric units in a typical organic polymer. In principle, comparison of the integral of the signals from carbons in the end-groups with that for carbons in the monomeric units should give the information being sought. Of course, the two types of signals must be resolved but, even if this requirement is met, there are other difficulties. In practice, the end-group signals are very small unless ^{13}C-enriched initiator is used in the preparation of the polymer and then it is necessary to have exact information on the level of enrichment (see section 3.1.2).

Consider, as an example, the case of poly(methyl methacrylate) prepared in such a way that there is one $Me_2C(CN)$– end-group, with enriched methyl carbons, for n monomeric units of formula $-CH_2 \cdot C(Me)(COOMe)-$; in a typical case, n might lie in the range 100–5000. Suppose that a is the abundance of ^{13}C in the enriched methyl groups and that b is the abundance at each of the sites in the monomeric units. If A is the total area under the ^{13}C NMR peaks for the two methyl carbons of the end-groups and B is the corresponding area for all five of the carbons of the monomeric units and, very importantly, if the spectral response of a ^{13}C nucleus is independent of its position in the polymer molecule (see section 3.1.2), then

$$\frac{A/a}{B/b} = 2/5n, \quad \text{so that } n = 2Ba/5Ab \tag{3.1}$$

It is usually supposed that the enrichment factor for the special carbons in the end-groups is the same as that for the corresponding carbons in the original initiator from which they were derived. The abundance of ^{13}C at the enriched sites in the initiator may be appreciably less than 100% and an isotope effect in the initiation process would mean that a different enrichment factor would be applicable to the end-groups. Any isotope effect is likely to be small for ^{13}C-labelled initiators although not for those enriched with deuterium.

Comparison of the numbers of monomeric units and end-groups has been used for determination of number-average molecular weights of polymers. For this purpose, it is necessary to know the number of the particular end-groups in the average polymer molecule. This number is not always two; it is smaller if end-groups of other types also are present and it can be larger if the macromolecules are branched. In general, it is desirable whenever possible to calibrate the method by the use of physical techniques such as osmometry.

Determination of the ratio of end-groups to monomeric units in systems of another type must be mentioned. Suppose that the end-group contains an element readily detected by NMR, such as fluorine or phosphorus, and that the bodies of the macromolecules are free from that element. Clearly, determination of, say, the fluorine content of the whole polymer could lead to a value of the required ratio. It is necessary to use as a standard a properly characterized fluorine-containing substance, preferably dissolved with the polymer in the NMR solvent, and to ensure identical responses of the ^{19}F nuclei in the standard and the unknown by careful choice of the conditions for recording the spectrum. An example of an application of this type is discussed in section 3.2.4.

In this chapter, there are frequent mentions of certain monomers and the polymers derived from them. Abbreviations are used for styrene (STY), methyl methacrylate (MMA), vinyl acetate (VAC), acrylonitrile (AN) and methacrylonitrile (MAN). Abbreviations for other materials are specified at the points at which they first appear.

3.1.2 Comments on experimental procedures

The principal experimental techniques involved in the examination of polymers by solution state NMR are considered in chapters 1 and 4 but certain points should be made here. This chapter contains examples of the use of substances highly enriched with ^{13}C. Most of them are not at present available commercially and so they need to be prepared specially; for this purpose, many compounds with 99 at% ^{13}C can be obtained. The case of azobisisobutyronitrile can be taken as an example; this substance is widely used as an initiator of radical polymerizations and fragments derived from it form end-groups in the resulting macromolecules (see section 3.2.2). The preparation of the azonitrile can be represented thus:

$$\text{acetone} + \text{hydrazine} + \text{sodium cyanide} \rightarrow Me_2C(CN){\cdot}NH{\cdot}NH{\cdot}C(CN)Me_2$$
$$\downarrow$$
$$Me_2C(CN){\cdot}N{:}N{\cdot}C(CN)Me_2$$

^{13}C-Enriched forms of acetone and sodium cyanide are available and so it is possible to prepare various types of ^{13}C-enriched initiator. Even the ^{13}C-enriched starting compounds are expensive and so the conversions need to be performed efficiently and on a small scale.

Frequently, it is possible to make effective use of initiators with about 50% ^{13}C at the selected site. Accurate information on the enrichment factor can be obtained by mass spectrometry and also, in some cases, by examination of the ^1H NMR spectrum of the ^{13}C substance. The signals due to protons attached to the ^{13}C atom are flanked by pairs of satellite lines arising from spin–spin coupling of the protons to the ^{13}C nucleus; measurements of their intensities can yield the required information [1].

Some use has been made in polymer systems of materials enriched with deuterium or with ^{15}N, followed by recording of ^2H or ^{15}N NMR spectra as appropriate. Many deuterated substances and a few enriched with ^{15}N are available and can be used in syntheses.

It should be mentioned that the ratio between the intensities of signals from 'abnormal' groups derived from a reagent, such as an initiator, and those from the much more numerous monomeric units can be improved by a factor of about 10 if the monomer is depleted in ^{13}C. It is possible to purchase some chemicals in which ^{13}C is present at about 0.1% instead of at its natural level of about 1% and depleted monomers might then be specially prepared. The procedure may be worth while in some exceptional cases but it is not generally applicable.

Quite commonly, the signals from isotopically enriched 'abnormal' units in a polymer are obscured by responses from the many 'normal' units. This problem can be overcome by the use of difference spectroscopy. A polymer is prepared with a reactant, say initiator, enriched with ^{13}C; a second polymer is prepared under identical conditions but with the reactant having ^{13}C at only its natural level. The two polymers are examined spectroscopically under precisely the same conditions. The only difference between the two spectra should be the considerable reinforcement of signals from the enriched sites in any of the ^{13}C-labelled reactant incorporated in the polymer. Subtraction of the 'unenriched' spectrum from the 'enriched' should yield a difference spectrum containing only the responses from the incorporated enriched reactant or ^{13}C-enriched fragments derived from it.

In practice, such subtractions may not be perfect because of slight differences between the actual constitutions of the two polymers and also imperfections during the spectroscopic examinations. Another point must be noted. The main polymer signals vanish as a result of the subtraction because they correspond to identical 'coherent' information but spectral noise on the two separate data sets is random and unrelated; therefore it will add. For this reason, the signal-to-noise ratio in the difference spectrum must be worse than for signals directly observed in a single measurement.

It is always necessary to confirm that signals, thought to be due to a reactant chemically incorporated in polymer, do not actually originate from occluded material of low molecular weight. A similar problem arises acutely when using reactants labelled with a radioactive isotope such as ^{14}C. Occluded substances usually give NMR signals which are sharper than those from substances chemically incorporated in high polymer; they can be reduced in intensity or even eliminated by the techniques, such as re-precipitation, commonly used to purify polymers. The chemical shifts for the reactant itself and for any low molecular byproducts from it generally differ from those for the structural unit derived from it; this feature means that NMR analyses for incorporated reactant can be highly specific and that rigorous purification of the polymer may not be required [1].

The possibility of observing spin–spin coupling must be considered for structural units containing two ^{13}C sites, as in the $Me_2C(CN)$– end-groups in polymers made using as initiator the azonitrile already referred to and prepared from $(^{13}CH_3)_2CO$. For the two bond ^{13}C–C–^{13}C unit in saturated systems, the splitting due to the coupling is typically less than 2 Hz and is frequently unresolvable. No homonuclear couplings have been found between the enriched methyl carbons in $Me_2C(CN)$– end-groups. In this connection, significant observations have been made on the ^{13}C signals from the initiator fragments in polymers of very low molecular weight made using unenriched azonitrile. For these end-groups, there is only a very small chance that both methyl groups contain ^{13}C and yet the signals cannot be distinguished from those from polymers made using initiator having both methyl groups enriched. It is significant also that the structures of the methyl signals derived from unenriched end-groups are the same as those for the corresponding signals from end-groups in which the nitrile groups are highly enriched with ^{13}C.

In the case of polyMMA with $Me_2C(CN)$– end-groups, the carbons of the end-group methyls have considerably longer T_1 relaxation times than the carbons of the α-methyl groups in the monomeric units (about 0.30 s as opposed to 0.05 s) [1] and it has been found for many other cases also that the terminal groups have longer T_1 than similar groups within polymer chains (see chapter 4). This feature can be useful in helping to decide whether particular signals arise from end-groups or from small numbers of certain in-chain units.

It is necessary now to return to the requirement (see section 3.1.1) that, for certain applications, the spectra should be recorded under conditions such that ^{13}C nuclei at all sites in a polymer molecule contribute equally to the spectrum. If this need is to be met, there must be no selective saturation of signals with rather large values of T_1; further, the nuclear Overhauser enhancements must be totally suppressed because their magnitudes vary with the position of ^{13}C in the macromolecule (see chapters 1 and 4). The various values of T_1 must therefore be determined with some accuracy. It is usual when recording spectra for quantitative purposes, to arrange for the interval between pulses to be 10 times the longest T_1 of relevance and to use a gated decoupling mode in which protons are irradiated only during signal acquisition so that all Overhauser effects are eliminated. This approach unfortunately extends the time needed to obtain spectra with satisfactory signal-to-noise ratio and usually it is necessary to accumulate up to 10^4 transients.

Rates of nuclear relaxations can be altered by the deliberate addition to the solutions used for NMR spectroscopy of species containing paramagnetic nuclei. The usual relaxation agent employed with substrates dissolved in organic solvents is the chelate chromium tris(acetylacetonate) ($Cr(acac)_3$), $[Me{\cdot}C\dot{O}{\cdot}CH{\cdot}C(\dot{O})Me]_3Cr$, i.e. chromium(III)-2,4-pentane-dionate. It acts to reduce values of T_1 via dipole–dipole interactions, leaving chemical shift values unaltered. $Cr(acac)_3$, when used at concentrations of about 0.05 M, substantially reduces values of T_1, particularly those for quaternary nuclei; simultaneously it removes the nuclear Overhauser enhancement from proton-decoupled ^{13}C and other resonances. Use of this reagent could therefore considerably shorten the times required for quantitative NMR measurements.

It is necessary, however, to consider some problems associated with the use of $Cr(acac)_3$. First, it is imperative that there are no chemical reactions or possibilities for formation of complexes involving the relaxation agent and the substrate. Even weak complexation would cause gross line-broadening for nuclei near the binding site. In the absence of specific interactions, generalized line-broadening may occur although not severely if $Cr(acac)_3$ concentrations are low. The requirement that the substrate should contain no specific binding sites in practice imposes considerable limitations on the use of relaxation agents. It is also possible that the relaxing ability becomes inefficient compared with fast proton dipolar processes in very large molecules. It must be noted that the use of a relaxation agent involves the 'contamination' of samples used for NMR. If further use is to be made of materials after spectroscopic examination, the need for re-purification must be considered.

3.2 End-groups

3.2.1 Introductory remarks

An important application of NMR techniques for study of abnormal groupings in polymers is concerned with the end-groups derived from reagents, such as initiators, used in the polymerization. Ordinarily there is only one such end-group for perhaps 1000 monomeric units and detection may be very difficult unless the end-group contains an 'unusual' element, e.g. fluorine, or it is enriched with an NMR-active isotope, such as ^{13}C. In some cases, the conditions for making the polymer can be adjusted so that oligomeric products are formed and the ratio of end-groups to monomeric units is not so unfavourable for detection. A modification of this procedure involves preparation of the polymer under 'ordinary' conditions followed by isolation of the low molecular weight fraction for examination. The sensitivity achievable in end-group analyses by NMR techniques cannot rival that in analyses involving radioactive isotopes, but the specificity is such that it is possible not only to identify the precise nature of the incorporated end-groups but also, in many cases, to obtain information about the monomeric units close to them in the attached macromolecular chains. Further, it is likely that the NMR signals from the end-groups can be distinguished readily from those due to unused initiator or waste products of low molecular weight.

An alternative in some cases to the use of ^{13}C-labelled initiator, or other source of end-groups, depends upon modification of unenriched end-groups in preformed polymers by means of ^{13}C-labelled reagents, so converting them into groups that can be studied by ^{13}C NMR; for example, –OH end-groups can be esterified with isotopically enriched reagents. The procedure could be used with reagents such as trifluoroacetic anhydride that give new end-groups containing nuclei easily detected by NMR. It appears that procedures of this type have not been applied extensively but they would have notable advantages including avoidance of the necessity to prepare special polymers and also the possibility of studying end-groups even in commercial polymers.

There is much interest in the preparation of special initiators to give end-groups of particular types. In radical polymerizations, organic peroxides are of little value in this connection because it is usual for the derived radical to be capable of dissociating further to give a radical of a second type and a stable molecule; this process leads to the presence in polymers of two distinct types of end-groups. Further, it must be noted that many organic peroxides are rather dangerous substances. Azo initiators are much more useful for the production of polymers with selected end-groups; thus *N*-[4,4′-azobis(4-cyanopentanoyl)]-bis-dibenz[*b*, *f*] azepine (**1**) has been used, although not in isotopically enriched form, to yield polymers with dibenz-azepine end-groups which can participate in cyclo-additions [2].

$$N{-}CO{-}CH_2{-}CH_2{-}C(CH_3)(CN){-}N{=}N{-}C(CH_3)(CN){-}CH_2{-}CH_2{-}CO{-}N$$

(1)

3.2.2 *Azo initiators*

3.2.2.1 General considerations. Many azo compounds, potentially useful as sources of radicals to initiate polymerizations, have been described; some of them are available commercially (see, for example, the data sheets issued by the Wako Company of Japan). The principal criteria governing the selection of the azo initiator for a particular case are: (i) the temperature at which the polymerization is to be performed; and (ii) the nature of the reaction medium, i.e. organic or aqueous. The best known azo initiator is azobis-isobutyronitrile (AIBN) but many other azonitriles of general formula $R_1R_2C(CN){\cdot}N{:}N{\cdot}C(CN)R_1R_2$ have been characterized [3]. AIBN gives 1-cyano-1-methylethyl radicals at a convenient rate at 60°C in organic systems, according to the equation

$$Me_2C(CN){\cdot}N{:}N{\cdot}C(CN)Me_2 \rightarrow 2\,Me_2\dot{C}(CN) + N_2. \qquad (3.2)$$

It is also used as a blowing agent for the production of materials such as expanded polymers. The extensive use of AIBN is taken as justification for the present detailed discussion of its chemistry.

Much use has been made of isotopically enriched AIBN in studies of end-groups in polymers made by radical polymerization. The 1-cyano-1-methylethyl radical can also be generated by the dissociation of the azoformamide 2-(carbamoylazo)-isobutyronitrile (AZOF), $Me_2C(CN)\cdot N{:}N\cdot CO\cdot NH_2$. This compound is appreciably soluble in water and it dissociates at a convenient rate at temperatures in the region of 100°C. ^{13}C-Enriched samples of AZOF have been used in much the same way as enriched AIBN; they have been used for study of the reactions of the 1-cyano-1-methylethyl radical at temperatures at which AIBN dissociates too quickly to be of use in polymerizing systems.

Objections have been raised to the use of AIBN because 'wasted' radicals can give rise to a toxic product, namely tetramethylsuccinodinitrile. For this reason, increasing attention is being paid to the azoester dimethyl 2,2′-azobis-isobutyrate (AZE), $Me_2C(COOMe)\cdot N{:}N\cdot C(COOMe)Me_2$ which dissociates at 60°C at about the same rate as AIBN. The first reports on applications of [^{13}C]AZE have been published [4, 5].

Other azo initiators have been used in ^{13}C-enriched forms, namely azobis-1,1′-phenylethane (APE), $Me\cdot CHPh\cdot N{:}N\cdot CHPh\cdot Me$, 1,1′-azobis(1,3-diphenylpropane), $Ph\cdot CH_2\cdot CH_2\cdot CHPh\cdot N{:}N\cdot CHPh\cdot CH_2\cdot CH_2\cdot Ph$ and 4,4′-azobis-(4-phenylbutyronitrile), $CH_2(CN)\cdot CH_2\cdot CHPh\cdot N{:}N\cdot CHPh\cdot CH_2\cdot CH_2\cdot CN$. These compounds are of little practical importance as initiators of polymerization but they give rise to radicals with structures that are of great interest for studying the factors influencing the reactivities of radicals towards monomers.

The choice of the site for ^{13}C-enrichment of AIBN may be important and use has been made of three types of material, enriched in the methyl groups ([β,β-^{13}C]AIBN), at the quaternary carbons ([α-^{13}C]AIBN) or in the nitrile groups (^{13}CN–AIBN). Fully deuterated AIBN and also material enriched in the nitrile groups with ^{15}N have been used to a limited extent [6].

The types of problems that may be solved by using isotopically enriched azo initiators, followed by consideration of the NMR spectra of the derived polymers, can be summarized thus:

1. identification of the initiator fragments in a polymer and determination of the number of monomeric units incorporated for each of the fragments;
2. characterization of the monomeric units attached to or near the end-groups derived from an initiator;
3. study of initiation in copolymerizations and comparison of the reactivities of monomers towards the initiating radicals.

3.2.2.2 Use in homopolymerizations. Examination by NMR of low conversion homopolymers prepared at 60°C with each type of isotopically enriched AIBN, in turn, has provided no evidence that the initiator gives rise to any end-groups other than $Me_2C(CN)$–; in particular, keteniminyl end-groups $Me_2C{:}C{:}N$– (or groups that might be derived from them) are absent. Experiments at 100°C with AZOF, ^{13}C-enriched in the methyl groups or in the nitrile group, have also indicated that the $Me_2\dot{C}(CN)$ radical gives rise only to the expected end-groups [7]. The reason for the interest in this topic lies in the fact that the compound $Me_2C(CN)\cdot N:C:CMe_2$ is produced in appreciable amounts when AIBN is decomposed in an inert solvent [8]; this product is not stable thermally and so it would be expected that any $Me_2C:C:N$– end-groups also would be labile.

When using ^{13}C-labelled initiator, it is necessary to discover whether any signals from incorporated initiator fragments are obscured by those arising from the monomeric units. A suitable check can be performed readily by difference spectroscopy (see section 3.1.2). In the case of [α-^{13}C]AIBN with STY, another procedure was applied, depending on the fact that initiator fragments were enriched at quaternary carbons [9]. Spectra were obtained using a pulse sequence leading to cancellation of all signals due to carbons carrying hydrogens [10]. It was confirmed that signals between 31.6 and 32.1 ppm were correctly attributed to the quaternary carbons in the end-groups and that no signals from those carbons were buried under signals due to methylene and methine carbons of the monomeric units.

It has been indicated already (see section 3.1.1) that, under some circumstances, NMR studies of end-groups can lead to values of number-average molecular weights of polymers. It has been found for polyMMA prepared using [β,β-^{13}C]AIBN[1] and for polySTY made with [α-^{13}C]AIBN[9] that satisfactory results can be obtained. The conditions for recording the ^{13}C spectra must be carefully selected and it is necessary to have information on the average number of initiator fragments per polymer molecule, i.e. on the relative importance of combination and disproportionation in the termination steps of the polymerizations and also on the frequencies of transfer reactions.

The chemical shift for an end-group derived from a ^{13}C-labelled initiator depends on the nature of the attached monomeric unit. In the case of AIBN, the dependence is expected to be most marked when [α-^{13}C]AIBN is used because the enriched site is one bond closer to the monomeric unit than for other forms of [^{13}C]AIBN but nevertheless [β,β-^{13}C]AIBN is preferred for many applications, notably those concerned with copolymers. The methyl-enriched initiator also has an advantage in some applications because it gives rise to considerably greater intensity of end-group signals as a result of nuclear Overhauser enhancement. Similar considerations apply to the azo ester AZE.

Figure 3.1 shows parts of the 25 MHz ^{13}C NMR spectra of polymers prepared using [β,β-^{13}C]AIBN. For each of the polymers and for those derived from many other monomers, there are two sets of methyl resonances covering equal areas. The presence of the two sets is due to the non-equivalence of the two methyl carbons in the $Me_2C(CN)$– end-groups, brought about by the presence in the adjoining monomeric unit of a chiral centre. Each set clearly contains more than one signal because the monomeric units near the initiator fragment can have various stereochemical arrangements. The values of T_1 are different for the two methyl carbons in the end-groups; they are, for example, 0.30 and 0.21 s for the end-groups in polyMMA. There are also

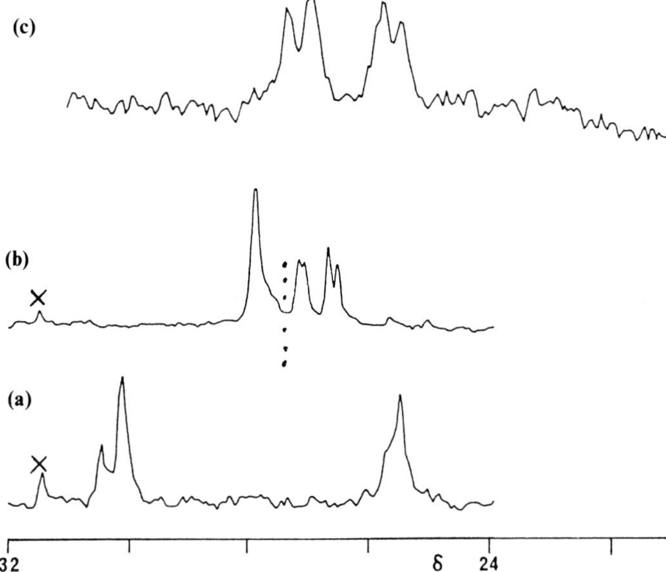

Figure 3.1 Parts of the 25 MHz ^{13}C NMR spectra of polymers made using [β,β-^{13}C]AIBN showing the methyl signals from $Me_2C(CN)$– end-groups. (a) PolyMMA; (b) polySTY; (c) poly(vinyl chloride). Peaks marked × in spectra (a) and (b) are due to residual hexane, used as precipitant for the polymers. The two sets of methyl resonances, referred to in the text, are well separated in spectra (a) and (c); in spectrum (b), the dotted line shows the division between the two sets.

differences between the values of T_2 relaxation times so that the two sets of end-group signals are unequal in area when the spectrum is recorded by the spin-echo technique (see chapter 4).

If the $Me_2C(CN)$– group derived from [β,β-^{13}C]AIBN is attached to a monomeric unit without a chiral centre, as for poly(vinylidene chloride) or the 1,4-unit in polybutadiene, there is only one set of signals from the methyl carbons. The same result is found for poly(ethyl acrylate) [11] even though the monomeric unit possesses a chiral site; in this case, the two sets must be almost coincident presumably because the terminal *meso* and *racemic* placements of the monomeric unit produce very similar shielding of the methyl groups in the initiator fragments.

The resonances for the enriched end-groups in polymers made with [^{13}C]AZE are consistent with the expectation that the end-groups have the structure $Me_2C(COOMe)$– [4, 5]. The spectra, recorded under suitable conditions, resemble those for the corresponding polymers with [^{13}C]AIBN; [β,β-^{13}C]AZE produces end-groups which give equal sets of signals if the monomeric units contain a chiral centre although again an alkyl acrylate seems anomalous [5]. In those cases where there is separation of the two sets of methyl resonances, it is greater when the end-groups are $Me_2C(COOMe)$– rather than $Me_2C(CN)$–.

The 75.5 MHz ^{13}C NMR spectra of low conversion polymers made from conjugated dienes, using [β,β-^{13}C]AIBN, have been analysed [12]. They differ from those in Figure 3.1 in not consisting of two well separated groups of equal area; the spectral complexities for the polymers of the dienes are due to the possibility of more than one type of structure for the unit attached to an initiator fragment. Polybutadiene made by radical polymerization consists mainly of 1,4-units and the principal end-group peak at 26.2 ppm is assigned to $Me_2C(CN)$– attached to such a unit; the presence of a shoulder indicates that the peak is composite because the 1,4-units can be of two types, namely *cis* and *trans*. The two minor peaks between 27 and 28 ppm are explained on the basis of about 8% of the initiator fragments being joined to 1,2-units; these units contain a chiral centre and the peaks are therefore expected to cover equal areas. Similar features are evident for poly(2,3-dimethyl-1,3-butadiene) prepared with [β,β-^{13}C]AIBN but greater complexity is found in the case of polyisoprene because the end-groups (R–) may be joined to monomeric units of four types as in the formulae

$$R\cdot CH_2\cdot CH{:}CMe\cdot CH_2–,$$

$$R\cdot CH_2\cdot CMe{:}CH\cdot CH_2–,$$

$$R\cdot CH_2\cdot \underset{\displaystyle CH{:}CH_2}{\underset{|}{CMe}}–, \quad R\cdot CH_2\cdot \underset{\displaystyle CMe{:}CH_2}{\underset{|}{CH}}–$$

The ^{13}C NMR spectra of polymers with enriched end-groups can yield much information about the stereochemical arrangements of the monomeric units near the ends of the macromolecular chain. The splitting of signals is very clear in the 25.5 MHz spectrum of polyMMA prepared with ^{13}CN–AIBN [1]. The two components have intensities in the ratio 3:1, as for the two components of the low-field signals for methyl-enriched end-groups (see Figure 3.1). The splitting is attributed to the presence of *meso* dyads at some chain-ends and *racemic* dyads at others. It appears that in no case does the stereochemistry of monomeric units close to the initiator fragments differ appreciably from that of units in-chain. Moad *et al.* [9] concluded that the 62.9 MHz ^{13}C NMR spectrum of polySTY, prepared using [α-^{13}C]AIBN, contains stereochemical information about at least eight adjacent monomeric units.

[^{13}C]APE has been used as a thermal initiator for homopolymerization at 100°C [13] and as a photo-initiator at 33°C [14]. The chemical shifts for the enriched sites in the end-groups depend on the nature of the monomer and usually lie in the range 34–39 ppm although, for the polymer from acenaphthylene, they are between 40 and 47 ppm. In most cases, difference spectroscopy is required to distinguish the end-group signals from those due to monomeric units. The 75.5 MHz spectra of homopolymers of conjugated dienes, prepared at 100°C with [^{13}C]APE [12], yield information comparable with that from similar studies involving [β,β-^{13}C]AIBN at 60°C. In the case of polybutadiene, end-groups attached to 1,4-units appear to give a single peak but measurement of T_1 shows it to be composite, presumably because of the occurrence of *cis* and *trans* forms of the unit. For the polymers of dienes, the proportions of the various types of units adjacent to the Me·CHPh– end-group are similar to those found when AIBN is used as initiator.

Studies with [α-^{13}C]AIBN and [β,β-^{13}C]AIBN have not provided evidence for the occurrence of head-addition of the $Me_2\dot{C}(CN)$ radical to any monomer. The presence of initiator fragments attached to the heads of VAC and vinyl formate units has however, been indicated by the presence of two distinct peaks in the 25 MHz spectra of low conversion polymers prepared with ^{13}CN–AIBN at 60°C or with ^{13}CN–AZOF at 100°C [15]. In the case of VAC, a peak at 124.2 ppm covers an area four times that corresponding to a peak at 122.8 ppm. The larger and smaller peaks are assigned to the groupings $Me_2C(CN){\cdot}CH_2{\cdot}CH(OAc)$– and $Me_2C(CN){\cdot}CH(OAc){\cdot}CH_2$–, respectively. This assignment is supported (but of course not firmly established) by consideration of the model compounds $Me_2C(CN){\cdot}CH_2{\cdot}CH(OAc){\cdot}Me$ and $Me_2C(CN){\cdot}CH(OAc){\cdot}Me$, the nitrile carbons of which give signals at 124.3 and 122.4 ppm, respectively. The proportion of head-initiation in the polymerization of VAC seems to be somewhat larger when the reaction is performed at comparatively low concentration of monomer and high concentration of initiator. The ratio of head- and tail-additions at a particular temperature, however, cannot be affected by the concentrations of these reactants; it must be concluded that initiation is not the sole source of head-groupings and that some of them arise from other processes, such as primary radical termination in which a radical derived directly from the initiator combines with a growing polymer radical.

3.2.2.3 Use in copolymerizations. Figure 3.2 shows part of the ^{13}C spectrum of a statistical copolymer of STY with MMA, prepared at 60°C with [β,β-^{13}C] AIBN [16]. There are four sets of signals, two corresponding to initiator fragments attached to MMA units (represented as $R{\cdot}M_1$–) and the other pair to end-groups adjacent to STY units and represented as $R{\cdot}M_2$–. The assignment is based on the spectra of the homopolymers and is supported by considering the dependence of the relative sizes of the peaks on the composition of the feed and therefore on that of the derived copolymer. Generally similar spectra are found for the STY/MMA system with [β,β-^{13}C]AZE as initiator [5]. The numbers of $R{\cdot}M_1$– and $R{\cdot}M_2$– end-groups can be compared by considering the areas under the appropriate peaks provided that the spectrum has been recorded under suitable conditions. For comparison of these areas, complete separation of the four sets of peaks is not essential. Referring to Figure 3.2, it is necessary only to be able to find the area (A) for one set and

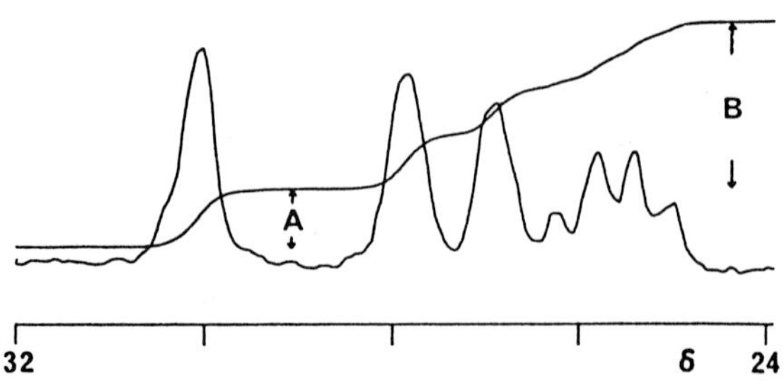

Figure 3.2 Part of the 25 MHz ^{13}C NMR spectrum of a copolymer of STY with MMA, made using [β,β-^{13}C]AIBN in a feed for which [STY]/[MMA] was 0.92. The peak near 30 ppm corresponds to half the $Me_2C(CN)$·MMA– end-groups; the other peaks correspond to the remaining $Me_2C(CN)$·MMA– and all the $Me_2C(CN)$·STY– end-groups.

the combined area (B) for the three other sets; the numbers of $R \cdot M_1$– and $R \cdot M_2$– groups are therefore in the proportions $2A$:(B-A). For systems such as that illustrated by Figure 3.2, the signals at approximately 30 ppm are well separated so that A and B can be found readily.

Figure 3.3 shows parts of the difference spectra of copolymers of MMA with 2,3-dimethylbutadiene (DMB), made using [β,β-^{13}C]AIBN [12]; the spectrum of a homopolymer of MMA, prepared similarly, is shown for comparison. As expected, the general pattern of the spectra differs from that of those in Figure 3.1 because the 1,4-unit derived from the diene does not possess a chiral centre. In Figure 3.3, the peaks near 27.3 ppm arise from initiator fragments adjacent to units derived from the diene and the other peaks from fragments joined to MMA units; the relative sizes vary in the expected manner with the composition of the feed. These spectra are discussed further in later paragraphs.

During the initiation of copolymerization, there is competition between the monomers for capture of the radicals derived from the initiator, according to the scheme

$$R\cdot + M' \rightarrow R\cdot M'\cdot, \quad \text{rate constant} = k'$$

$$R\cdot + M'' \rightarrow R\cdot M''\cdot \quad \text{rate constant} = k''$$

so that

$$\frac{k'[M']}{k''[M'']} = \frac{\text{no. of } R\cdot M'\text{– groups}}{\text{no. of } R\cdot M''\text{– groups}} = \frac{\text{spectral area for } R\cdot M'\text{–}}{\text{spectral area for } R\cdot M''\text{–}} \tag{3.3}$$

where [M′] and [M″] are the concentrations of the monomers in the feed. Application of this equation allows comparison of the rate constants for the reactions of the monomers with the initiating radical. MMA is particularly suitable as a reference monomer in comparisons involving the radicals $Me_2\dot{C}(CN)$ and $Me_2\dot{C}(COOMe)$, derived from AIBN (or AZOF) and AZE,

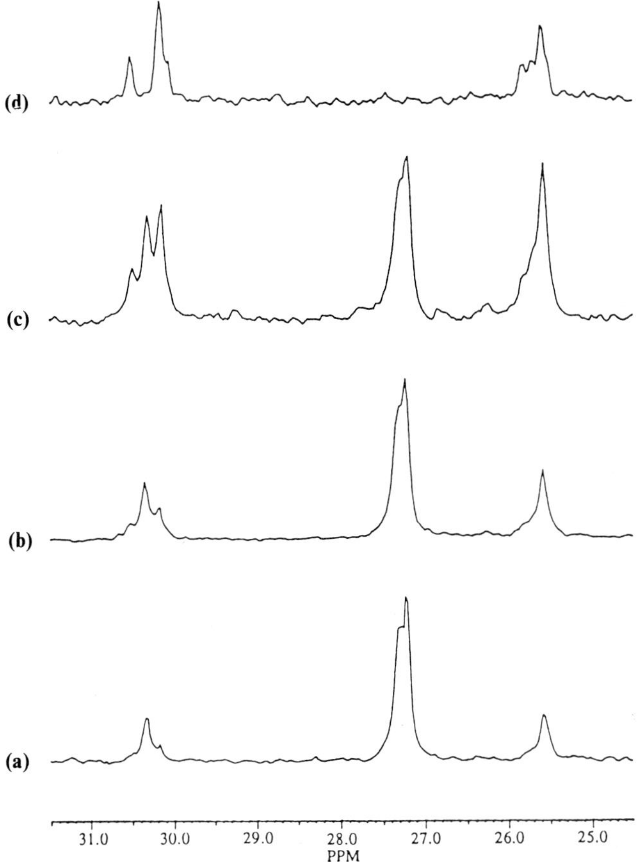

Figure 3.3 Difference spectra for copolymers of MMA with 2,3-dimethylbutadiene (DMB), made using [β,β-^{13}C]AIBN. Spectra were recorded on a spectrometer operating at 75 MHz. [DMB]/[MMA] in the feed had values (a) 0.58; (b) 0.40; (c) 0.17. Spectrum (d) refers to the homopolymer of MMA and can be compared with spectrum (a) in Figure 3.1, recorded at 25 MHz.

respectively, because of the wide separation between the two sets of resonances for the enriched methyl carbons in the initiator fragments attached to MMA units. It is also helpful that MMA copolymerizes readily with many monomers and that the processes are well characterized. The treatment has been applied to many monomers.

Comparisons have also been made of the reactivities of monomers towards the 1-phenylethyl radical, using [^{13}C]APE to initiate copolymerizations [13] and following the general procedure applied for the $Me_2\dot{C}(CN)$ radical. In this case, acenaphthylene was selected as the reference monomer because the chemical shifts for the end-groups attached to it are quite different from those for the end-groups joined to monomeric units of other types.

The signals from end-groups R·M′–, where R– is derived from a ^{13}C-labelled initiator in a homopolymer of monomer M′ cannot be identical to those from the same end-groups in a copolymer of monomers M′ and M″. Differences must arise from effects of the second and subsequent monomeric units which may be either M′ or M″ in a copolymer containing comparable numbers of the two types of monomeric units but can only be M′ in a homopolymer. The spectral differences are usually not evident in 25 MHz spectra but can be clearly seen in spectra recorded at higher fields, say 75 MHz.

In Figure 3.3, the downfield signals for $Me_2C(CN)$·MMA– end-groups are composite; the relative sizes of the components vary systematically with the composition of the feed and therefore with that of the copolymer. The peaks at approximately 30.2 and 30.6 ppm grow relative to that at approximately 30.4 ppm with increasing proportion of MMA in the feed and the central peak is absent for the homopolymer of MMA. The outer peaks are therefore assigned to end-groups R·MMA·MMA– and the central peak to R·MMA·DMB–; reasonable comparisons of areas can be made. Application of an equation analogous to (3.3) makes possible comparison of the rate constants for the reactions of 2,3-dimethybutadiene (DMB) and MMA with the radical R·MMA·; it is found that k_{DMB}/k_{MMA} is about 2.9. A procedure of this type was first used for comparisons of the reactivities of MMA and AN towards the radical Me·CHPh·AN·, produced during the copolymerization of those monomers with APE as initiator [17].

The comparisons of the reactivities of monomers towards the radicals $Me_2\dot{C}(CN)$ and Me·$\dot{C}$HPh have contributed to a re-assessment of a fundamental assumption in the usual treatment of radical copolymerization, namely the absence of so-called penultimate group effects in growth reactions. The reactivity of a growing radical is supposed to be governed solely by the nature of the monomeric unit last added, i.e. there is no influence of units further from the reactive site.

The dependence of the compositions of copolymers on those of the mixtures of monomers from which they are formed can provide evidence for penultimate group effects. Measurements of sequence lengths in copolymers by analysis of their NMR spectra, as discussed in chapter 2, are, however, rather more satisfactory for detecting the effects [18]. Both approaches show that the nature of the penultimate group in a polymer radical may affect the ratio of the rate constant for the competing growth reactions ($P_A^{\cdot} + M_A \rightarrow$) and ($P_A^{\cdot} + M_B \rightarrow$), i.e., selectivity in the reactions of the radical. If the two rate constants are affected to a similar extent, the effects of penultimate groups may be revealed only by measurements of rates of copolymerizations, as in the case of STY with MMA [19]. Determination of the tacticity of polyMMA, found by analysis of the carbonyl signals in the ^{13}C NMR spectrum of the polymer (see chapter 1), suggests that the growth reaction in the radical polymerization at 60°C is significantly influenced by the penpenultimate unit in the growing radical [20].

The initiating radicals derived from AIBN, AZE and APE can be regarded as models for the growing radicals of MAN, MMA and STY, respectively; in each case, the model radical has a hydrogen atom in place of a polymer chain. In the absence of effects of non-terminal groups, it would be expected that the ratio of the rate constants ($k_{M'}/k_{M''}$) for the reactions of two monomers with, say, the $Me_2\dot{C}(CN)$ radical, would be close to that for their reactions with the polyMAN radical, as deduced from studies of copolymerizations and values of monomer reactivity ratios.

Generally there is fair agreement between selectivity in the reactions of a polymer radical with monomers and that in the corresponding reactions of the appropriate model radical. This feature can be well illustrated by considering the case of STY and MMA with the radicals $Me{\cdot}\dot{C}HPH$ and $Me_2\dot{C}(COOMe)$; as expected from studies of copolymerizations, MMA is about twice as reactive as STY towards the radical regarded as a model for the polySTY radical whereas it is about half as reactive towards the radical representing a polyMMA radical [5].

Divergences between results for models and macroradicals are greatest in those cases where penultimate group effects are known to occur (or could reasonably be expected) in copolymerization. Discrepancies are evident also for monomers known to have quite low ceiling temperatures in their polymerizations [21]. The study of penultimate groups was extended by examining the reactivities of STY and AN towards the initiating radicals 1-(3-cyano-1-phenylpropyl) ($CH_2CN{\cdot}CH_2{\cdot}\dot{C}HPh$) and 1-(1,3-diphenylpropyl) ($Ph{\cdot}CH_2{\cdot}\ CH_2{\cdot}\dot{C}HPh$) which can be regarded as models for the polySTY radical with penultimate AN and STY units, respectively [22]. The radicals were generated from the symmetrical azo compounds referred to already, enriched with ^{13}C at position 1; the procedure depended on the application of an equation similar to (3.3). If penultimate group effects are ignored, the two radicals would be expected to show the same selectivity in their reactions with monomers; it was shown, however, that the values of k_{STY}/k_{AN} differed by a factor of about 2.5.

3.2.3 Peroxides

At one time, benzoyl peroxide (BPO) was the favoured initiator for radical polymerizations in non-aqueous media. It is still widely used and it is also important as a component of redox initiating systems; it is discussed in some detail in this section. BPO is an example of a diaroyl peroxide and there are many other types of organic peroxides used as initiators of polymerizations, giving end-groups of various types; extensive lists are available, e.g. the catalogues and pamphlets issued by AKZO and by the Nippon Oil and Fats Company.

There have been few studies by NMR techniques of end-groups derived from peroxides other than BPO and a fluorinated derivative. Di-*tert*-butyl diperoxyoxalate has been used as a source of *tert*-butoxy radicals giving end-groups Me_3CO– in polySTY [23]. The ^1H and ^{13}C NMR signals from the end-groups in polymers of moderately high molecular weight were seen as a result of application of the spin-echo T_2 discrimination technique. End-groups in polybutadiene made with hydrogen peroxide as initiator have been examined. The numbers of the resulting –OH end-groups attached to the three types of butadiene unit (**2**, **3** and **4**) were compared [24] through differences between the chemical shifts for the protons in the adjacent methylene groups.

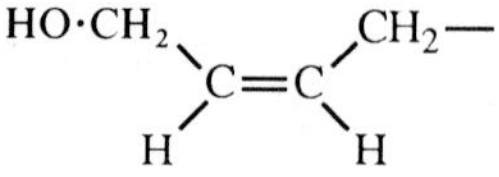

$HO{\cdot}CH_2(H)C{=}C(H)CH_2{-}$ **(2)**

$HO{\cdot}CH_2(H)C{=}C(H)CH_2{-}$ 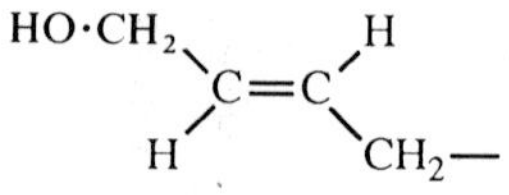 **(3)**

$HO{\cdot}CH_2{\cdot}CH(CH{:}CH_2){-}$ **(4)**

The first account of the use of benzoyl-carbonyl-[^{13}C]-peroxide ([^{13}C]BPO), given by Moad *et al.* [25], referred to benzoate end-groups in polySTY. The 62.5 MHz ^{13}C NMR spectrum of the polymer contains a signal at approximately 166.2 ppm, attributed to the expected end-groups $Ph{\cdot}CO{\cdot}O{\cdot}CH_2{\cdot}CHPh$– formed by attachment of the benzoyloxy radical to the tail of a molecule of the monomer. The two components, not fully resolved but clearly of approximately equal intensities, correspond to *d* and *l* configurations about the chiral site in the monomeric unit. Further splitting can be expected because of stereochemical arrangements for the second and subsequent monomeric units in the polymer chain. The spectrum contains other signals derived from benzoate groups, covering in total an area about 10% of that corresponding to the main signal from the incorporated initiator fragments. It was supposed, with support from the consideration of the carbonyl resonances of appropriate model compounds such as 1-phenylethyl benzoate, that these additional resonances are due to end-groups $Ph{\cdot}CO{\cdot}O{\cdot}CHPh{\cdot}CH_2{\cdot}CH_2{\cdot}CHPh$– and $Ph{\cdot}CO{\cdot}O{\cdot}CHPh{\cdot}CH_2{\cdot}CHPh{\cdot}CH_2$– in which the benzoate group is joined to the head of a monomeric unit. End-groups of the first of these types arise from head-initiation followed by the 'normal' addition of a second molecule of monomer; end-groups of the other type can be formed by primary radical termination. Benzoate groups joined to the heads of monomeric units can also result from transfer to initiator. The results were consistent with those of trapping studies [26].

Similar experiments indicated about 5% head-initiation with MMA [27]. There seem as yet to be no reports on the application of the ^{13}C NMR procedure to the polymers derived from other monomers with BPO.

The use of [^{13}C]BPO for a detailed study of the initiator fragments in polymers has a limitation in that it gives no direct information on phenyl end-groups derived from the peroxide; these groups are formed because some of the benzoyloxy radicals undergo decarboxylation to give phenyl radicals that can react with monomer in an initiation process or with growing polymer radicals in primary radical termination. The balance between benzoate and phenyl end-groups depends upon the nature of the monomer and conditions prevailing during polymerization (concentration of monomer, temperature and pressure). Under some circumstances, phenyl groups form a high proportion of the incorporated initiator fragments. For study of both benzoate and phenyl end-groups, the use of peroxide appropriately labelled with radioactive isotopes (^{14}C and/or ^{3}H) is effective but it cannot distinguish between end-groups attached to the heads of monomeric units and those joined to the tails [28]. It is possible that useful results could be obtained by the use of peroxide enriched with ^{13}C in the rings although there might be great complexities unless the initiator could be enriched only at particular sites in the phenyl rings [1].

When *p*-fluorobenzoyl peroxide is used as initiator for MMA, ^{19}F NMR signals from fluorobenzoate and fluorophenyl end-groups in the polymer can be clearly recognized; their intensities can be compared to give the relative numbers of the two types of initiator fragments in the polymer [29]. Of course, the introduction of the electron-withdrawing fluorine atom into the benzoyloxy radical modifies its reactivity towards monomers and also its tendency to undergo decarboxylation.

The ^{13}C NMR spectra of copolymers of MMA with STY, made using [^{13}C]BPO, contain the signals expected for benzoate groups joined to MMA units and also those expected for the groups attached to STY units; the relative intensities depend upon the composition of the feed [27]. The two types of signals are resolved at 62.5 MHz but not at 25 MHz. The peaks are somewhat different from those in the spectra of the two homopolymers because the second and subsequent monomeric units in copolymers may be derived from either monomer. In the spectra of copolymers, the ratios of the areas covered by signals arising from the groups $Ph{\cdot}CO{\cdot}O{\cdot}CH_2{\cdot}CHPh$– and $Ph{\cdot}CO{\cdot}O{\cdot}CH_2{\cdot}C(Me)(COOMe)$– are consistent with the results obtained with radioactively labelled peroxide, showing that at 60°C STY is about eight times as effective as MMA in capturing benzoyloxy radicals.

[^{13}C]BPO has been used in extensive studies of the participation in radical polymerizations of unsaturated substances not usually regarded as comonomers. Much of the work has involved stilbene and similar substances that show unexpectedly high reactivities towards the benzoyloxy radical. These additives become incorporated in polyMMA, for example, mainly at sites very close to the end-groups; the resulting structures may be useful in subsequent attempts to modify the polymers.

The spectra of polymers of MMA, prepared using [^{13}C]BPO with stilbene-like substances present at low concentrations, show the signals at approximately 166 ppm expected for benzoate groups attached to MMA units (see Figure 3.4). They are accompanied by other benzoate signals having intensities depending upon the concentration and nature of the additive in the feed

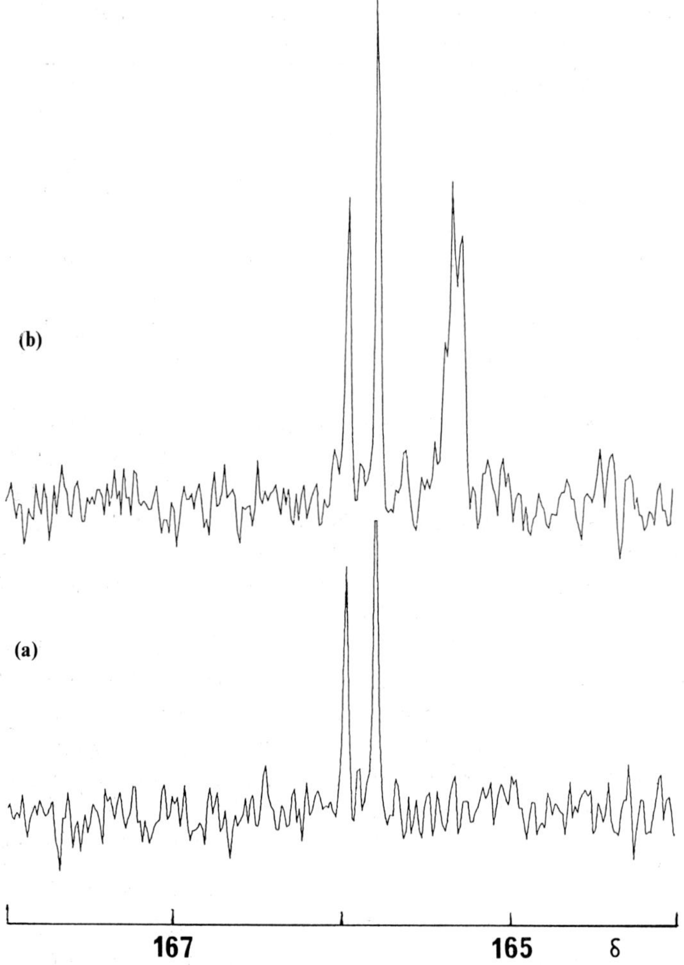

Figure 3.4 Parts of the 100 MHz ^{13}C NMR spectra of samples of polyMMA made using [^{13}C]BPO, showing the carbonyl signals from benzoate end-groups. (a) Polymer prepared from MMA without additive; (b) polymer prepared from monomer containing the *trans*-isomer (*E*-isomer) of 4-(2-phenylethenyl)-1,1′-biphenyl, referred to as Biph–STL, $C_6H_5{\cdot}CH{:}CH{\cdot}C_6H_4{\cdot}C_6H_5$; [Biph–STL]/[MMA] = 1.06×10^{-2}. These spectra were recorded at the University of Warwick through the SERC NMR Service, by courtesy of Dr. O.W. Howarth.

[30, 31]. These signals cover the much smaller peaks attributed to benzoate groups joined to the heads of MMA units; it must be supposed that they arise from benzoate groups adjacent to units derived from the unsaturated additive. Integration of the spectra and application of equation (3.3) lead to evaluation of k'/k where k' and k are the rate constants for the additions of the initiating radical to an additive and MMA, respectively. The high reactivities of stilbene, its derivatives and its analogues towards the benzoyloxy

Table 3.1 Reactivities of some stilbene-like compounds (Ph·CH:CH·R) towards the benzoyloxy radical (see [33] for detailed references)

Group R in general formula	Relative reactivity at 60°C
Phenyl	48
p-Chlorophenyl	90
4-Biphenyl	103
1-Naphthyl	113
9-Phenanthryl	145
4-Pyridyl	8
2-Thienyl	290
Methyl methacrylate	1 (standard)
Styrene	8

radical are indicated by the information presented in Table 3.1. Lower but nevertheless significant reactivities towards the benzoyloxy radical have been found for various other α, β-disubstituted monomers such as β-methylstyrene [32]. In all these cases, very little of the additive is incorporated as a result of 'normal' copolymerization; the additive chemically bound in the polymer is very largely confined to sites adjacent to benzoate end-groups. The topic has been considered in detail in a review [33].

Experiments with 4-fluorostilbene provided confirmation of the presence near benzoate end-groups of units derived from stilbene-like compounds [34]. The additive was used in polymerizations of STY initiated by BPO and by AIBN. Figure 3.5 shows the 93.7 MHz ^{19}F NMR spectra of polymers prepared at 60°C from similar mixtures of STY with 4-fluorostilbene. The product made using AIBN exhibits signals derived from incorporated 4-fluorostilbene units flanked on both sides by STY units. The complexity of the spectrum is due in part to stereochemical factors and in part to the presence of two forms of the 'foreign' units, namely $-CH_2 \cdot CHPh \cdot \underline{CHPh \cdot CHX} \cdot CH_2 \cdot CHPh-$ and $-CH_2 \cdot CHPh \cdot \underline{CHX \cdot CHPh} \cdot CH_2 \cdot CHPh-$ where X represents C_6H_4F; the two forms correspond to head- and tail-additions of the additive to the polySTY growing radical. Particular attention must be paid to the additional signals in the spectrum of the polymer made using BPO as initiator; the difference spectrum in Figure 3.5 shows clearly that, in the polymer made using BPO, some of the units derived from 4-fluorostilbene have an environment quite different from that of the others. This finding is interpreted as confirming that there is a substantial number of units attached at one end to a benzoate end-group and at the other to a unit derived from STY. One of the two sets of peaks in the difference spectrum corresponds to $Ph \cdot CO \cdot O \cdot CHPh \cdot CHX-$ and the other set to $Ph \cdot CO \cdot O \cdot CHX \cdot CHPh-$. This explanation could be tested by examination of model compounds and also by repetition of the study using the symmetrical 4,4′-difluorostilbene as additive so that, in the formulae, Ph would be changed to X and only one set of peaks would be expected in the difference spectrum.

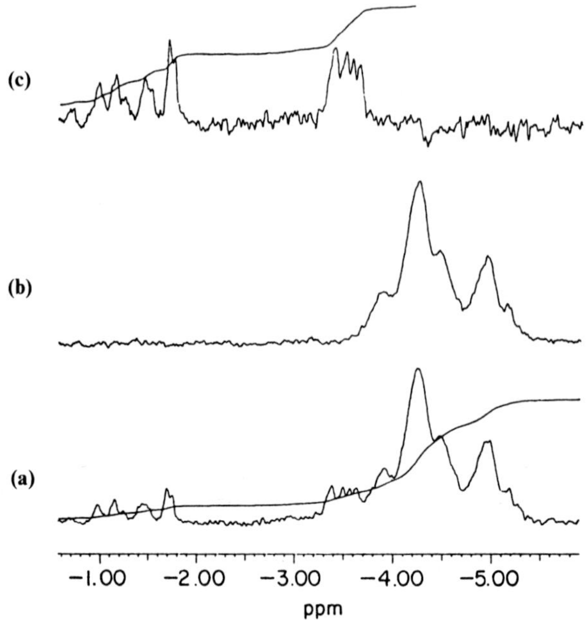

Figure 3.5 93.7 MHz ^{19}F NMR spectra of polymers made from STY with 4-fluorostilbene (FSTL). (a) BPO used as initiator with [STY]/[FSTL] = 6.17; (b) AIBN used as initiator with [STY]/[FSTL] = 7.10; (c) difference spectrum constructed from spectra (a) and (b).

Substances of certain other classes resemble stilbene in possessing high reactivity towards the benzoyloxy radical; they include 1,4-diarylbuta-1,3-dienes [35, 36]. New principles are not involved in these studies but one point merits attention. There is great complexity in the ^{13}C NMR signals originating from the benzoate groups attached to units derived from one of these additives because the units may correspond to 1,2-, *trans*-1,4- or *cis*-1,4-placements. Similar effects are found when 1,2,3-triphenylcyclopropene is used as an additive [37].

In each of the cases so far examined experimentally, the *trans* isomer of the stilbene-like compound is appreciably more reactive than the *cis* isomer. This feature can be illustrated by consideration of the *trans,trans*, *trans,cis* and *cis,cis* isomers (i.e. the *EE*, *EZ* and *ZZ* isomers, respectively) of 1,4-distyrylbenzene, Ph·CH:CH·C_6H_4·CH:CH·Ph, which can be regarded as a derivative of stilbene with a styryl group at the *para* position in one of the phenyl groups. Each of the isomers has two sites for reaction with the benzoyloxy radical. Figure 3.6 shows parts of the ^{13}C NMR spectra of polymers of MMA, made using [^{13}C]BPO and an isomer of distyrylbenzene; the signals near 166 ppm arise from benzoate end-groups attached to MMA units and the others from the initiator fragments adjacent to units formed from the additive. Using the procedure already described, the reactivities

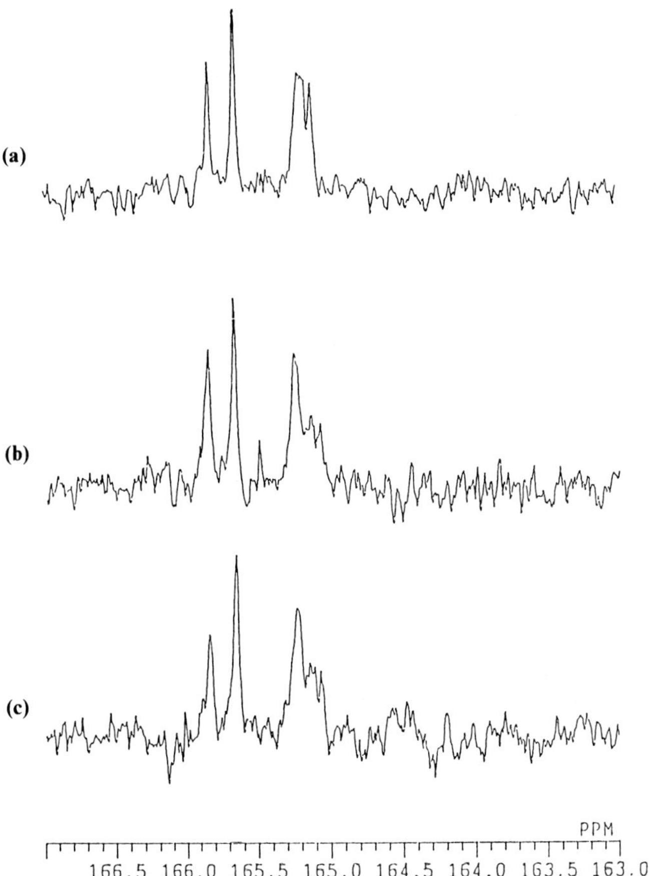

Figure 3.6 Parts of the 100 MHz ^{13}C NMR spectra of samples of polyMMA, made using [^{13}C]BPO in the presence of 1,4-distyrylbenzene (DSB); the spectra show the carbonyl signals from benzoate end-groups. (a) *EE*-isomer of DSB, [DSB]/[MMA] = 0.26×10^{-2}; (b) *EZ*-isomer, [DSB]/[MMA] = 0.63×10^{-2}; (c) *ZZ*-isomer, [DSB]/[MMA] = 0.99×10^{-2}.

towards the benzoyloxy radical (relative to that of MMA) are about 430, 170 and 100 for the *EE*, *EZ* and *ZZ* isomers, respectively [38]. It is possible to explain the similarity between the spectra for polymers prepared using the *EZ* and *ZZ* isomers and the difference between them and the spectrum for the polymer made with the *EE* isomer. The *EZ* isomer is expected to react preferentially at the *E*-styryl group so that most of the resulting units in the polymer resemble those formed from the *ZZ* isomer in the sense that, in both cases, the unreacted styryl groups have the *Z* configuration; on the other hand, an *E*-styryl groups remains in each unit derived from the *EE* isomer.

3.2.4 Other initiators

An application of NMR for study of end-groups in polymers produced by anionic polymerization involved use of the spin-echo technique [23]. It was applied for detection of *tert*-butyl end-groups in polyMMA made using unenriched di-*tert*-butyl magnesium as initiator.

Very early and notable applications of ^{13}C-enriched materials in initiating systems referred to polymerizations of the Ziegler–Natta type [39, 40]. Propylene and other α-olefins may be polymerized using catalysts based upon mixtures of aluminium alkyls and transition metal halides; the resulting polymers generally are highly isotactic. The mechanism of stereochemical control in such systems has been much debated with some arguments in favour of a mechanism (equation (3.4)) whereby control arises from the repeated stereospecific insertion of the olefin into a transition metal–alkyl bond at the catalyst surface. In an alternative proposal (equation (3.5)), metallocarbene and metallocyclobutane intermediates participate. If the latter mechanism were to apply, stereochemical control might be exercised by the growing chain itself rather than by the catalyst.

$$\text{Mt–CH}_3 \xrightarrow{C_3H_6} \text{Mt–CH}_2\text{–CHMe}_2 \xrightarrow{C_3H_6} \text{Mt–CH}_2\text{–CHMe–CH}_2\text{–CHMe}_2 \quad \text{(etc.)} \tag{3.4}$$

$$\text{Mt–CH}_3 \longrightarrow \overset{\text{H}}{\dot{\text{M}}}\text{t}=\text{CH}_2 \xrightarrow{C_3H_6} \begin{array}{c}\text{H}\\ \dot{\text{M}}\text{t}-\text{CH}_2\\ |\qquad|\\ \text{CH}_2\text{–CH–Me}\end{array}$$

$$\swarrow \qquad \searrow$$

$$\begin{array}{c}\text{Mt}\quad\text{Me}\\ |\qquad|\\ \text{CH}_2\text{–CH–Me}\end{array} \qquad \begin{array}{c}\text{Mt–CH}_2\\ |\\ \text{Me–CH–Me}\end{array} \tag{3.5}$$

The uncertainty about the mechanism has been resolved very elegantly by Zambelli *et al.* [39] who prepared isotactic polypropylene using catalysts in which the initial active centre consisted of Ti–^{13}CH$_3$ groups. The ^{13}C NMR spectra of the polymers gave evidence for ^{13}C-enrichment only in the terminal methyl groups of the polymer (**5**) with no enrichment at a methylene site. The mechanism shown in equation (3.5) envisages isomerization involving methyl and methylene groups and so it would correspond to enrichment at sites of both types. It must be noted in connection with (**5**) and (**6**) that ^{13}C can be present at only one site in any particular chain and the structures shown represent averages for assemblies of chains. Similar techniques and arguments were applied to the polymerization of 1-butene [40].

$$\sim\text{CH}_2\text{—CH}\begin{array}{l}\diagup {}^{13}\text{CH}_3\\ \diagdown {}^{13}\text{CH}_3\end{array} \qquad \sim{}^{13}\text{CH}_2\text{–CH}\begin{array}{l}\diagup {}^{13}\text{CH}_3\\ \diagdown {}^{13}\text{CH}_3\end{array}$$

(5) **(6)**

Maleimide and derivatives such as N-phenylmaleimide can be polymerized with triphenylphosphine as initiator [41] to give products with the repeat unit (**7**). The ^{31}P NMR spectrum of the polymer dissolved in d_6-dimethylsulphoxide contained signals at -28.3 and -13.4 ppm (referenced to H_3PO_4) attributed to phosphonium salts and ylids, respectively, thought to be present as end-groups (**8**) and (**9**).

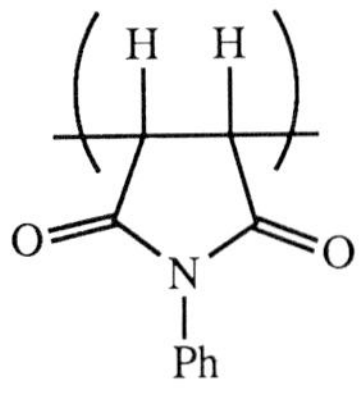

(7) 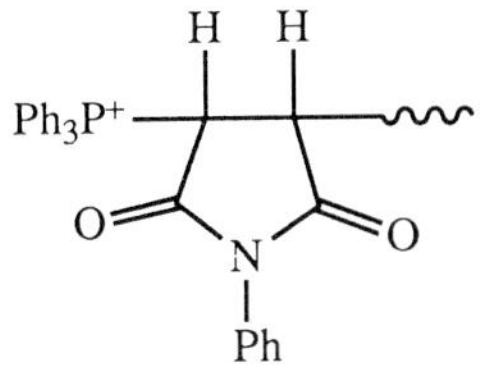 **(8)** 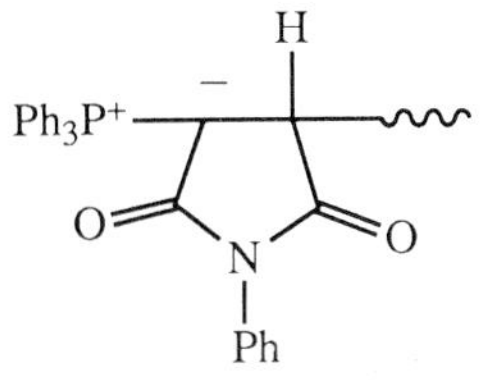 **(9)**

The signals were complex probably because of the occurrence of alternative configurations of the monomeric units near the end-groups. Addition of diazabicyclononene to the solution under examination caused disappearance of the signals due to phosphonium salt and reinforcement of those due to ylid. The change was reversed quantitatively by subsequent addition of trifluoroacetic acid. The sensitivity of ^{31}P NMR is such that large signals were obtained for materials containing about 0.2% phosphorus by weight, corresponding to about 100 phenylmaleimide units for each phosphorus atom.

Maleic anhydride also can be polymerized with triphenylphosphine as initiator but the product is somewhat complex. The ^{31}P NMR spectrum of the polymer of maleic anhydride and the effects on it of treatments with diazabicyclononene and trifluoroacetic acid suggest that all the phosphorus atoms in the polymer are present as phosphonium salts [41].

There is interest in potassium perphosphate as an initiator of radical polymerization, either by itself or as a component of redox systems. Polymers of acrylamide, made with various systems involving perphosphate and subsequently purified by dialysis against water, showed no ^{31}P NMR signals; this result suggests that phosphate radical-ions are not incorporated to an appreciable extent in the final polymers [42]. IR spectroscopic evidence was previously cited [43] as showing the presence of phosphate groups in poly-MMA made using perphosphate with sodium thiosulphate. The polymers were purified by precipitation techniques. It is notoriously difficult to remove all traces of inorganic material from polymers during their precipitation from solutions in organic solvents. It is possible that the IR absorption attributed to phosphorus-containing end-groups was actually due to occluded perphosphate or material derived from it.

It is convenient to consider at this point the case of poly(aryl ether ether ketone) (PEEK), prepared by step-growth polymerization involving a fluoroketone and resorcinol; the first stage is represented thus

$$F{\cdot}C_6H_4{\cdot}CO{\cdot}C_6H_4{\cdot}F + HO{\cdot}C_6H_4{\cdot}OH \longrightarrow F{-}C_6H_4{-}CO{-}C_6H_4{-}O{-}C_6H_4{-}OH + HF \quad (3.6)$$

The polymerization does not require an initiator of one of the types so far considered in this chapter but use of a slight excess of the ketonic reactant leads to the polymer having a fluorine atom at each chain-end. An accurate value of the fluorine content of the polymer would permit determination of its number-average molecular weight.

The analysis has been performed by means of ^{19}F NMR with the sodium salt of trifluoroacetic acid as internal standard for both chemical shift and fluorine content [44]. The notable advantage of the NMR method is the avoidance of interference from fluorine-containing impurities in the polymer. Any such impurities would give rise to ^{19}F signals quite distinct from those originating from the fluorines at the chain-ends. The T_1 relaxation times for the end-of-chain fluorines and those in the standard are 0.3 and 1.1 s, respectively, and so to ensure valid comparison between peak areas and fluorine contents for standard and polymer, the delay between scans was extended to 5.6 s.

3.2.5 Transfer agents and retarders

Valuable information about the growing centres in ionic polymerizations can be acquired by using reagents that terminate growth and give end-groups that can be examined by NMR. The particular interest lies in the precise nature and stereochemistry of the monomeric units near the end-groups. Hogen-Esch and co-workers [45–47] have made extensive use of $^{13}CH_3I$ in anionic polymerizations of monomers including 2- and 4-vinylpyridine and MMA. The general reaction is represented as

$$P_n^- + {}^{13}CH_3I \rightarrow P_n{\cdot}{}^{13}CH_3 + I^- \quad (3.7)$$

Analysis of the ^{13}C NMR spectrum of a polymer made under these conditions can give information about the stereochemistry of two or three of the dyads at the end of the chain. When methyl iodide terminates the anionic polymerization of MMA, the resulting terminal unit is $-CH_2{\cdot}CMe_2{\cdot}COOMe$ which resembles that produced when a particular azo ester is used as a radical initiator for the monomer (see section 3.2.2).

Diphenylchlorophosphonate terminates anionic polymerizations thus

$$P_n^- + Cl{\cdot}P(O)(OPh)_2 \rightarrow P_n{\cdot}P(O)(OPh)_2 + Cl^- \quad (3.8)$$

In the ^{31}P NMR spectra of the resulting polymers, the chemical shifts for the end-groups depend upon the nature of the attached chain [48]; for ethylene oxide and styrene oxide, they are respectively −11.5 and −13.0 ppm referenced to H_3PO_4. In the case of styrene oxide, comparison of the spectrum with those of model compounds indicated that propagation most probably occurs through a secondary alcoholate ion, meaning that ring-opening occurs by β-scission

$$\text{(styrene oxide: H, H, Ph, H; }\beta\text{, O, }\alpha\text{)} \xrightarrow{\beta\text{ - scission}} \text{- CH}_2\text{·CHPh-O}^- \qquad (3.9)$$

and not by α-scission which would produce $-CHPh·CH_2-O^-$ (see section 3.3.1). A similar method of working was applied to the cationic polymerization of tetrahydrofuran and other cyclic ethers, using *tert*-phosphines as terminating agents [49].

The value of NMR for study of transfer reactions in radical polymerizations can be illustrated by considering a case involving VAC. Transfer reactions for that monomer occur more readily than for most other monomers; the reactions include some which lead to structural branching in the polymer (see section 3.4.3). Industrially, large quantities of polyVAC are hydrolysed to the water-soluble poly(vinyl alcohol) (see chapter 2) which, for some applications, needs to be of comparatively low molecular weight and has improved properties if the ends of the molecules have quite large hydrophobic groups. These groups might be introduced into the precursory polyVAC by choice of an appropriate initiator for the polymerization but reduction in the average chain length of the polymer and the introduction of the required end-groups can be achieved simultaneously by the use of a suitable transfer agent. Such a substance is 2-mercaptoethanol [50] for which the dominant reactions are

$$P_n^{\cdot} + HS·CH_2·CH_2·OH \rightarrow P_n·H + {}^{\cdot}S·CH_2CH_2·OH \qquad (3.10)$$

$$HO·CH_2·CH_2·S^{\cdot} + M \rightarrow HO·CH_2·CH_2·S·M^{\cdot}. \qquad (3.11)$$

The presence of hydroxyethylmercapto end-groups in the derived poly(vinyl alcohol) was confirmed by ^1H NMR for a sample of polymer with a number-average degree of polymerization of about 80. Similar experiments involving *n*-alkyl mercaptans up to $C_{18}H_{37}SH$ gave comparable results [51].

^{29}Si NMR has been used in conjunction with ^{13}C NMR for study of transfer to monomer in the cationic cyclopolymerization of diallylmethylphenylsilane [52]. The process is represented in equation (3.12) but the number-average degree of polymerization ($\overline{DP}_n$) of the product is low and is remarkably insensitive to the concentration of the monomer in the polymerizing system and to other experimental conditions. It is supposed [53] that transfer to monomer occurs readily according to equation (3.13)

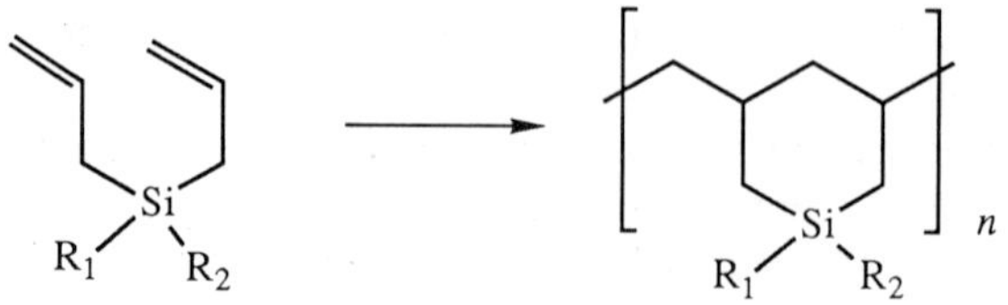

(3.12)

(3.13)

in which the first product corresponds to an uncharged chain-end and the second to a cation that may re-initiate polymerization. The dominant group of resonances in the ^{29}Si NMR spectrum of the polymer corresponds to 74% of the total signal intensity; for the polymer under consideration, the value of $\overline{DP}_n$ is eight and so these signals can be attributed to the silicons of in-chain units. The remaining signals are associated with the silicons in terminal units. The ^{13}C NMR spectrum of the polymer is consistent with these conclusions.

3.2.6 *Main-chain scission*

End-groups can be formed by scission of the main chains of polymer molecules and ^{13}C NMR has been used in this connection with ethylene-propylene copolymers exposed to γ-irradiation [54]. Crosslinking also occurs so that linewidths increase with radiation dose because of decrease in molecular mobility and decrease in the spin–spin relaxation time (T_2) (see chapter 4). The total intensity of the NMR signals decreases because of the immobilization caused by crosslinking and the resulting strong dipole–dipole coupling. Nevertheless, resonances due to new methyl end-groups were detected and attributed to cleavage of C–C bonds in the main chains; it was deduced that cleavage occurs preferentially at bonds adjacent to units derived from propylene.

There is great interest in the preparation of telechelic oligomers to be used in the formation of block copolymers and in reactive processing. One procedure, giving α,ω-functional oligomers, involves the use of high polymers having small contents of comonomers which give units susceptible to main chain cleavage. The presence of particular end groups in the products has in many cases been confirmed by NMR studies which have also been used to determine the actual contents of the end-groups. Examples of the procedure include ozonolysis of the in-chain double bonds resulting from the incorporation of conjugated dienes in polymers of alkyl acrylates and methacrylates [55]. The nature of the end-groups in the resulting oligomers depends on the type of diene used as comonomer and on the procedure used to work up the ozonide.

Reductive treatment of the ozonides formed from copolymers of MMA with butadiene leads to oligomers with –OH end-groups; their presence was confirmed by the appearance of ^{1}H NMR signals between 3.9 and 4.1 ppm and also ^{13}C NMR signals at 62.3 and 64.6 ppm from the adjacent methylene groups. Acceptable values of $\bar{M}_n$ for the oligomers were obtained from comparison of the areas of the proton peaks of these methylene groups with those of the methoxyl protons in the MMA units.

The ozonides derived from copolymers can also be subjected to oxidative work-up. In the case of butyl acrylate with 2,3-diphenylbuta-1,3-diene, the spectra of the derived oligomers showed the ^{1}H peaks around 7.4 ppm, expected for the protons in phenylketone end-groups, and the ^{13}C peaks at 197.2 and 198.8 ppm for the carbonyl carbons in those end-groups. The following formulae represent (a) the cleavage of the diene in a 1,4-placement and (b) the end-groups in the resulting oligomer of butyl acrylate

$$-CH_2-CPh\,\vdots\!=\!\vdots\,CPh-CH_2- \quad \textbf{(a)} \qquad Ph\cdot CO\cdot CH_2-[CH_2\cdot CH(COOBu)]_n\cdot CH_2\cdot CO\cdot Ph \quad \textbf{(b)}$$

^{19}F NMR was used to characterize fluorophenyl ketone end-groups in oligomers produced by ozonolysis of polymers of MMA and STY, containing units derived from 2,3-di(4-fluorophenyl)buta-1,3-diene [56].

In all the cases examined, the diene is very largely incorporated as 1,4-units but the pendant double bonds of the few 1,2-units can cause complications by giving rise to additional end-groups. In some cases at least, the difficulty can be avoided by replacing the diene by a suitable acetylenic compound as the comonomer used to furnish the in-chain double bonds [57]. For example, phenylacetylene has been used instead of 2,3-diphenylbutadiene to produce oligomers of MMA, having phenyl ketone and carboxylic acid end-groups as indicated by the overall equation

$$\underset{\displaystyle\text{COOMe}}{-CH_2-\underset{|}{C}Me}-CH{=}CPh-CH_2-\underset{\displaystyle\text{COOMe}}{\underset{|}{C}Me}- \xrightarrow[\text{KI/AcOH}]{\text{O}_3\text{/CHCl}_3} -CH_2-\underset{\displaystyle\text{COOMe}}{\underset{|}{C}Me}-COOH + O{=}CPh-CH_2-\underset{\displaystyle\text{COOMe}}{\underset{|}{C}Me}- \tag{3.14}$$

Another promising method for the preparation of oligomers of MMA uses the photolysis of high polymers containing a limited number of units bearing carbonyl groups. These units can be introduced by using methyl vinyl ketone as a comonomer. It is well established that absorption of a UV quantum by a carbonyl group can result in the scission of a carbon–carbon bond in the vicinity of the chromophore. In the cases under discussion, the ^1H and ^{13}C NMR spectra of the oligomeric products are entirely consistent with the view that the end-groups formed as a result of the photolysis can be represented as $CH_2{=}CH{\cdot}CH_2$– [58] and not as $CH_2{=}C(COOMe){\cdot}CH_2$– as had been suggested previously [59], i.e. the predominant unsaturated end-groups are not acrylic in nature.

3.3 In-chain units

3.3.1 Regioselectivity in growth reactions

Growth during the radical polymerization of a vinyl monomer, $CH_2{:}CHX$, occurs almost exclusively by head-to-tail additions so that the reactive centre is represented as $P_n{\cdot}CH_2{\cdot}\dot{C}HX$. Occasional head-to-head addition produces a centre $P_n{\cdot}CH_2{\cdot}CHX{\cdot}CHX{\cdot}\dot{C}H_2$ but the next addition is likely to be tail-to-tail to regenerate the normal growing radical $P_n{\cdot}CH_2{\cdot}CHX{\cdot}CHX{\cdot}CH_2{\cdot}CH_2{\cdot}\dot{C}HX$. The resulting disturbance in the structural regularity of the macromolecular chain may have important consequences in connection with crystallinity of the polymer and also with its thermal stability. The termination of growing radicals by combination leads to a structure such as $P_n{\cdot}CH_2{\cdot}CHX{\cdot}CHX{\cdot}CH_2{\cdot}P_m$ containing a head-to-head grouping but no adjacent tail-to-tail grouping.

Head-to-head placements in polyVAC have been examined using a chemical method [60, 61]. The polymer was converted to poly(vinyl alcohol) so that any head-to-head groupings became vicinal diglycols which could then be cleaved by periodic acid with resulting decrease in the average molecular weight of the polymer. Measurements of molecular weight before and after the cleavage lead directly to the number of head-to-head placements in the average polymer molecule. The method is very sensitive since one cleavage per polymer molecule leads to the halving of the number-average molecular weight. The results of the pioneering work have been confirmed and extended by NMR studies. It must be noted that the conversion of polyVAC to the alcohol may affect the skeleton of the macromolecule in the sense that some branches may become detached and there may be some degradation (see section 3.4.3); it may be necessary to allow for this effect.

If head-to-head and tail-to-tail additions occur during the polymerization of VAC, then a chain grown from left to right can be represented thus when it has been converted to poly(vinyl alcohol)

$$-CH_2{\cdot}\underset{x}{CH(OH)}{\cdot}CH_2{\cdot}CH(OH){\cdot}CH(OH){\cdot}\underset{y}{CH_2}{\cdot}\underset{z}{CH_2}{\cdot}CH(OH){\cdot}CH_2{\cdot}CH(OH)-$$

^1H and ^{13}C NMR have failed to give firm evidence for the presence of 1,2-glycols in samples of the polymer. Tests on poly(vinyl alcohol) made from copolymers of VAC with vinylene carbonate (units derived from which must give rise to 1,2-diglycols) showed that it is possible to determine the vicinal diglycol units by NMR techniques when their number is about 15 for every 100 monomeric units but the sensitivity is insufficient for the method to be useful for poly(vinyl alcohol) prepared from ordinary samples of polyVAC.

Ovenall [62] used a different approach; he showed that the adjacent methylene carbons (*y* and *z* in the formula), resulting from the tail-to-tail addition, gave clear signals in a 100 MHz ^{13}C spectrum, with chemical shifts 28.0 and 33.2 ppm. The chemical shift for methylene carbon *x* was 38.8 ppm. The number of reversed growth reactions was found to be 1.9 for 100 monomeric units, in good agreement with the results obtained by the physico-chemical method for similar polymers.

Head-to-head growth is comparatively frequent during the polymerizations of vinyl fluoride and vinylidene fluoride. The polymers have been studied by 470 MHz ^{19}F NMR with proton-decoupling [63, 64]. The spectrum for poly(vinylidene fluoride) is simpler than that for poly(vinyl fluoride) because the former polymer contains no asymmetric carbons.

A section of a poly(vinylidene fluoride) molecule is shown in the formula

$$-\underset{}{CH_2}\cdot\underset{a}{CF_2}\cdot CH_2\cdot\underset{b}{CF_2}\cdot\underset{c}{CF_2}\cdot CH_2\cdot CH_2\cdot\underset{d}{CF_2}\cdot CH_2\cdot CF_2-$$

for growth from left to right. Here a sequence of head-to-tail reactions is followed by a head-to-head and then a tail-to-tail addition before resumption of head-to-tail processes. The fluorines *a* to *d* have different environments and the 188 MHz ^{19}F NMR spectrum contains four peaks; comparison of their areas indicates that a typical commercial polymer has a structure resulting from one head-to-head and the following tail-to-tail process for about 30 head-to-tail reactions.

The assignment of resonances was originally based on comparison of the chemical shifts with those for appropriate fluorocarbons. Confirmation has been provided by a procedure involving the preparation of polymers by a method leading to materials with various proportions of inverted monomeric units. Copolymers of $CH_2{:}CF_2$ with low contents of $CF_2{:}CHCl$ were prepared. The presence of the large chlorine atom in the molecule of the comonomer ensures that it enters the copolymer in such a way that the structure $-CH_2\cdot CF_2\cdot CF_2\cdot CHCl\cdot CH_2\cdot CF_2-$ is produced. The copolymers were treated with tri-*n*-butyltin hydride so that the comonomer units were converted to those that would be formed by head-to-head addition of a molecule of vinylidene fluoride.

A section of the chain of poly(vinyl fluoride) can be represented thus

$$-\underbrace{CH_2\cdot CHF\cdot CH_2\cdot CHF}_{a}\cdot\underbrace{CHF\cdot CH_2\cdot CH_2\cdot CHF}_{c}\cdot CH_2\cdot CHF-$$

(with b spanning $CH_2\cdot CHF\cdot CHF\cdot CH_2$ and a spanning $CH_2\cdot CHF\cdot CH_2\cdot CHF$ on the lower line) where, for growth from left to right, units *a*, *b* and *c* correspond to head-to-tail, head-to-head and tail-to-tail additions, respectively. The proton-decoupled 188 MHz ^{19}F spectrum contains many peaks because of the presence of asymmetric carbons in the chain but the signals can be placed into two sets, namely large signals between 178 and 185 ppm for $-CH_2{\cdot}CF_2{\cdot}CH_2-$ and smaller signals between 188 and 200 ppm for $-CH_2{\cdot}CF_2{\cdot}CF_2{\cdot}CH_2-$ groups. Integration of the spectrum allows comparison of the numbers of units of the two types.

The assignment was confirmed by examination of poly(vinyl fluoride) free from inverted units. This material was made by reductive dechlorination of poly(1-fluoro-1-chloroethylene) by means of tri-*n*-butyltin hydride; the bulky chlorine atom in the monomer effectively prevents head-to-head growth. The ^{19}F spectrum of poly(vinyl fluoride) prepared by this indirect route contained only the signals in the range 178–185 ppm.

Considerable attention has been paid to regioselectivity during the ring-opening polymerizations of olefin oxides; polymers and copolymers (both block and statistical) of ethylene oxide, propylene oxide, 1,2-butylene oxide and styrene oxide have been examined from this point of view using ^{13}C NMR [65–67]. In principle, either of the carbon–oxygen bonds in the three-membered ring can be cleaved during the polymerization; this feature and the lack of stereospecificity in many of the polymerizations lead to complex microstructures for the chains and to NMR spectra with many peaks. Interpretations of the spectra have been greatly assisted by the use of spectral editing techniques, such as DEPT (see chapter 1), so that CH and CH_2 subspectra could be obtained. Resolution enhancement has also been valuable.

The molecular weights of the polymers were not high, corresponding to average chain lengths of the order of 100, and so it was possible to identify both primary and secondary hydroxyl end-groups in some of the polymers. Application of a simple first-order Markov model (see chapter 2) indicated that a primary chain-end adds preferentially to the CHPh head of a molecule of styrene oxide but that a secondary chain-end is more likely to react with the CH_2 tail. Identification of the end-group signals was confirmed by the variation in intensity with the molecular weight of the polymer. The growth reaction in the polymerization of styrene oxide has also been studied by another method (see section 3.2.5).

3.3.2 Abnormal units

PolyAN readily discolours on exposure to base or heat; the phenomenon has been studied by IR and also by ^{1}H and ^{13}C NMR [68–70]. Coloured samples of polyAN, produced either by heating polymer at 100°C in vacuo or by treatment of a solution of polymer in dimethylformamide with 10% aq. NaOH, show a small ^{1}H NMR peak at 2.7 ppm in addition to the usual methine and methylene proton signals at 2.0 and 3.1 ppm, respectively [69].

This small peak probably arises from the methine and methylene protons in sequences of cyclized monomeric units (see equation (3.15)). The ^1H spectrum of the coloured polymer also contains a small peak at 8.31 ppm, probably due to the imino protons at the ends of the cyclized sequences. The base-initiated colouration of polyAN probably proceeds as shown in equation (3.15). This mechanism would explain why polymers made using anionic initiators based on alkali metals, especially at relatively high temperatures, tend to be discoloured.

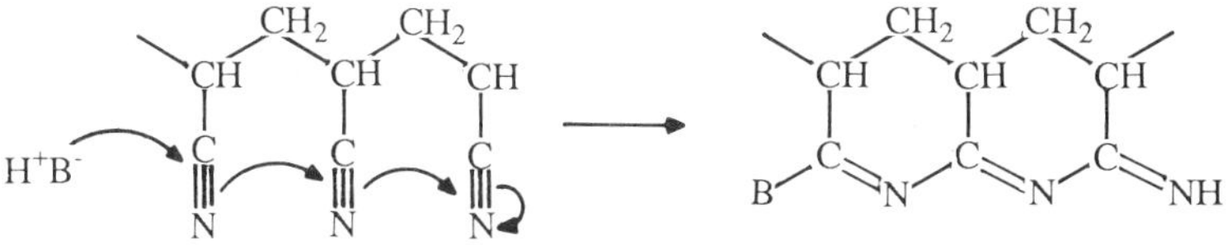

(3.15)

Coloured samples of polymer, prepared using anionic initiators, display several small ^{13}C signals between 12 and 16 ppm. It is likely that they are associated with chain branching rather than with the cyclized structures.

Other defects that may occur in polyAN include acrylamide (**10**) and acrylic acid (**11**) units, resulting from adventitious hydrolysis of AN units, and also derived cyclic structures such as anhydrides (**12**), imino-amides (**13**) and imino-esters (**14**). These structures give rise to characteristic additional peaks in the ^1H and ^{13}C NMR spectra of the polymer and they also contribute to the purely aliphatic regions of the spectra [68–70].

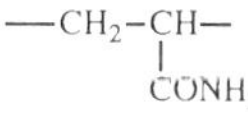

$—CH_2—CH(CONH_2)—$ (**10**)

$—CH_2—CH(COOH)—$ (**11**)

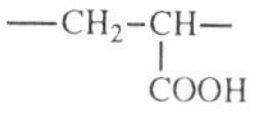

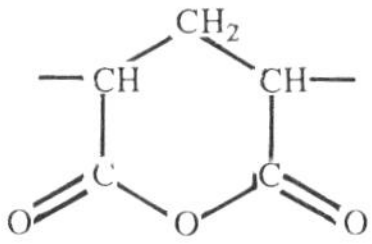

(**12**)

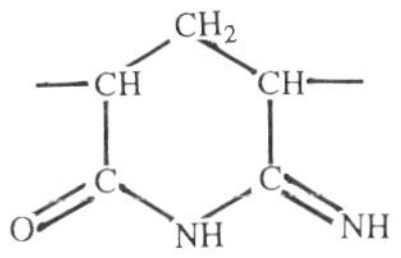

(**13**)

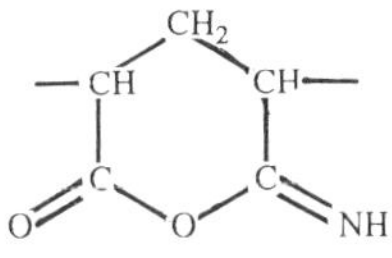

(**14**)

Under some circumstances foreign units can arise at a detectable level during the making of polymers with AIBN as initiator. Polymerization of STY initiated by a comparatively high concentration of [α-^{13}C]AIBN and carried to high conversion (about 85%) gave a polymer containing units formed by copolymerization of [^{13}C]MAN; there was one MAN unit for just over 2000 units of STY [9]. It appears that the enriched monomer had come from initiator 'wasted' in reactions such as

$$2Me_2\dot{C}(CN) \rightarrow Me_2C(CN)H + CH_2{:}C(CN)Me \qquad (3.16)$$

and that it had been produced steadily although slowly throughout the polymerization. The chemical shifts for the carbons in MAN units, as found from examination of authentic copolymers of STY with MAN, were found to be distinctly different from those of the carbons in $Me_2C(CN)$– end-groups. The relaxation time T_1 for the ^{13}C in the in-chain units was, as expected, appreciably less than the T_1 for the corresponding carbon in end-groups derived from the initiator. It is conceivable that 'foreign' units could arise during polymerizations involving the azo ester AZE (see section 3.2.2) as initiator; in this case, 'wasted' radicals might give MMA by interaction according to the equation

$$2Me_2\dot{C}(COOMe) \rightarrow Me_2CH{\cdot}COOMe + CH_2{:}C(COOMe)Me \quad (3.17)$$

and it would subsequently enter polymer.

The cationic polymerization of an olefin, other than the simplest, may lead to a product containing several types of units because of hydride and methide shifts in the growing centres; high field ^{13}C NMR makes it possible to detect the various types of unit and then to determine their proportions although it may be necessary to apply semi-empirical rules to assign signals [71]. In the case of 4-methyl-1-pentene, $-CH_2{\cdot}CH_2{\cdot}C(Me)(Et)-$ units are present in the polymer at 1–2% [72]. The occurrence of units of several other types has been demonstrated but $-CH_2{\cdot}CH(CH_2{\cdot}CHMe_2)-$, corresponding to 1,2-addition, has not been found although it could have been expected to be the most abundant unit. The absence of the 1,2-units was established by the use of the DEPT technique; signals previously assigned to methyl and methine carbons in 1,2-units must actually originate from methylene groups.

3.4 Branching

3.4.1 General remarks

Branching is a structural feature often arising during polymerization; it is particularly likely to develop during radical processes and is a consequence of attack by a growing chain on a monomeric unit already incorporated in the polymer. An important cause of branching is the process of 'transfer to polymer'. The reaction can occur intramolecularly in which case it is commonly termed 'back-biting' and leads to short branches which are evident even in the earliest stages of polymerization. Branching caused by intermolecular transfer to polymer is most pronounced in polymers produced in reactions carried to high conversions; there is a distribution of branch lengths but the average is comparable with that for the main chains and so the branches are described as long.

Branching can also occur by another mechanism for dienes, conjugated and otherwise. When the monomer first enters the polymer, it forms a unit containing a double bond. In some cases, this double bond very soon becomes incorporated in the same macromolecular chain by a process referred to as cyclopolymerization. In other cases, it reacts at a later stage, giving first a branch and usually later crosslinking develops. Step-reaction polymerization involving a monomer with functionality greater than two leads first to branching and then to crosslinking and products which are insoluble and infusible. Lightly crosslinked polymers swell extensively in those liquids which are solvents for the corresponding polymers without crosslinks. Frequently, the techniques of solution NMR can be successfully applied to highly swollen gels because there is still sufficient molecular mobility (see chapter 4).

3.4.2 Low density polyethylene

The archetypical example of a branched polymer is low density polyethylene (LDPE), the product of radical polymerization at high temperature and pressure. That LDPE is significantly branched was first suspected because of the influences of polymerization conditions upon the crystallinity of the polymer and upon its rheological behaviour in the melt and in solution. Confirmation was provided by IR analysis which indicated a considerable excess of methyl groups; ideally, the maximum number of such groups would be two per molecule, corresponding to the end-groups of a linear alkane.

The advent of high resolution ^{13}C NMR has allowed branching in LDPE to be studied in detail [73–75]. The spectrum of a typical polymer of this type, recorded for the molten material or close to its melting point in a high boiling solvent such as *o*-dichlorobenzene, displays a strong signal at 30.0 ppm, attributed to methylene carbons in long linear sequences. There are also methylene signals in the range 22.9–33.4 ppm, methyl signals between 8.1 and 14.2 ppm, methine signals at 37.1 and 38.0 ppm and even a quaternary carbon signal at 44.5 ppm. Agreement seems to have been reached on the detailed assignments of these signals to carbons within, or in the vicinities of, various types of short and long branches. The assignments have been made with the aid of spectra of low molecular weight linear and branched hydrocarbons and also by considering model polymers such as copolymers of ethylene with 1-butene.

A typical LDPE contains 18.9 branches for 1000 backbone carbons [74], made up thus: methyl, 0; ethyl, 0.7; *n*-propyl, 0; *n*-butyl, 13.1; *n*-amyl, 2.9; *n*-hexyl and longer, 2.2. The preponderance of C_4 branches is consistent with the view that back-biting occurs mainly through the thermodynamically favoured six-membered cyclic transition state as indicated in equation (3.18).

$$\begin{array}{l} \qquad\qquad\quad \cdot CH_2 \\ \qquad\quad H \qquad \diagdown CH_2 \\ \qquad\quad | \qquad\quad | \\ -CH_2-CH \qquad .CH_2 \\ \qquad\qquad\quad \diagdown CH_2 \end{array} \longrightarrow -CH_2\cdot\dot{C}H\cdot CH_2\cdot CH_2\cdot CH_2\cdot CH_3 \qquad (3.18)$$

There is evidence that some of the ethyl branches occur in 1,3-pairs and that some are attached to quaternary carbons, i.e. they are present at double-branch points. It appears that some samples of LDPE contain a few 2-ethyl-*n*-hexyl branches.

The influence of branching upon the crystallinity of LDPE, and hence upon its physical and mechanical properties, is of great importance for the applications of the polymer. This topic has been studied in two main ways, both involving the use of ^{13}C NMR. In the first procedure, the amorphous regions of various partially crystalline copolymers of ethylene with propylene, 1-butene, 1-hexene, 1-octene, 1-dodecene or 4-methyl-1-pentene, all synthesized by Ziegler–Natta polymerization, were degraded oxidatively by fuming HNO_3 [76]. The residual portions were entirely crystalline and were examined by solution-state ^{13}C NMR. The probabilities of incorporating side-chains from the olefin comonomer into the crystal lamellae were in the order Me > Et > *n*-butyl = *n*-hexyl = *n*-decyl > iso-butyl, i.e. the probability falls with increasing bulk of the side-chain. The alternative method for probing the distribution of side-chains between amorphous and crystalline phases makes use of solid-state ^{13}C NMR, often with two distinct pulse sequences, one selected to favour crystalline regions and the other to deal with amorphous regions [70–79] (see chapter 5). This second procedure has shown that there is a tendency for branches and other minor structural features, such as end-groups, to be excluded from the crystalline regions.

3.4.3 Polymers from other vinyl monomers

There is convincing evidence, much of it from study of solution properties, for branching in many samples of polyVAC; it has been supplemented by the results of NMR examination of polymers and of the samples of poly(vinyl alcohol) derived from them; the exact identities and distributions of the branches have not, however, been established [80–82]. An indication that abstraction of the hydrogen atom from a methine carbon of a VAC unit can occur, leading to branches, comes from the finding that the intensity of the methine carbon signal is, for many polymers, reduced relative to that of the methylene signal by up to 15% [81]. Similar reductions in intensity of methine signals have been found for some ethylene/VAC copolymers, suggesting that in these materials there may also be branches attached to VAC units [83]. A type of branching clearly present in polyVAC is ‘hydrolysable’ branching but its presence has yet to be confirmed by NMR studies. Branching of this type is believed to arise from abstraction of a hydrogen atom from the acetoxy methyl of a VAC unit. If this transfer to polymer occurs intermolecularly, the branch is on average long and it is attached to the main chain through an ester linkage; if the transfer is intramolecular, the branches are short but an ester linkage is built into the main chain of the polymer. The existence of ‘hydrolysable’ branching is indicated by comparison of the solution properties of the original polyVAC with those of the material made by re-acetylation of the poly(vinyl alcohol) prepared by hydrolysis of the original polymer.

Branching in poly(vinyl chloride) (PVC) is complicated and cannot be studied satisfactorily by direct NMR examination of the polymer. Several studies [84, 85] have been made on samples of PVC which have been reductively dechlorinated with tri-*n*-butyltin hydride or by tri-*n*-butyltin deuteride; the resulting polymers are essentially hydrocarbons and they have been examined by solution-state ^{13}C NMR. Monodeuteration of a ^{13}C atom converts its proton-decoupled NMR signal from a singlet to a triplet; the signal and those of the neighbouring carbons are moved upfield. Comparison of the ^{13}C NMR spectrum of the hydrogenated polymer with that of the corresponding deuterated polymer makes it possible to identify the points of chlorine attachment in the original PVC. This procedure has shown that a typical sample of PVC contains 2,4-dichloro-*n*-butyl and 2-chloroethyl branches, presumably arising from back-biting, and also long branches originating from intermolecular transfer to polymer. The dominant minor structure in PVC is, however, the chloromethyl branch; it is thought to be formed as a result of a head-to-head addition of monomer to a growing radical, followed by a 1,2-shift of a chlorine atom and subsequent head-to-tail growth according to the scheme

$$\begin{array}{l} -CH_2\cdot CHCl\cdot CHCl\cdot \dot{C}H_2 \rightarrow -CH_2\cdot CHCl\cdot \underset{\displaystyle CH_2Cl}{\underset{|}{\dot{C}H}} \xrightarrow{\text{monomer}} \\ \qquad -CH_2\cdot CHCl\cdot \underset{\displaystyle CHCl}{\underset{|}{CH}}\cdot CH_2\cdot \dot{C}HCl \end{array} \tag{3.19}$$

Minor tertiary fluorine signals have been found for samples of poly(vinyl fluoride) [86]; they are associated with CF groups at branch points, either $-CH_2\cdot \overset{|}{C}F\cdot CH_2\cdot CHF-$ (^{19}F signal at -147 ppm) or $-CH_2\cdot \overset{|}{C}F\cdot CHF\cdot CH_2-$ (^{19}F signal at -162 ppm), the latter being a unit originally formed by head-to-head addition (see section 3.3.1). There is one branch for between 80 and 200 monomeric units, depending upon reactor conditions. It appears that branching develops as a result of transfer to polymer, a growing radical $-CH_2\cdot \dot{C}HF$ abstracting hydrogen either inter- or intramolecularly; a small ^{19}F signal at -220 ppm provides evidence for the resulting $-CH_2\cdot CH_2F$ end-groups. Additional evidence for these saturated end-groups was provided by an ingenious differential decoupling experiment in which ^{1}H NMR spectra were recorded with or without selective irradiation at the frequency of the weak ^{19}F signal; small signals at 2.0 and 4.5 ppm were found for the methylene and fluoromethyl protons of these end-groups.

3.4.4 Amino and phenolic resins

Branching occurs early in the reactions of aldehydes, such as formaldehyde, with polyfunctional amines and amides such as urea and melamine (2,4,6-triaminotriazine). Systems of this type have been extensively studied by NMR, with the aim of identifying various linear and branched structures and the effects of reaction conditions such as temperature, ratio of monomers and pH [87–94]. Inevitably, ^{13}C NMR provides more detailed information than ^{1}H NMR because of the wider range of chemical shifts for the former; this point is particularly important when attempting to distinguish between structures that are only subtly different. ^{13}C spectra of urea/formaldehyde condensates, isolated before the gel points, contain a variety of signals due to methylene carbons (from formaldehyde) and carbonyl carbons (from urea) [88–90]. The methylene signals are especially useful because comparison of their areas allows determination of the relative amounts of formaldehyde incorporated as methylene links, methylene ether links and as methylol groups. Further, methylene links in linear portions of the chains can be distinguished from those at branch points; the same consideration applies to methylene ether links. Some assignments are given in Table 3.2.

Similar ^{13}C NMR studies have been performed on formaldehyde/melamine condensates. Signals from methylene, methylene ether and methylol groups occur for these materials also [91–93] but the azine carbon signals give additional information about the patterns of substitution in the triazine rings and hence about the various types of branching [92]. For amino resins, useful information can be obtained from natural abundance ^{15}N NMR since the chemical shifts of tertiary amino nitrogens, i.e. those bearing two substituents (methylol groups, methylene links or methylene ether links) are approximately 20 ppm downfield from those of secondary amino nitrogens, i.e. those carrying a single substituent, which in turn are approximately 25 ppm downfield from those of the unsubstituted primary amino nitrogens [94]. There are also small but useful differences of chemical shift according to the nature of any substituents attached to a nitrogen atom.

Analysis of amino resins at stages much beyond the gel point requires the

Table 3.2 Methylene carbon assignments in ^{13}C NMR spectra of urea/formaldehyde resins

Assignment	Chemical shift (ppm)
$-NHCH_2NH-$	47.7
$-NHCH_2N<$	54·3
$>NCH_2N<$	58·9
$-NHCH_2OH$	65·5
$>NCH_2OH$	72.3
$-NHCH_2OCH_2NH-$	70·0
$>NCH_2OCH_2NH-$	76.5

use of CPMAS solid-state ^{13}C and ^{15}N NMR (see chapter 5). Several such examinations have been performed; the general assignment principles are as for the solution studies [93–96].

There have been accounts of the use of NMR to characterize phenol/aldehyde condensates. These condensates may show extensive branching before the gel point is reached but ultimately crosslink to give insoluble thermosets. A recent study was concerned with a detailed ^{13}C NMR analysis of cresol/formaldehyde novolak compositions, i.e. materials made under acidic conditions with cresol in excess over formaldehyde and intended for use as moulding powders [97]. The study focused on the identification of structures in condensates of *m*-cresol with formaldehyde. Commercial products are generally based on a mixture of the isomers of cresol but, even when only one isomer is used, the number of structural possibilities increases rapidly as condensation proceeds. For example, six methylene-linked dimers are possible if two molecules of *m*-cresol condense with one of formaldehyde. These dimers give rise to six methylene carbon, nine methyl carbon and nine phenolic carbon resonances. These signals have been assigned with information from ^{13}C–^{13}C J correlation two dimensional (2-D) experiments (INADEQUATE), as have resonances for some branched tetrameric model compounds; they have been used to assign signals in the ^{13}C spectra of condensates of higher molecular weight. In these condensates, it is possible to distinguish between signals from end-groups (the chemical shifts are similar to those of the dimers), those from in-chain units and those arising from units at branch-points (they resemble the tetrameric models). The fractions of rings at branch-points can be determined most easily from the pattern of methyl carbon resonances between 12 and 22 ppm since the methyl carbon is particularly sensitive to the pattern of substitution around the ring. Branch densities in these novolaks, having molecular weights around 4000, are typically about 15%.

3.4.5 Polysiloxanes

The hydrolysis of tetra-alkylsilanes gives rise not to perfect dense silica networks, as would be expected for complete reaction, but to linear and branched polymers having a siloxane backbone and SiOH and SiOR side-groups. The early stages of the reaction between tetramethoxysilane and water have been followed by means of ^{29}Si NMR with ^{1}H–^{29}Si polarization transfer [98]. The polarization transfer (DEPT experiment) boosts the signals from Si to an extent governed by the number of attached protons; this feature helps in the assignment of the various Si peaks. In the spectra of the products of hydrolysis, the signals can be placed into four sets corresponding to Si atoms in monomer units in end-groups, at in-chain sites, at branch-points and in fully crosslinked units. The changes in the signals with time have been used to follow the kinetics of the reaction. At low ratios of alkoxide to water, the hydrolysis is incomplete and the degree of condensation is therefore limited; at high ratios, the hydrolysis is fast and it is the condensation process that limits the rate.

3.5 Chemical modification of polymers

3.5.1 Synthetic polymers

Chemical modification of polymers can occur casually as a result of reactions with atmospheric oxygen, moisture or more aggressive agents such as ozone. The changes may have profound effects on structures and properties; some of the reactions have interesting features, thus the auto-oxidation of polySTY is surprisingly slow compared with that of the model compound isopropylbenzene. Efforts continue to discover efficient stabilizers for incorporation in polymers to prevent undesirable reactions. The additives are used at low levels and some are chemically attached to the macromolecules in order to stop loss by diffusion; clearly NMR techniques have applications in this field. This section is devoted, however, to the deliberate chemical modification of polymers and its study by the methods of NMR. The changes are performed in order to prepare new materials with particular properties. Excluded from attention are important topics such as reactions of end-groups leading to chain-extension or to the formation of block copolymers, the scission of main chains to give oligomers with functional terminal groups (see section 3.2.6), the joining of separate chains to give crosslinked structures and the preparation of graft copolymers.

Chemical modification of aromatic polymers, e.g. polystyrene and polyacenaphthylene, is used to obtain materials containing functional groups; many such products have application as, for example, polymer-supported reagents for synthesis. In practice, the reagents are prepared and used as insoluble beads, monomers such as divinylbenzene being incorporated to provide the necessary crosslinking; the actual reactions involved in the modifications are, however, better studied with soluble polymers. Commonly, it is desirable that the reactions should occur at specific sites in the aromatic rings of the parent polymer. This requirement would be met automatically if the functionalized polymer were prepared by direct polymerization or copolymerization of the appropriate monomer but frequently these processes are impracticable because of severe problems either in making the required monomer or in polymerizing it; thus difficulties invariably arise in attempts to polymerize a monomer having a group such as –OH, $-NO_2$ or $-NH_2$ attached directly to an aromatic ring because those groups cause great retardation or even inhibition of polymerization. The case of poly(*p*-hydroxystyrene) [99] can be cited; this polymer is of importance in connection with photolithography. It is possible to polymerize *p*-acetoxystyrene without difficulty and then the MeCOO– groups in the polymer can be converted to –OH; processes of this type can be monitored in a straightforward manner by use of NMR spectra.

IR can usually be applied for following changes in polymers but examination by NMR can be particularly useful. The bromination of polystyrene can be quoted as an example. The reagent and conditions can be selected so that substitution occurs exclusively in the rings and it has been confirmed, by ^{1}H and ^{13}C NMR studies, that regioselectivity can be very high with at least 95% of the introduced halogen being at the *para* positions. Polymers of *p*-bromo- and *o*-bromostyrene and copolymers of those monomers with STY were used as reference materials [100].

When a polymer is altered by chemical reaction and the process is not carried to completion, the modified units are not necessarily distributed at random through the macromolecules. The partial conversion of polyVAC to poly(vinyl alcohol) gives a product in which the residual VAC units tend to occur in blocks [82] (see chapter 2). This conclusion was reached from ^{13}C NMR studies of the materials dissolved in D_2O. The methylene carbon resonances were used to determine the mean lengths of sequences of VAC and of vinyl alcohol units.

3.5.2 *Natural polymers*

NMR can be employed effectively for examination of many modification processes for cellulose. Studies on acetylation have been reviewed by Usmanov [101]. The quantitative analysis of ^{1}H NMR data relies upon information derived from specifically modified model analogues. Similarly, ^{13}C chemical shift data for various cellulose oligomers permit the analysis of the spectra of samples of cellulose acetate. Quantitative analyses for the distribution of acetate groups in cellulose acetates of various types have been achieved.

Many other derivatives of cellulose have been studied using NMR methods (see chapter 2). Takahashi *et al.* [102] investigated the distributions of ring substituents from examination of the ring-carbon signals. For trityl, tosyl and methyl cellulose and also cellulose formate, the reactivity was normally C-6 > C-2 > C-3, where the numbering refers to the carbon positions on the glucose ring. For cellulose S–Me xanthate, however, the C-3 site was the most reactive. The behaviour of cellulose xanthate in solution in $NaOH/D_2O$ was examined as a function of time using ^{13}C NMR and DEPT methods. In this way, structural differences for xanthates prepared by various methods were examined. Substitution patterns in ethyl hydroxyethyl celluloses have been investigated through ^{13}C NMR studies of their hydrolysates, together with those of hydroxyethyl and ethyl cellulose [103]. The high mobility of the hydrolysates led to much better spectral resolution than that for solutions of the corresponding intact polymers. Specific substitution at the C-2 and C-3 sites could be determined. In highly substituted samples, it was possible to find directly the fraction of unsubstituted C-6 sites.

The nitration of cellulose can be monitored by NMR methods. It has been shown [104] that reaction for cotton cellulose occurs simultaneously at all three OH sites but at different rates; reaction at C-6 is fastest. Mono-, di- and tri-substituted chains co-exist in macromolecular samples. More detailed examination [105] has revealed the distribution of mono-, di- and tri-substituted monomer units, leading to the conclusion that the relative reactivities are 30:2:1 for C-6, C-2 and C-3, respectively. Nitration of cellulose fibrils has been examined in detail [106]. For reactions at -25 to $+22°C$, an equilibrium degree of substitution (DS) of 1.2 was established rapidly, yielding mostly mononitrocellulose after 1 h. When nitrated cellulose was de-nitrated and then reprocessed, high-resolution ^{13}C data suggested that values of DS were high after 1 h. At 48 h, values of DS were similar for initially de-nitrated and for normal materials but the ^{13}C spectra revealed significant differences in microstructure. There has been a study using ^{15}N NMR after reaction with a nitration mixture enriched with ^{15}N; three clearly resolved signals were found, with chemical shifts in the order N-6 > N-2 > N-3 [107].

There have been investigations on the modification of dextrans, α-D-(1,6)-linked glucose polymers. In one case, a ^{13}C NMR study of the oxidation of dextran T 10 with aqueous bromine and a comparison with model glucosiduloses showed [108] that keto groups were produced, mainly at C-2 and C-4; over-oxidation afforded a small amount of acidic ring-cleavage products. Dextran and chitosan have been nitrated and the structures of products elucidated using ^{13}C NMR [109]; assignments of ^{13}C signals from trimethylated dextran have been given [110]. A combined ^{1}H and ^{13}C NMR examination of dextran and its acetylated and benzylated derivatives showed that, in all cases, (1,6)-linked glucopyranoside units were present in the $^{4}C_1$ chair conformation [111]. In an investigation of processes to produce materials suitable as matrices for the immobilization of proteins and enzymes and subsequent slow release *in vivo*, the naturally derived polymers dextran, glycogen, hydroxyethyl starch and maltodextrin were derivatized at high pH with the glycidyl ester of acrylic acid [112]. The extent of derivatization was assessed using ^{1}H NMR which proved to be more precise and convenient than an assay involving bromination. Partly modified dextrans with acetyl groups were prepared by reaction with either acetyl chloride or acetic anhydride under homogeneous conditions. Differences in the distributions of substituents in the products were found from the ring-carbon spectra [113]. For the products obtained using the acid chloride, the distributions corresponded to the reactivities being C-2 > C-4 > C-3; for the products made with the anhydride, the deduced reactivities were C-2 ≃ C-3 > C-4. The results were explained in terms of both steric and intramolecular hydrogen bond effects. It has been shown [114] that ^{13}C NMR spectra can yield information about the characteristics of crosslinked dextran (Sephadex-G) hydrogels.

3.6 Concluding comments

It has been shown in this chapter that NMR techniques can be of great value in solving many problems connected with minor structures in macromolecules. The possibilities are certainly not exhausted and many interesting systems remain to be examined. If one outstanding problem were to be mentioned, it could well be branching in polymers derived from vinyl monomers; this important structural feature has been examined thoroughly in only a few cases.

The developments and improvements in NMR spectrometers lead to increased sensitivity and better resolution in spectra. It may become possible, in work on end-groups for example, to use initiators and other reagents that are not highly enriched with ^{13}C and do not contain nuclei that are very sensitive in NMR. It is necessary, however, to point out that valuable results can come from spectrometers operating at comparatively low fields, say 25 MHz for ^{13}C; the information deduced from spectra, such as those in Figures 3.1 and 3.2, illustrates this feature. It is unlikely that NMR techniques will ever have sensitivities to match those achievable by the use of radioactive isotopes but the specificity associated with NMR is incomparably better. There are also great practical advantages resulting from the avoidance of the hazards and problems associated with the use of radioactive materials.

It seems certain that two-dimensional NMR techniques will be very effective in work on 'abnormal' structures in polymer molecules. Mention has been made [4] of the value of two-dimensional NMR for providing unambiguous assignments for end-groups in copolymers made using azo initiators. Early results on ^{13}C–^{1}H shift correlations for resonances from $Me_2C(CN)$– end-groups have been published [115]. Preliminary reports have been given [116] of two-dimensional NMR studies of minor structures in polyAN. A notable feature of this development is the likelihood of achieving good resolution of signals at fields lower than those necessary when using the more conventional one-dimensional techniques.

References

1. J.C. Bevington, J.R. Ebdon and T.N. Huckerby, *Eur. Polym. J.* **21** (1985) 685–694.
2. C.H. Bamford, A. Ledwith and Y. Yagci, *Polymer* **19** (1978) 354–356.
3. C. Walling, *Free Radicals in Solution*, Wiley, New York (1957) 512–513.
4. J. Krstina, G. Moad and D.H. Solomon, *Eur. Polym. J.* **28** (1992) 275–282.
5. J.C. Bevington, R.A. Lyons and E. Senogles, *Eur. Polym. J.* **28** (1992) 283–286.
6. J.C. Bevington, T.N. Huckerby and N.W.E. Hutton, *Eur. Polym. J.* **18** (1982) 963–965.
7. J.C. Bevington, S.W. Breuer and T.N. Huckerby, *Polymer Commun.* **25** (1984) 260–261.
8. M. Talât-Erben and S. Bywater, *J. Am. Chem. Soc.* **77** (1955) 3710–3711.
9. G. Moad, D.H. Solomon, S.R. Johns and R.I. Willing, *Macromolecules* **17** (1984) 1094–1099.
10. M.R. Bendall, D.T. Pegg, D.M. Doddrell, S.R. Johns and R.I. Willing, *J. Chem. Soc., Chem. Commun.* (1982) 1138–1140.

11. J.C. Bevington, T.N. Huckerby and N.W.E Hutton, *Eur. Polym. J.* **20** (1984) 525–528.
12. J.C. Bevington, B.F. Bowden, D.A. Cywar, R.A. Lyons, E. Senogles and D.A. Tirrell, *Eur. Polym. J.* **27** (1991) 1239–1249.
13. J.C. Bevington, D.A. Cywar, T.N. Huckerby, E. Senogles and D.A. Tirrell, *Eur. Polym. J.* **26** (1990) 41–46.
14. D.A. Cywar and D.A. Tirrell, *Macromolecules* **19** (1986) 2908–2911.
15. J.C. Bevington, S.W. Breuer, E.N.J. Heseltine, T.N. Huckerby and S.C. Varma, *J. Polym. Sci., Polym. Chem. Ed.* **25** (1987) 1085–1092.
16. J.C. Bevington, T.N. Huckerby and N.W.E. Hutton, *J. Polym. Sci., Polym. Chem. Ed.* **20** (1982) 2655–2660.
17. D.A. Cywar and D.A. Tirrell, *Eur. Polym. J.* **25** (1989) 657–664.
18. R.G. Farmer, D.J.T. Hill and J.H. O'Donnell, *J. Macromol. Sci., Chem.* **14** (1980) 51–68.
19. T. Fukuda, Y.-D. Ma and H. Inagaki, *Macromolecules* **18** (1985) 17–26.
20. G. Moad, D.H. Solomon, T.H. Spurling, S.R. Johns and R.I. Willing, *Aust. J. Chem.* **39** (1986) 43–50.
21. J.C. Bevington, D.A. Cywar, T.N. Huckerby, E. Senogles and D.A. Tirrell, *Eur. Polym. J.* **26** (1990) 871–875.
22. D.A. Cywar and D.A. Tirrell, *J. Am. Chem. Soc.* **111** (1989) 7544–7553.
23. S.R. Johns, E. Rizzardo, D.H. Solomon and R.I. Willing, *Makromol. Chem., Rapid Commun.* **4** (1983) 29–32.
24. L.S. Bresler, E.N. Barantsevich, V.I. Polyansky and S.S. Iantchev, *Makromol. Chem.* **183** (1982) 2479–2489.
25. G. Moad, D.H. Solomon, S.R. Johns and R.I. Willing, *Macromolecules* **15** (1982) 1188–1191.
26. G. Moad, E. Rizzardo and D.H. Solomon, *Aust. J. Chem.* **36** (1983) 1573–1588.
27. G. Moad, E. Rizzardo, D.H. Solomon, S.R. Johns and R.I. Willing, *Macromolecules* **19** (1986) 2494–2497.
28. J.C. Bevington and J.R. Ebdon, in *Developments in Polymerisation—2*, ed. R.N. Haward, Applied Science Publishers, London (1979), pp. 1–43.
29. J.C. Bevington, T.N. Huckerby and N. Vickerstaff, *Makromol. Chem., Rapid Commun.* **4** (1983) 349–352.
30. J.C. Bevington and T.N. Huckerby, *Macromolecules* **18** (1985) 176–178.
31. J.C. Bevington, S.W. Breuer and T.N. Huckerby, *Macromolecules* **22** (1989) 55–61.
32. C.A. Barson, J.C. Bevington and T.N. Huckerby, *Polym. Bull.* **22** (1989) 131–135.
33. J.C. Bevington, *Angew. Makromol. Chem.* **185/186** (1991) 1–10.
34. J.C. Bevington, T.N. Huckerby, N. Vickerstaff and C.A. Barson, *Polymer* **27** (1986) 1823–1825.
35. C.A. Barson, J.C. Bevington and T.N. Huckerby, *Polymer* **31** (1990) 145–149.
36. J.C. Bevington, S.W. Breuer, T.N. Huckerby, R.F.D Jones and C.A. Barson, *J. Polym. Sci., Polym. Chem. Ed.* **28** (1990) 3271–3278.
37. C.A. Barson, J.C. Bevington and T.N. Huckerby, *Polym. Bull.* **28** (1992) 657–662.
38. C.A. Barson, J.C. Bevington, S.W. Breuer and T.N. Huckerby, *Makromol. Chem., Rapid Commun.* **13** (1992) 97–101.
39. A. Zambelli, P. Locatelli, M.C. Sacchi and E. Rigamonti, *Macromolecules* **13** (1980) 798–800.
40. A. Zambelli, M.C. Sacchi, P. Locatelli and G. Zannoni, *Macromolecules* **15** (1982) 211–212.
41. P. Hodge, E. Khoshdel and A.A. Naim, *Polym. Commun.* **27** (1986) 322–323.
42. D.B. Staniforth, Ph. D. Thesis, University of Lancaster (1988).
43. P.L. Nayak, S. Lenka and M.K. Mishra, *J. Polym. Sci., Polym. Chem. Ed.* **19** (1981) 839–842.
44. J. Devaux, D. Daoust, R. Legras, J.M. Dereppe and E. Nield, *Polymer* **30** (1989) 161–164.
45. S.S. Huang, A.H. Soum and T.E. Hogen-Esch, *J. Polym. Sci., Polym. Lett.* **21** (1983) 559–563.
46. A.H. Soum and T.E. Hogen-Esch, *Macromolecules* **18** (1985) 690–694.
47. R. Volpe and T.E. Hogen-Esch, *Macromolecules* **23** (1990) 4196–4199.
48. S. Sosnowski, A. Duda, S. Słomkowski and S. Penczek, *Makromol. Chem., Rapid Commun.* **5** (1985) 551–557.
49. K. Matyjaszewski and S. Penczek, *Makromol. Chem.* **182** (1981) 1735–1742.
50. T. Sato and T. Okaya, in *Polyvinyl Alcohol*, ed. C.A. Finch, Wiley, Chichester (1992), pp. 112–114.

51. T. Sato and T. Okaya, in *Polyvinyl Alcohol*, ed. C.A. Finch, Wiley, Chichester (1992), pp. 119–125.
52. R.H. Cragg, R.G. Jones and A.C. Swain, *Eur. Polym. J.* **27** (1991) 785–788.
53. R.G. Jones, R.H. Cragg and A.C. Swain, *Eur. Polym. J.* **28** (1992) 651–655.
54. J.H. O'Donnell and A.K. Whittaker, *Polymer* **33** (1992) 62–67.
55. L.R. Dix, J.R. Ebdon, N.J. Flint and P. Hodge, *Eur. Polym. J.* **27** (1991) 581–588.
56. L.R. Dix, J.R. Ebdon and P. Hodge, *Polymer*, **34** (1993) 406–411.
57. J.R. Ebdon and N.J. Flint, *A.C.S. Polym. Prepr.* **33(1)** (1992) 972–973.
58. S.G. Bond and J.R. Ebdon, *Polym. Commun.* **32** (1991) 290–292.
59. Y. Amerik and J.E. Guillet, *Macromolecules* **4** (1971) 375–379.
60. P.J. Flory and F.S. Leutner, *J. Polym. Sci.* **3** (1948) 880–890.
61. R.L. Adelman and R.C. Ferguson, *J. Polym. Chem., Polym. Chem. Ed.* **13** (1975) 891–911.
62. D.W. Ovenall, *Macromolecules* **17** (1984) 1458–1464.
63. R.E. Cais and J.M. Kometani, in *NMR and Macromolecules*, ed. J.C. Randall, ACS Symposium Series No. 247, American Chemical Society (1984), pp. 153–166.
64. R.E. Cais and J.M. Kometani, *Macromolecules* **18** (1985) 1354–1357.
65. M. Sepulchre, A. Kassamaly, M. Moreau and N. Spassky, *Makromol. Chem.* **189** (1988) 2485–2501.
66. C. Campbell, F. Heatley, G. Holcroft and C. Booth, *Eur. Polym. J.* **25** (1989) 831–837.
67. F. Heatley, G.-E. Yu, W.-B. Sun, E.J. Pywell, R.H. Mobbs and C. Booth, *Eur. Polym. J.* **26** (1990) 583–592.
68. J. Lovy, V. Janout, H. Hrudkova, *Coll. Czech. Chem. Commun.* **49** (1984) 506–512.
69. V.R. Pai Verneker and B. Shaha, *Macromolecules* **19** (1986) 1851–1856.
70. T.C. Hunter, PhD Thesis, University of Lancaster (1991).
71. L.A. Lindeman and J.Q. Adams, *Anal. Chem.* **43** (1971) 1245–1252.
72. A. Mizuno and H. Kawachi, *Polymer* **33** (1992) 57–61.
73. H.N. Cheng, F.C. Schilling and F.A. Bovey, *Macromolecules* **9** (1976) 363–365.
74. J.N. Hay, P.J. Mills and R. Ognjanovic, *Polymer* **27** (1986) 677–680.
75. D.C. Bugada and A. Rudin, *Eur. Polym. J.* **23** (1987) 809–818, 847–850.
76. S. Hosoda, H. Nomura, Y. Goto and H. Kihara, *Polymer* **31** (1990) 1999–2005.
77. F. Laupretre, L. Monnerie, L. Barthelemy, J.P. Vairon, A. Sauzeau and D. Roussel, *Polym. Bull.* **15** (1986) 159–164.
78. E. Perez and D.L. Vanderhart, *J. Polym. Sci., Part B, Polym. Phys.* **25** (1987) 1637–1653.
79. M.F. Grenier-Loustalot, *Polym. Commun.* **31** (1990) 329–332.
80. Y. Morishima, Y. Irie, H. Iimuro and S. Nozakura, *J. Polym. Sci., Polym. Chem. Ed.* **14** (1976) 759–766, 1267–1275, 1277–1282.
81. A.S. Dunn and S.R. Naravane, *Br. Polym. J.* **12** (1980) 75–77.
82. D.C. Bugada and A. Rudin, *Polymer* **25** (1984) 1759–1766.
83. D.C. Bugada and A. Rudin, *Eur. Polym. J.* **28** (1992) 219–227.
84. W.H. Starnes, Jr., G.M. Villacorta, F.C. Schilling, I.M. Plitz, G.S. Park and A.H. Saremi, *Macromolecules* **18** (1985) 1780–1786.
85. M.-F. Llauro-Darricades, N. Bensemra, A. Guyot and R. Petiaud, *Makromol. Chem., Macromol. Symp.* **29** (1989) 171–184.
86. D.W. Ovenall and R.E. Uschold, *Macromolecules* **24** (1991) 3235–3237.
87. M. Chiavarini, R. Bigatto and N. Conti, *Angew. Makromol. Chem.* **70** (1978) 49–58.
88. A.J.J. de Breet, W. Dankelman, W.G.B. Huysmans and J. de Wit, *Angew. Makromol. Chem.* **62** (1977) 7–31.
89. J.R. Ebdon and P.E. Heaton, *Polymer* **18** (1977) 971–974.
90. B. Tomita and S. Hatano, *J. Polym. Sci., Polym. Chem. Ed.* **16** (1978) 2509–2525.
91. B. Tomita and H. Ono, *J. Polym. Sci., Polym. Chem. Ed.* **17** (1979) 3205–3215.
92. J.R. Ebdon, B.J. Hunt and W.T.S. O'Rourke, *Br. Polym. J.* **19** (1987) 197–203.
93. J.R. Ebdon, B.J. Hunt, W.T.S. O'Rourke and J. Parkin, *Br. Polym. J.* **20** (1988) 327–334.
94. J.R. Ebdon, P.E. Heaton, T.N. Huckerby, W.T.S. O'Rourke and J. Parkin, *Polymer* **25** (1984) 821–825.
95. G.E. Maciel, N.M. Szeverenyi, T.A. Early and G.E. Myers, *Macromolecules* **16** (1983) 598–604.
96. I. Chuang, B.L. Hawkins, G.E. Maciel, and G. Myers, *Macromolecules* **18** (1985) 1482–1485.
97. L.E. Bogan, Jr., *Macromolecules* **24** (1991) 4807–4812.
98. F. Brunet, B. Cabane, M. Dubois and P. Perly, *J. Phys. Chem.* **95** (1991) 945–951.

99. K. Nakabayashi, R. Schwalm and W. Schnabel, *Angew. Makromol. Chem.* **195** (1992) 191–204.
100. M. Camps, J.P. Monthéard, F. Benrokia, J.M. Camps and Q.T. Pham, *Eur. Polym. J.* **26** (1990) 53–59.
101. T.I. Usmanov, *Polymer Science USSR* (Eng. Transl.) **33** (1991) 611–635.
102. S.-I. Takahashi, T. Fujimoto, B.M. Barua, T. Miyamoto and H. Inagaki, *J. Polym. Sci., Polym. Chem. Ed.* **24** (1986) 2981–2993.
103. P. Zadorecki, T. Hjertberg and M. Ardwisson, *Makromol. Chem.* **188** (1987) 513–515.
104. G.L. Ryzhova and N.V. Novikova, *Vysokomol. Soedin., Ser. A* **27** (1985) 155–158.
105. N.A. Azancheev, E.N. Sérgeev, V.N. Sopin and G.N. Marchenko, *Vysokomol. Soedin., Ser. B.* **30** (1988) 296–298.
106. A.H.K. Fowler, H.S. Munro and D.T. Clarke, in *Cellulose and its Derivatives*, ed. J.F. Kennedy, Horwood, Chichester (1985), pp. 345–347.
107. G.N. Marchenko, V.F. Sopin, V.I. Kovalenko, N.M. Azancheev, E.N. Sergeev and E.M. Belova, *Vysokomol. Soedin., Ser. B* **30** (1988) 295–296.
108. O. Larm, K. Larsson, E. Scholander, B. Meyer and J. Thiem, *Carbohydr. Res.* **91** (1981) 13–20.
109. V. Panov, V.D. Spichak and V.D. Dubina, *Vysokomol. Soedin. Ser. A* **23** (1981) 412–421.
110. H. Friebolin, G. Keilich, N. Frank, U. Dabrowski and E. Siefert, *Org. Magn. Reson.* **12** (1979) 216–222.
111. D. Gagnaire and M. Vignon, *Makromol. Chem.* **178** (1977) 2321–2333.
112. M. Lepistö, P. Artursson, P. Edman, T. Laakso and I. Sjöholn, *Anal. Biochem.* **133** (1983) 132–135.
113. F. Arranz and M. Sanchez-Chaves, *Polymer* **29** (1988) 507–512.
114. R. Scherrer, R. Lundin, M. Benson and S.C. Witt, *Carbohydr. Res.* **124** (1984) 151–155.
115. J.C. Bevington and T.N. Huckerby, *Polymer* **33** (1992) 1323–1325.
116. T.C. Hunter and J.R. Ebdon, in *Macromolecular Preprints '92*, preprints from *Functional Polymers and Biopolymers*, RSC/ACS Conference, University of Kent (1992), pp. 101–102.

4 Liquid state NMR studies of polymer dynamics and conformation

O.W. HOWARTH

4.1 Introduction

Nuclear magnetic resonance experiments can reveal not only the chemical structures of polymers, together with their defects, but also a wealth of detail concerning their conformations and their motions. The local conformations combine to determine the bulk coiling properties of the polymer, such as the radius of gyration and hydrodynamic volume. They may be treated in a simplified way as a finite number of rotationally isomeric states. The presence and the weighting of these states may be approximately deduced either from computer modelling, or from fits to coiling properties, or from a number of theories which relate the local conformation to chemical shift, in a way that depends upon local tacticity. Further information about them may be available from spin–spin (J) couplings. This is because the size of three-bond and longer range spin–spin couplings depends on the dihedral angles between the bonds to the nuclei under observation. Nuclear Overhauser enhancements (NOEs) and other relaxation-related parameters depend in a complex way upon the distances from the observed nucleus to the nearest neighbour protons. Although these distances are fixed for a protonated carbon atom, they depend on the local conformation when an unprotonated carbon is under observation. NOEs between vicinal protons depend on local dihedral angles in a similar way.

The flexibility of polymer chains very largely arises from transitions between accessible conformational states. This is true both for overall chain flexibility and for more local motions. If no distinct alternative conformations are available within thermal energies, then the result is an unmalleable polymer such as polyphenylene. A second possibility is that alternative, distinguishable local states are available, but that the transitions between them are slow. An example might be crystalline polypropylene at room temperature. In this case, carbon atoms that would on average be equivalent in the dissolved polymer remain inequivalent in the solid, and thus give distinct resonances. Information of this kind arises when the separation of the chemical shifts, measured in Hertz, exceeds the rate of interconversion of the conformations and is illustrated in chapter 5. If the motions are more rapid than this then only averaged chemical shifts will be observed, although the conformational transitions may still be appreciably slower than in solution. The motional properties will be an average for the various conformations.

The normal case with a molten or dissolved polymer is that its conformational jumps are quite rapid, compared with the tumbling and looping motions of entire chains. A typical barrier to a conformational jump in a vinyl polymer chain without external constraints is only approximately 12–20 kJ mol^{-1}, and is therefore readily surmounted at room temperature, typically at least 10^{10} times per second. In this case, the local motions will contribute to the NMR relaxation parameters of the polymer, such as T_1, T_2 and NOE. Similar motions will occur in the gel state and in the flexible solid, above T_g.

However, the relaxation parameters, in particular T_2 and NOE, are also sensitive to the slower molecular motions, such as chain reptation, and also to chain–chain interactions. It is not a trivial matter to disentangle the contributions of different motions to each overall relaxation parameter, but this separation can be achieved if sufficient data are available. A full relaxation and conformational analysis of this type can yield a wealth of motional and averaged structural information. This chapter sets out the relationships between all these properties in a simplified way, which nevertheless relates to physical fundamentals.

The experimental data which the NMR experiment can readily obtain are as follows.

T_1, the spin–lattice relaxation time. This may be measured for any resolved spin type, e.g. ^{1}H or ^{13}C, within one tactic sequence, and over a range of temperatures and spectrometer frequencies as introduced in chapter 1. The term lattice is to be understood as summarising the entire surroundings of any one spin under investigation, including nearby like spins. T_1 is the average time taken by the spin in question to gain or lose attraction (e-1/e) of its energy to these surroundings. It is therefore measured by perturbing the spin (group) away from its Boltzmann equilibrium state, and then, via NMR, observing its return to equilibrium (see section 4.7). Figure 4.1 shows an example. Of course, T_1 is only rigorously definable if this return is exponential. Fortunately, this is the case, within experimental error, for all proton-decoupled ^{13}C spectra of molecules undergoing reasonably isotropic motions. It is less true in some more subtle investigations involving other, coupled spins, or for rapidly rotating methyl groups attached to a much more slowly moving chain. As will be explained below, T_1 values are most sensitive to motions at or near the NMR frequency.

T_2, the spin–spin relaxation time. With many homogeneous polymers above their glass transition temperature, this simply equals $1/(\pi \times$ width at half peak height in Hz) (see chapter 1) and it is easily measured. But when the linewidth is a composite of contributions from both relaxation and chemical shift variation, then T_2, the contribution from relaxation, must be measured

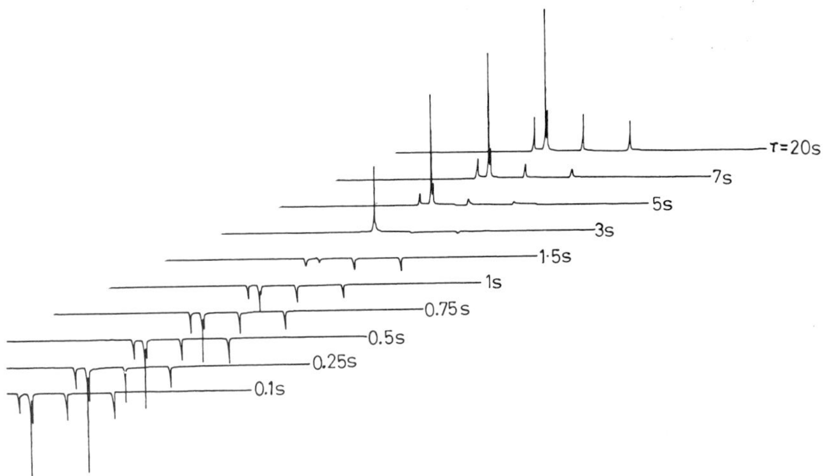

Figure 4.1 Inversion-recovery proton-decoupled ^{13}C NMR spectra of *n*-hexadecane. τ is the variable waiting time between the π and $\pi/2$ pulses. Note that the methyl resonance at the right recovers its equilibrium magnetisation (top spectrum) more slowly than the tall resonance arising from the eight central methylene carbons.

by one of a range of spin–echo techniques. Different measures of linewidth are appropriate below the glass-transition temperature, and these are discussed in chapters 6 and 7. The value of T_2, like T_1, depends on variables such as temperature, observation frequency and method. There is a closely related variable $T_{1\rho}$, which can be obtained from a variant observation method which involves the spin-locking technique (see chapter 7). Unlike T_1, T_2 values are particularly sensitive to slow molecular motions.

Internuclear Overhauser enhancements. T_1 and T_2 are relatively crude measures of the transfer of energy of a given spin. An alternative and equally crude measure is to detect the deviation of the spin, i.e. of its signal intensity, from Boltzmann equilibrium when all significant nearby spins are irradiated, and hence themselves perturbed from Boltzmann equilibrium. This occurs as a matter of course during the acquisition of proton-decoupled carbon NMR spectra as described in chapter 1. The resulting measurement is necessarily heteronuclear, for otherwise the irradiation itself would perturb the spin under investigation. The resulting fractional intensity change is called the nuclear Overhauser enhancement, or NOE. Although it depends upon the same underlying motional variables as do T_1 and T_2, it has a different dependence upon the NMR frequency, and so provides a cheap alternative to variable-frequency NMR. It is generally insensitive to slow molecular motions when these occur, in the presence of more rapid motions.

Much more informative Overhauser enhancements may also be observed between like (or unlike) nuclei of different chemical shift, especially protons, provided the irradiation is selective, and does not affect the spins under observation. (Selective T_1 contributions can be measured similarly, but less conveniently or accurately). As will be seen below, the presence and size of such pairwise NOEs is a valuable source of averaged structural information.

The practical details of these measurements are outlined in the final section of this chapter. But our first aim must be to develop their relationship to the fundamentals of molecular motion. The logic of the present treatment is as follows. The simplest model for the angular motion of molecules is the unrestrained rigid rotor, for we shall see below that this can be usefully described by one single characteristic time parameter, even though the angular motion is random. Therefore we must start with this model. It would apply literally to a spherical polymer micelle, mobile in a solvent but at a sufficiently low temperature for its internal motions to be essentially frozen. An example of this is given later. It has also been shown experimentally that it offers an acceptable, although ill-defined approximation to the slower, bulk chain motions, such as reptation, of a normal polymer above T_g. Thus, it is a practical as well as a necessary theoretical basis for the description of polymer relaxation.

Of course, most dissolved polymers are also internally flexible, and so the rigid rotor theory is extended below, to recognise the presence of superimposed, and generally more rapid local molecular motions. Finally, we discuss a more sophisticated model for the slow, bulk chain motions, which relates these to, for example, chain length and solvent penetration.

4.2 NMR relaxation theory

A nucleus in an external magnetic field is just as subject to Newton's first law as is any other object. It can only change its motion, i.e. its spin state, through the agency of a force, or equivalently of a spatially varying potential V. The magnetic component of radiation provides one such force, which can bring about magnetic resonance provided it varies in an appropriate way with time. In most polymers, the only other dominant time dependent forces arise from the interaction of the nuclear magnetic dipole under study with other nearby dipoles. However, other forces can be present in some cases, e.g. deuterated polymers, and, so we will initially treat these 'lattice' forces in a general way. They normally bring about magnetic relaxation.

The effect of these forces on the nucleus essentially follows Newton's second law, which is re-expressed in quantum mechanics as the 'time-dependent' wave equation [1]. They appear in this equation as a time-dependent potential $V(t)$, superimposed upon the much stronger, static potential from the main magnetic field B_0. The effect of this potential $V(t)$, as will be seen in more detail in following sections, is to create a transition probability W between any pair of Zeeman states. More precisely, it leads to slightly different probabilities W_+ and W_-, for upward and downward transitions, respectively. Because the lattice is always able to receive energy given out by a downwards transition, but very occasionally unable to supply energy for an upwards transition, it follows that W_- is always slightly greater than W_+. This contrasts with the extra transition probability brought about by radiation, which is shown below to be strictly the same for both upwards jumps, i.e. absorption, and downwards jumps, i.e. stimulated emission.

4.2.1 The flow of spin populations

A surprising amount of insight concerning nuclear spin relaxation can be obtained simply by treating the various available spin states as analogous to chemical states linked by kinetics. Although the spin transition probabilities such as W_+ remain to be determined either by experiment or by quantum mechanical theory, as do kinetic rate constants, nevertheless the flow of spins obeys essentially the same kinetic laws as does any other equilibrating system.

The simplest spin system available for magnetic resonance has a single spin, I, with two energy states, or Zeeman 'levels', as in Figure 4.2. The upper state has a spin population n_u^I and the lower state n_l^I. We may first consider the case of thermal equilibrium. At equilibrium, the net upwards flow of spins must equal the net downwards flow, so that

$$W_+^I n_l^I = W_-^I n_u^I \tag{4.1}$$

Now the ratio

$$W_+^I / W_-^I = \exp(-\gamma\hbar B_0/kT) \tag{4.2}$$

because the energy separation of the spin states is $\gamma\hbar B_0$, (see chapter 1), and

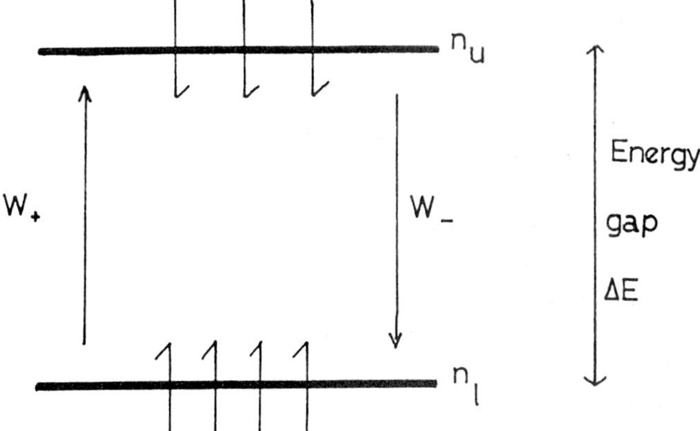

Figure 4.2 Two-state system, for a single spin type, as described by equations (4.1)–(4.4).

because Boltzmann's law describes the availability of lattice motions having the energy to stimulate upwards transitions, relative to their availability with zero energy. It follows that n_u^l/n_l^l must also obey Boltzmann's law, once thermal equilibrium is attained.

Equation (4.2) may also be used to study the approach to thermal equilibrium. Thermal non-equilibrium can arise either from deliberate perturbation of the spins, as described later, or simply as the sample is placed into the main magnetic field at the outset of the experiment. In general, for any spin,

$$\begin{aligned} \mathrm{d}n_2/\mathrm{d}t = \mathrm{d}n_1/\mathrm{d}t &= n_1 W_+ - n_2 W_- \\ &= N W_+ - n_2(W_+ + W_-) \end{aligned}$$

where $N = n_1 + n_2 = \text{constant}$. This rearranges to the simple first order 'kinetic' equation

$$\frac{\mathrm{d}}{\mathrm{d}t}\left(n_2 - \frac{NW_+}{W_+ + W_-}\right) = -\left(n_2 - \frac{NW_+}{W_+ + W_-}\right)(W_+ + W_-) \qquad (4.3)$$

which means an exponential decay of n_2 to its thermal equilibrium value of $NW_+/(W_+ + W_-)$ with a rate constant

$$(W_+ + W_-) = 1/T_1 \qquad (4.4)$$

Alternatively, one may say that the population difference $(n_2 - n_1)$, which determines the strength of the NMR signal, reverts to its Boltzmann value $N(W_- - W_+)/(W_+ + W_-)$.

A further possibility for obtaining a non-Boltzmann equilibrium arises when the spins receive fairly strong irradiation at their resonance frequency. This adds a further transition probability W_{irr} to both W_+ and W_- above. In the extreme case where $W_{irr} \gg W_+, W_-$, it follows that $n_1 = n_2$. As nuclear magnetic resonance depends upon the spin population difference $n_1 - n_2$, it follows that the resonance disappears. It is said to be saturated. Even modest saturation must be avoided in quantitative NMR. However, it is useful both for the selective suppression of resonances with long T_1, such as solvent peaks, and for the nuclear Overhauser experiments described below.

Unfortunately, it is necessary to consider a more complex spin system in order to explain relaxation in the presence of other spins, the behaviour of a carbon atom in a methine group being a good example. This is because the dominant relaxation process in this case arises from the magnetic dipole of the bonded proton, so that the possibility of proton spin transitions must also be included. Figure 4.3 shows the simplest such system. The proton spin, conventionally labelled S, has its projection along B_0 represented by the longer arrow. In this case, one must expect not only the one-spin transitions of probabilities W_I and W_S, but also concerted spin transitions of probabilities W_2 ('flip-flip') and W_0 ('flip-flop'). In each case, W_x will imply separate upwards and downwards rates of flow, in the appropriate Boltzmann ratio, and may also include an added transition rate W_{irr} due to irradiation.

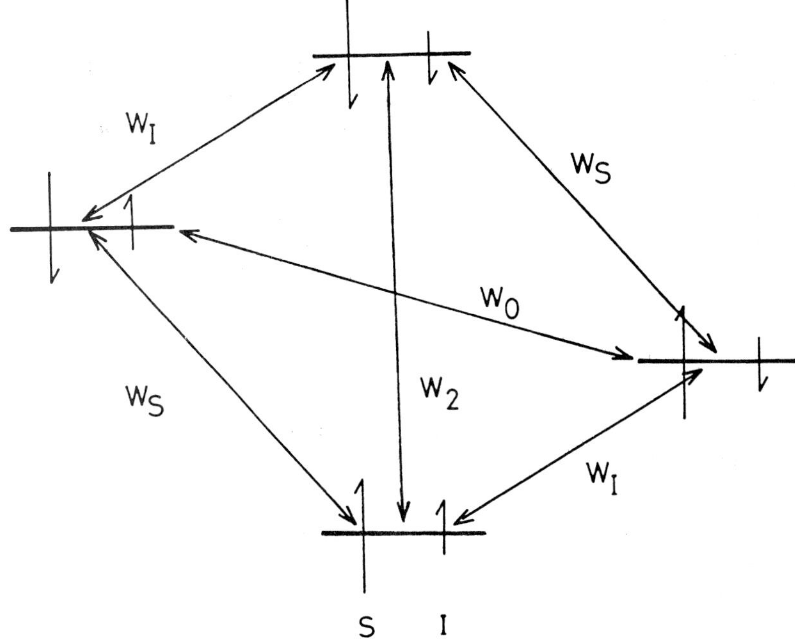

Figure 4.3 Four-state system, for two linked spin types I and S, as described by equations (4.5)–(4.7).

If 'kinetic' analysis is applied to the spin system in Figure 4.3, then four equations may be obtained for the rates dn_{1-4}/dt, of the type

$$dn_1/dt = n_2 W_I^- - n_1 W_I^+ + n_3 W_S^- - n_1 W_S^+ + n_4 W_2^- - n_1 W_2^+ \qquad (4.5)$$

One of these, however, is not independent of the other three. The four equations can describe a wide variety of relaxation behaviour, not all of which is exponential. Here we consider only one case, where S, typically the proton spin, is saturated via continuous irradiation. Other possibilities are treated elsewhere [2].

If S is saturated, then at all time $n_1 = n_3$ and $n_2 = n_4$. The algebraic pattern of the problem then reduces to the one considered above, but now,

$$1/T_1^I = 2W_I + W_2 + W_0 \qquad (4.6)$$

and the driven equilibrium population difference or signal strength $(n_2 - n_1)$ is given by

$$1 + \eta = \frac{\text{driven equilibrium signal}}{\text{Boltzmann equilibrium signal}} = 1 + \frac{\gamma_S}{\gamma_I}\left(\frac{W_2 - W_0}{2W_I + W_2 + W_0}\right) \qquad (4.7)$$

The magnetogyric ratios γ_S and γ_I arise in this equation from the different Boltzmann ratios appropriate to the different transitions. The relaxation remains exponential, if more rapid than before.

The ratio $(1+\eta)/1$ in equation (4.7) is called the nuclear Overhauser enhancement or NOE. Alternatively, η is called the nuclear Overhauser enhancement factor, or NOEF. One should not be surprised that $\eta \neq 0$, because energy is being continuously supplied to the spins.

In a later section we will indicate how the rates W_I, W_S, W_2 and W_0 may be derived from first principles. At this stage, it may be useful to anticipate these conclusions with the following summary.

1. When the molecular tumbling is rapid, and the carbon relaxation is dominated by the C–H dipolar interaction, then
$$W_2 : W_I : W_0 = 12:3:2 \tag{4.8}$$
so that the carbon NOE, $1+\eta = 2.998$.
2. Even though each W, and hence $1/T_1$, depends linearly upon the number of nearby protons, and also strongly upon the rate of the molecular tumbling motion, their ratio, and hence the NOE is essentially independent of both, provided that tumbling remains rapid.
3. When molecular tumbling is very slow, then the above ratios become much more nearly equal, and the NOE then drops to 1.153.
4. If another relaxation pathway is available for the spins I, such as the presence of nearby paramagnetic centres, or of other, unirradiated nearby protons, then W_I alone increases. If $W_I \gg W_2, W_0$ then the NOE = 1, i.e. no enhancement is observed. The proton spins have become irrelevant.
5. The same equations apply to a pair of protons which dominate each other's relaxation, as in an isolated methylene group. However, η_I now equals η_S, so that the NOE with rapid tumbling is now 1.5 (i.e. 50% enhancement), and with very slow tumbling it is -1, so that the resonance becomes inverted. The presence of other, unirradiated protons will always reduce the magnitude of these factors, as in case (4) above. It may also lead to non-exponential relaxation, because the other protons also act as stores for spin energy.

If several protons affect the carbon atom in question, and also undergo mutually correlated motions, as in a rotating methyl group, then they become even less independent of each other. Their resulting relaxation behaviour is complex, and is now non-exponential even when all the protons are simultaneously irradiated [2]. For this reason, particular care must be taken in treating methyl group relaxation in polymers.

4.2.2 *Microscopic relaxation theory*

It is now necessary to consider the derivations of the transition probabilities in more detail. The following sketch of the theory is treated in more detail elsewhere [1, 3]. The 2-level spin system of Figure 4.1 has individual energy states ψ_u and ψ_l, and must therefore have the overall wave function

$$\psi = C_l\psi_l \exp(-i\varepsilon_l t/\hbar) + C_u\psi_u \exp(-i\varepsilon_u t/\hbar) \tag{4.9}$$

where $\varepsilon_u - \varepsilon_l = \Delta E_{ul} = \hbar\omega_{ul}$ and C_l, C_u are the respective weighting factors. Insertion of (4.9) into the time-dependent wave equation, including the time-dependent perturbation $V(t)$ described in the introduction of section 4.2, enables one to deduce the absorption rate

$$\mathrm{d}|C|_u^2/\mathrm{d}t = W \tag{4.10}$$

$V(t)$ can be either, or both, the effect of radiation and that of spin–lattice interactions. In each case, it may be written as the product of a static interaction between u and l, V, and a time-dependent part $f(t)$, i.e. $V(t) = Vf(t)$. One finds that

$$C_u = (V_{ul}/i\hbar)\int_0^t f(t')\exp(i\omega_{ul}t')\,\mathrm{d}t' \tag{4.11}$$

in the simple case where all spins start in the lower level. The same equation must therefore also apply to C_l by simple reversal of the labels u and l. In the case of radiation alone, $V_{ul} = V_{lu} = \int w_u^* V w_l \,\mathrm{d}\tau = \int w_l^* V w_n \,\mathrm{d}\tau$. For spin lattice interactions, however, they are not exactly equal, but are instead in the correct Boltzmann ratio.

In equation (4.11), t' is a variable, and t is a fixed interval of time, during which spins transfer to the upper state. One may also differentiate (4.11) with respect to the interval t' in order to find the incremental contribution to C_u over the time interval dt. One obtains

$$\mathrm{d}C_u/\mathrm{d}t = (V_{ul}/i\hbar)f(t)\exp(i\omega_{ul}t) \tag{4.12}$$

Hence

$$W = \mathrm{d}|C_u|^2/\mathrm{d}t = C_u\,\mathrm{d}C_u^*/\mathrm{d}t + C_u^*\,\mathrm{d}C_u/\mathrm{d}t$$

$$= (|V_{ul}|^2/\hbar^2)\left\{\int_0^t f(t')f(t)\exp[i\omega_{ul}(t-t')]\,\mathrm{d}t' + \int_0^t f(t')f(t)\exp[-i\omega_{ul}(t-t')]\,\mathrm{d}t'\right\} \tag{4.13}$$

Now if we let $t' = t + \tau$, then the term in curled brackets becomes

$$\left\{\int_{-t}^0 f(t+\tau)f(t)\exp(-i\omega_{ul}\tau)\,\mathrm{d}\tau + \int_{-t}^0 f(t+\tau)f(t)\exp(i\omega_{ul}\tau)\,\mathrm{d}\tau\right\} \tag{4.14}$$

The first definite integral in equation (4.14) may be rewritten as

$$\int_t^0 f(t-\tau)f(t)\exp(i\omega_{ul}\tau)\,\mathrm{d}(-\tau) = \int_0^t f(t-\tau)f(t)\exp(i\omega_{ul}\tau)\,\mathrm{d}\tau$$

simply by altering τ to $-\tau$.

We now suppose that $f(t)$ fluctuates in some random manner, such that the fluctuations retain their general character in any given interval of time. This would be true for, e.g. polymer motions averaged over successive intervals of 1 s, but not true if the temperature varied. Under these conditions, $\overline{f(t-\tau)f(t)}$ will equal $\overline{f(t)f(t+\tau)}$, (where the bar indicates an average value). Both are simply the extent to which the value of $f(t)$ correlates with its value τ seconds earlier, or later. The function $f(t)f(t+\tau)$ is called the autocorrelation function of $f(t)$. It is clearly independent of the instant of measurement, t, and is conventionally written as $G(\tau)$ to emphasise this independence.

This concept of an autocorrelation function is central to the understanding of polymer motions. A homely analogy may make it more accessible. A telephone directory is (in theory) an accurate list of numbers and addresses at the moment its editing ceases. In other words, it correlates precisely with the truth, which makes $G(\tau) = 1$ at $\tau = 0$. As τ, the age of the directory, increases, this autocorrelation decays, with a half-life of a few years, and must eventually dwindle to near zero because of the finite lifespan of both humans and institutions. Exactly the same would be true in reverse if some time-traveller were able to obtain a copy of a future directory.

In the above case, the equality of $f(t-\tau)\ f(t)$ with $f(t)\ f(t+\tau)$ reduces expression (4.14) to

$$\int_{-t}^{t} f(t+\tau)f(t)\exp(i\omega_{\text{ul}}\tau)\,\mathrm{d}\tau \tag{4.15}$$

Furthermore (4.15) becomes independent of t once t exceeds the value beyond which $f(2t)$ no longer correlates significantly with $f(t)$. In this case, the integral limits extend to $\pm\infty$, and thus the entire integral becomes simply the Fourier transform of $G(\tau)$ evaluated at the Larmor frequency ω_{ul}. It is normally written as $J(\omega_{\text{ul}})$, the 'spectral density' of lattice motions at this frequency. Finally, (4.13) becomes

$$W = |V_{\text{ul}}|^2 J(\omega_{\text{ul}})/\hbar^2 \tag{4.16}$$

$J(\omega_{\text{ul}})$ may be quite a complex function of ω_{ul}, especially in the presence of motions of limited angular extent, for which $G(\tau)$ cannot decay all the way to zero. It may also depend upon which states u and l are under consideration, in more complicated cases such as in Figure 4.3. Initially, we consider the simplest case, that of an essentially rigid molecule tumbling isotropically. This could be, for example an unreacted monomer molecule, or a very short oligomer in which the overall tumbling is considerably faster than the internal motions.

Rigid, isotropic rotor. If we suppose that the rotor makes random jumps to new, random orientations then its orientational autocorrelation function will decay in the same first-order way as radioactive fission of nuclei. Therefore, in this case

$$G(\tau) = \exp(-|\tau|/\tau_R) \tag{4.17}$$

The same exponential decay can also be shown to arise from diffusional motion built up from smaller, random rotations. For such a decay, one can define the single rotational correlation time τ_c. It is the time during which $G(\tau)$ decays by a factor of 1/e. τ_R may also be approximately regarded as the mean time necessary for a major rotational jump or substantial angular diffusion to occur. For a very small molecule in a non-viscous solvent, τ_R will be of the order of the reciprocal of the rotational frequency, typically 0.1 ps. It will lengthen to 10–100 ps for a typical organic molecule, and its exact value will depend on solvent viscosity and on temperature. It lengthens further when the molecule interacts strongly with the solvent, as do aqueous ions, sugars and most biomolecules. Thus a decapeptide in water at room temperature has $\tau_R \approx 1$ ns, rising to approximately 20 ns for a small protein. Larger polymers than this are often floppy, which means that they display a range of internal motions, usually of restricted extent, in addition to their overall motions. It is therefore less appropriate to analyse them simply as rigid rotors.

Equation (4.17) may be combined with the definition of the spectral density $J(\omega)$ to give

$$J(\omega) = \int_{-\infty}^{\infty} \exp(|-\tau|/\tau_R)\exp(i\omega\tau)\,d\tau$$

$$= \left(\frac{2\tau_R}{1+\omega^2\tau_R^2}\right) \tag{4.18}$$

Thus $J(\omega)$ has a Lorentzian dependence upon the value of τ_R, just as an exponentially decaying free induction decay gives rise to a Lorentzian line-shape, i.e. $f(x) = 1/(1+x^2)$.

In equation (4.18), ω is the NMR frequency in radians per second, i.e. typically 2×10^9 Hz. Hence $J(\omega)$, W and $1/T_1$ rise with increasing τ_R until τ_R reaches the value $1/\omega$, i.e. 5×10^{-10} s. Beyond this value, $J(\omega)$ falls again. Qualitatively, those lattice motions available to stimulate W, first become more concentrated at the Larmor frequency and below, so that W rises, but then as τ_R rises further, they concentrate below the Larmor frequency, so that W falls.

4.2.3 Relaxation mechanisms

Intramolecular dipolar interactions. One may now proceed to calculate T_1 and NOE quantitatively. First, the underlying dipolar interaction V has to be expressed as a quantum-mechanical operator, in order to calculate its different but related contributions to the various transitions shown in Figure 4.2. One thus deduces [4] that

$$W_0 = 2a/[1 + (\omega_I - \omega_S)^2\tau_R^2] \tag{4.19}$$

$$W_1 = 3a/[1 + \omega_I^2 \tau_R^2] \tag{4.20}$$

$$W_2 = 12a/[1 + (\omega_I + \omega_S)^2 \tau_R^2] \tag{4.21}$$

Here, $a = n\mu_0^2 \gamma_I^2 \gamma_S^2 \hbar^2 \tau_R S(S+1)/240\pi^2 r^6$ and ω_I, ω_S are in rad s^{-1} as before. $S = 1/2$ for a proton, and μ_0 is the permittivity of the vacuum. For a CH_n group, r is normally taken as 0.109 nm, which makes $a/n\tau_R$ equal in this case to 1.0742×10^{-9}. Both T_1 and the NOE can be deduced directly from the above three equations and from equations (4.6) and (4.7). The resulting dependence of the calculated carbon nT_1, and of the NOE upon τ_R is shown in Figure 4.4, along with one example of nT_1 when a single, restricted internal motion is also included. The partially separated lines refer to different carbon NMR frequencies. Similar curves may be deduced for the slightly different equations that apply to proton–proton dipolar relaxation [3], although in this case the distance r will be a proton–proton distance, and separate contributions to r^{-6} must be added for each nearby proton. This is also necessary in the case of an unprotonated carbon resonance, for which T_1 will generally be about 20 times the protonated case, but it is not of importance with a protonated carbon because in this case the bound protons dominate the r^{-6} sum.

Figure 4.4 also includes the spin–spin relaxation time T_2. The linewidth at half height equals $(n\pi T_2)^{-1}$ Hz. T_2 arises in part from the same inter-level jumps as T_1, but also from direct perturbation of the levels u and 1 by $V(t)$. This latter contribution is analogous to the broadening of the averaged resonance that arises from near-rapid chemical exchange. As the exchange rate decreases, the resonance broadens. At low values of τ_R, this rather slow perturbation contributes half of the linewidth, but at high τ_R it dominates. This is the main reason why polymer linewidths are often frustratingly large. In the nomenclature of equations (4.19)–(4.21),

$$1/T_2 = 1/2T_1 + 6a/(1 + \omega_S^2 \tau_R^2) + 4a \tag{4.22}$$

A very similar equation [5] applies to $T_{1\rho}$, the relaxation rate in the rotating frame.

Other relaxation mechanisms. The above principles may be applied much more widely than to CH and HH pairs. A few examples are relevant to polymer relaxation.

Intermolecular dipolar relaxation. Nearby protons on other molecules, including solvent, will not contribute significantly to the relaxation of protonated carbons, because of the r^{-6} dependence deduced above. They do have a detectable influence on unprotonated carbons, however, typically contributing 0.1 Hz to $1/T_1$. They can also influence the proton resonances, especially if they are held fairly close to the proton in question by, for example, some polymer entanglement. In this case $f(t)$ will include not only any angular

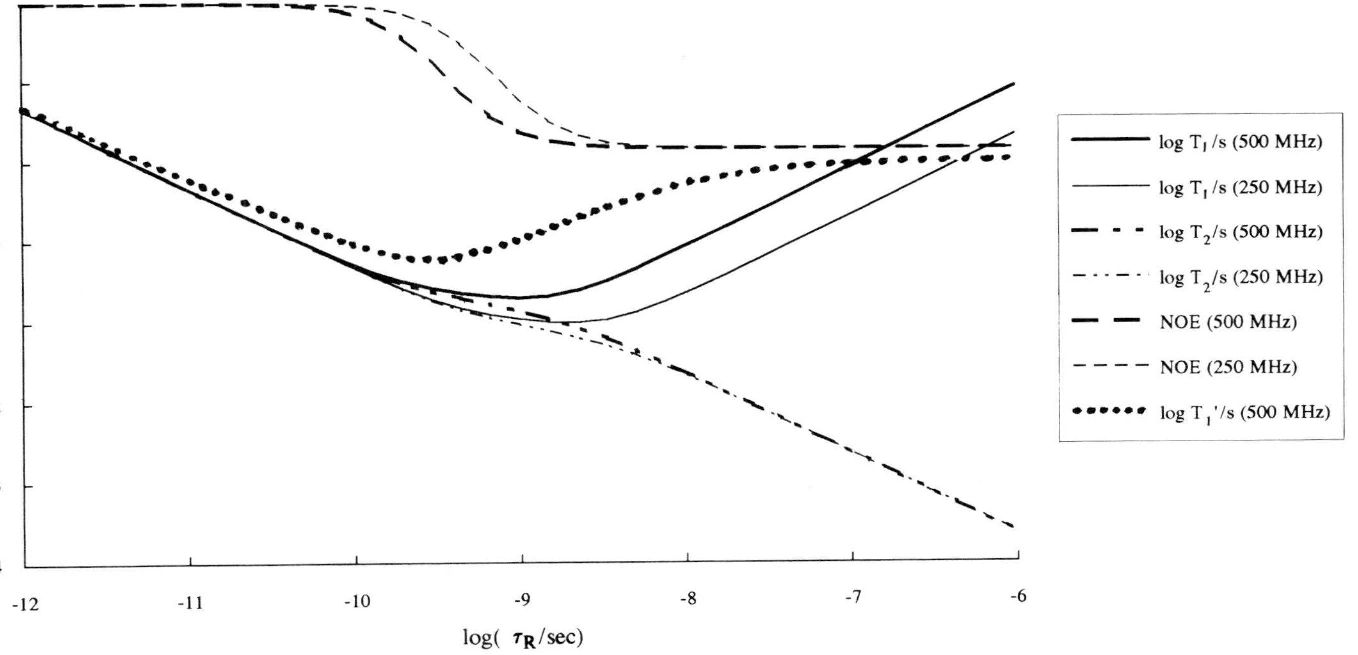

Figure 4.4 Dependence of NOE, nT_1 and nT_2 upon rotational correlation time τ_R, for a proton-decoupled ^{13}C resonance in a rigid rotor, in spectrometers with proton resonance frequencies of 250 and 500 MHz. The dotted curve represents nT'_1 for a non-rigid rotor with an internal motion which rapidly halves the angular order parameter, S^2. The curves for other values of S^2 may be estimated by interpolation.

motion of the H–H vector, but also its variation as the H–H distance fluctuates. Experimentally, one does not normally observe such interactions with solvent molecules, because these are deuterated. This reduces γ_S^2 by a factor of 42.4. However, one can observe them between, for example, different regions of block polymers or blends, typically at concentrations above 40% by volume. Below this concentration one may assume that solvent molecules intervene and separate the protons under study, and also that local motions become more rapid and extensive, thus reducing $J(\omega)$.

Quadrupolar interactions. One sometimes needs to observe the resonances of a quadrupolar nucleus such as ^{2}H, which has been introduced into a polymer, as described in chapter 8. In this case $J(\omega)$ will remain the same, but V will now be dominated by the interactions of the nuclear quadrupole with the local electric field gradient that arises from the bond by which the deuterium atom is attached. Thus no NOE will be observed, and the value of T_1 and T_2 in Figure 4.4 will be decreased, for a typical C–D bond, by a factor of about 100. The resulting resonances are therefore very broad. Further splittings will arise when the molecule is partially oriented, as described below.

Shift anisotropy. If the chemical shift depends upon molecular orientation, as it does in an aromatic ring, then this will also contribute to $V(t)$ as a result of motionally induced fluctuations in the electron shielding field. The contribution increases with B_0^2, and typically contributes 0.02 Hz to $1/T_1$, in high-field spectrometers. Again, if it arises in partially oriented molecules then it affects lineshape, this time asymmetrically. This is discussed further in chapters 5 and 7.

4.3 The dynamics of flexible molecules and macromolecules

All the above theory also applies to flexible molecules, except for the assumption that $G(\tau)$ decays exponentially. Let us now consider a somewhat more sophisticated motional model in which the relevant internuclear vector, e.g. the C–H bond, jumps freely within a cone [6] of semi-angle X, but is prevented from further movement either by other parts of the molecule, or by, for example, neighbouring polymer chains. This will turn out to be a surprisingly good approximation for local motions in a dissolved or molten polymer. The added motion fairly rapidly reduces $G(\tau)$ from 1 to

$$S^2 = \{(\cos X - \cos^3 X)/2(1 - \cos X)\}^2 \qquad (4.23)$$

One may note that S^2 varies from unity at $X = 0$ to zero at $X = \pi/2$. It is one kind of order parameter for the final state. The decay of $G(\tau)$ by random angular jumps is once again exponential, as in the previous case where X was equal to π. However, its time constant τ_G may well be far shorter than the previous τ_R, because the motions are librations, involving only a few atoms at a time, rather than bulk rotations. They represent a first, and deliberately non-specific step towards the description of local internal flexibility.

If the molecule also undergoes slow isotropic rotation as before, then this will also tend to reduce $G(\tau)$ towards zero, but much more slowly than the timescale for libration. (The same will often turn out to be true for the looping motions of a polymer chain). Thus,

$$G(\tau) = \{S^2 + (1 - S^2)\exp(-|\tau|/\tau_G)\}\exp(-|\tau|/\tau_R) \qquad (4.24)$$

This model can also be extended to further levels of librational motion, each with its own order parameter and correlation time, and has been underpinned more formally in the 'three-τ' case of two librational levels plus rotation [7]. It is clearly of a general nature. One may indeed choose to abandon relationship (4.23) and to re-interpret S^2 in (4.24) as a more general order parameter. One thereby achieves a 'model-free' theory, which has advantages over alternative, more specific models for polymer motion (see chapter 6), when the motional details are not clear.

Intuitively, one would expect S^2 to decrease near to the end of a flexible chain or side chain. This is borne out by ^{2}H NMR measurements on oriented polymers (see below and in chapter 8) where the deuterium resonance is a broad doublet, due to quadrupole coupling, whose separation is proportional to S^2. It is also qualitatively evident in Figure 4.1, where the fastest relaxation occurs at the centre of the 16-carbon chain.

One may now use (4.24) to obtain

$$J(\omega) = S^2\left(\frac{2\tau_R}{1 + \omega^2\tau_R^2}\right) + (1 - S^2)\left(\frac{2\tau_1}{1 + \omega^2\tau_1^2}\right) \qquad (4.25)$$

where $\tau_1 = 1/(1/\tau_R + 1/\tau_G)$. In the case where τ_G is extremely short, this simply reduces both $1/T_1$ and the linewidth by the factor S^2 over the entire range of τ_R. More realistic values of τ_R lead to somewhat smaller reductions of relaxation rate, especially when τ_R is less long, and also to counterintuitive increases of NOE at very long τ_R.

The above 'two-τ' calculations have been applied with considerable success to explain both the main-chain and the side-chain relaxation parameters in poly *n*-butyl and poly *n*-hexyl-methylacrylate in solution [6, 8]. A more detailed analysis shows that equations very similar to the above also describe the relaxation of an anisotropic but otherwise unrestricted rotor, such as an oligomer of a liquid-crystalline polymer, and that of a rotating methyl group attached to an isotropically tumbling molecule or molecular segment. Because the theory presented so far is almost model-free, it can therefore be applied with some success even to tangled polymer chains, even though the underlying physical meaning of the various motional parameters remains to be explained. As an example, consider the T_1 data [9] on a natural, lightly crosslinked rubber shown in Figure 4.5. The curves give the temperature dependence of nT_1

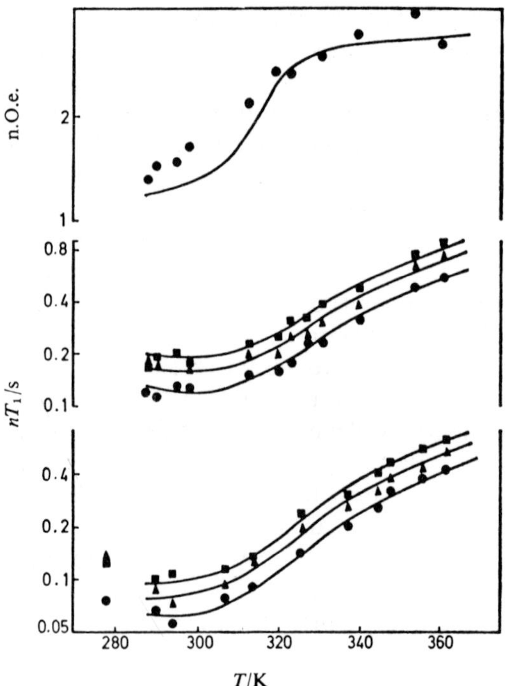

Figure 4.5 Proton-decoupled ^{13}C nT_1 values of a natural rubber, observed at carbon frequencies of 45.2 MHz (central set) and 22.6 MHz (lower set) plus NOE values at 45.2 MHz (upper curves). The labelling is $\{CH_2(\blacktriangle)–C(Me)=CH(\bullet)–CH_2(\blacksquare)\}_n$ and the solid curves are calculated from the '3-τ' theory described in the text.

for =CH–$\underline{C}H_2$–, –$\underline{C}H_2$–C(Me) and =$\underline{C}$H–carbon types at two field strengths. All three carbons have a minimum at each field at the same temperature, where $\omega \approx \tau_R$, as is predicted by equation (4.25). The minima differ, however, because the extent of local motion, S^2 or X in the theory above, decreases with increasing steric restriction of the atom in question. One may deduce maximum libration angles of 0.73, 0.65 and 0.5 rad, respectively. Similar angles are deduced from the analysis of the methacrylate solution data [6].

However, the nature of the motion that corresponds to the overall tumbling of an isolated, polymer molecule remains unclear. It clearly has a relatively slow component, because the linewidths are considerably greater than one would deduce from equations (4.22) and (4.25). They may be approximated by the more complex model invoking three levels of motion, mentioned above, provided that the slowest motion has τ_R as long as 18 ns. One also notes empirically that the slowest level of motion, in the bulk state, becomes more dominant when, for example, the rubber is blended with particles of an inert, solid filler. This results in broader resonances, beyond what would be expected anyway from local field inhomogeneities. Similar observations have been made concerning gels with adjustable crosslinking [10].

Whole-chain motions. A much more elaborate model is required for any further interpretation of overall chain motions. Its derivation is well beyond the scope of this chapter, but may be followed in [11]. The authors of this review concentrate on the Doi–Edwards reptation model [12] for tangled polymer chains, either in relatively concentrated solutions or in mobile bulk states. Such a chain portion is represented in a simplified way in Figure 4.6. There are essentially four levels of angular motion available to the chain. The normal modes of molecular vibration constitute the most rapid motions, but their angular extent, a few degrees, is insufficient to contribute significantly and directly to the correlation function.

The next most rapid type of motion is identified with deliberate imprecision as 'defect diffusion'. It is closely similar to the fairly rapid vibrational excursions of limited angular extent that are considered in the theory presented above. Motions of this type can arise either from cumulative low energy distortions of bond angle within one set of potential energy minima, or from thermally accessible local changes of the rotationally isomeric states of a few bonds. The latter changes will probably be correlated, in order to satisfy the

Figure 4.6 Simplified model for polymer reptation, showing the 'tight tube' created by contingent polymer chains plus possible plasticiser or solvent molecules.

overall geometrical constraints on the polymer chain. They serve to give the chain sufficient flexibility for reptational motions.

Kimmich *et al.* [11] approximate this rapidly decaying contribution to the overall correlation function, $A(\tau)$ in the usual way, i.e. by an exponential decay with correlation time τ_S to a lower, constant value. Thus

$$A(\tau) = a_1 \exp(-\tau/\tau_S) + a_2 \tag{4.26}$$

The exponential term here is a simple consequence of assuming random jumps within the librational solid angle. More complex models give similar functions in this time range. Typical experimental values for τ_S are of the order of 1 ns, but it is not a well-defined parameter. The residual order parameter a_2 is found to decrease when the polymer is swollen by, for example, a solvent. No doubt this swelling effectively relaxes the constraint of the 'tube', and hence permits wider librational excursions.

The next type of motion in the above theory corresponds to the overall reptational (sliding) motion of the polymer chain within its tube-like constraint. This motion affects the orientation of bonds, not only because the constraining tube is itself curved, but also because in places it may be sufficiently large to accommodate tight loops of chain. As a particular bond is pulled past the extremity of a loop, it may reverse its direction completely. The loss of angular correlation $B(\tau)$ which results is not exponential, and is therefore not strictly describable by a correlation time. Instead,

$$B(\tau) = \exp z(\tau)(\operatorname{erfc}\sqrt{z(\tau)}) \tag{4.27}$$

Here, erfc $\sqrt{z(\tau)}$ is the complementary error function, i.e.

$$\operatorname{erfc}\sqrt{z(\tau)} \equiv 1 - 2/\sqrt{\pi}\int_0^{\sqrt{z(\tau)}} \exp(-u^2)\,\mathrm{d}u$$

The function $z(\tau)$ is half the mean square displacement of any point along the chain, relative to a curvilinear coordinate along it. Calculation of reptation shows that

$$z(\tau) = D_1^K(\tau + 2\sqrt{\pi\tau T_d})/l^2 \tag{4.28}$$

where $T_d = N_K^2 b^2/\pi^2 D_d$. In these equations, D_1^K is the curvilinear diffusion coefficient of the overall (Kuhn) chain, and therefore also closely relates to the linear diffusion rate of the polymer. D_d is a defect diffusion coefficient, l is the correlation length of the 'tight tube', i.e. of the actual polymer periphery, N_K is the number of statistical segments in the (Kuhn) chain model and b is the r.m.s. length of each segment in this chain model. Thus $N_K b$ is approximately the chain length. This gives some dependence of $B(\tau)$ upon chain length at longer time values. However, if τ is small, then $B(\tau)$ is exponential in $\tau^{1/2}$, and if T_d is small, then $B(\tau)$ reduces to a simple exponential, somewhat steepened by the erfc term. The calculation shown in Figure 4.7(b) involves

(a)

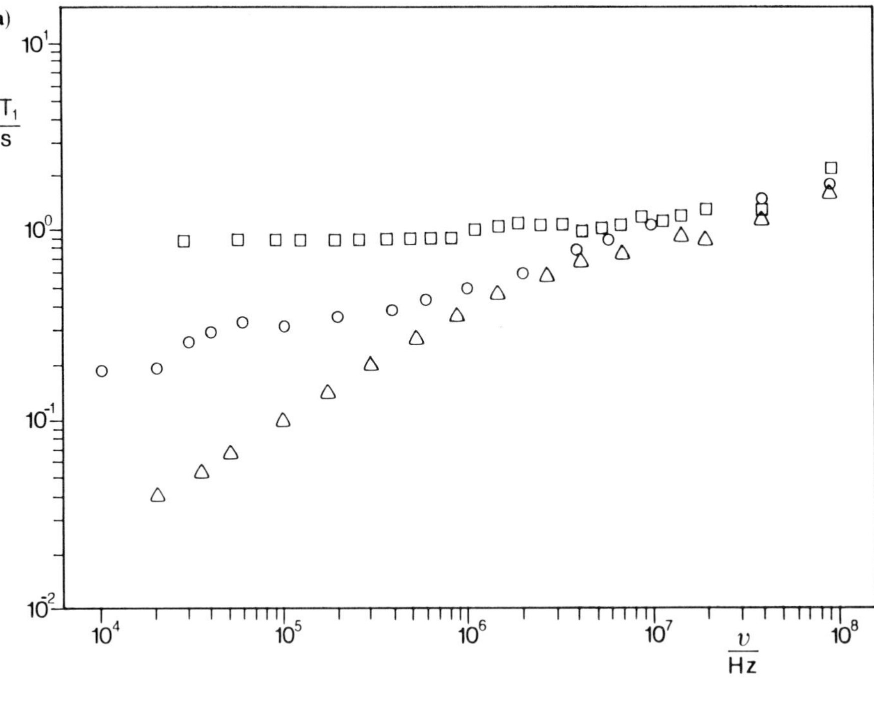

(b)

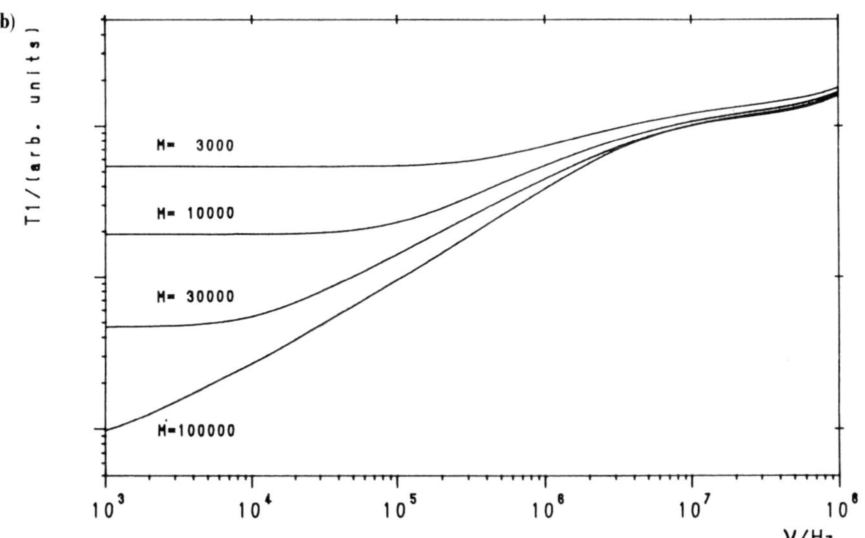

Figure 4.7 (a) Variable-field proton T_1 data at 200°C for three polyethylene fractions with molecular weights: (□) 2440; (○) 7200; (△) 125 000. (b) T_1 values, calculated as outlined in the text. From Kimmich *et al.* [11]. Reproduced by permission of Pergamon Press.

this assumption, and allows a convincing fit with the experimental data in Figure 4.7(a). Fits of this type permit one to estimate a characteristic although imprecisely defined decay time for these in-tube reptational motions. It depends upon molecular weight, and lies typically between 1 and 30 ns. Thus the reptational motions can occur as rapidly as the defect diffusional motions, and may be related to them, in part.

The slowest motion considered is that of the 'tube' itself. The resulting correlation function $C(\tau)$ is complex, for its derivation differs between the central part of tube, and the ends, where the possibility of complete temporary disengagement of the chain must be considered. It may be approximated by

$$C(\tau) = (\mathrm{erf}\sqrt{\tau_r/\tau})^2$$

which is an non-exponential decay with a characteristic time similar to T_d but steeply dependent upon the molecular mass M. A fit for data similar to that in Figure 4.7(a), but for the mean polyisoprene proton T_2, was obtained using

$$\tau_R = 3.5 \times 10^{-16} M^3 (1 - 30M^{-1/2})^2 \text{ s}$$

which equals 0.88 μs for $M = 2500$ and 0.35 s for $M = 100\,000$. This is a very steep dependence on M, which will be particularly evident in any relaxation process that depends on low frequencies, such as the linewidth.

The overall correlation function is then

$$G(\tau) = A(\tau)\,B(\tau)\,C(\tau) \tag{4.29}$$

It may very crudely be considered as a sequence of exponential decays, as in the earlier '3-τ' model, with only the final, M-dependent one being complete.

The above theory can be made to fit further experimental data, such as the variation of proton linewidth with temperature. An example of such data, for the net proton T_2 of polyisobutene, is shown in Figure 4.8. A typical activation enthalpy, applied to all the motional constants such as D_1^K, was 56.5 kJ mol^{-1}. The dependence on M, the molecular weight, is most evident at higher temperatures.

In some respects, the rather elaborate motional model presented above is complementary to the preceding, almost model-free approach. It is much more informative about the slower motions, and their dependence upon molecular weight in concentrated polymer solutions and melts. But it is less informative about the more rapid motions, and in particular about the way these vary from atom to atom within a dyad unit, or between dyads of different tacticity.

This same limitation was apparent in a number of earlier theories of polymer relaxation, which must nevertheless be mentioned for completeness. Bloembergen's rigid-rotor theory [13] was initially modified by replacing τ_R by a rather wide distribution of correlation times, such as a log χ^2 distribution [14]. This broadens the curve of log T_1 versus τ_R, now an average, and raises

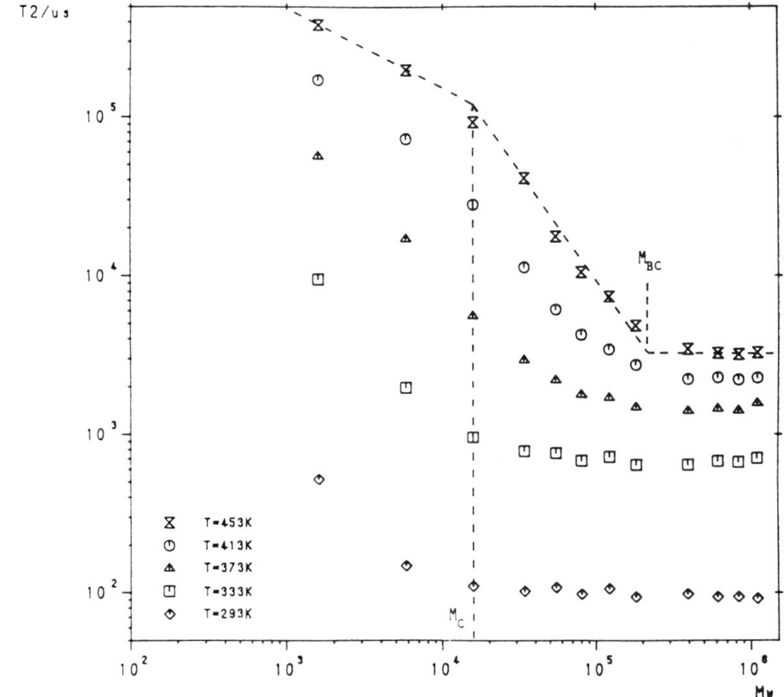

Figure 4.8 Molecular weight dependence of the mean proton T_2 values at various temperatures, for polyisobutene. From Kimmich *et al.* [11]. Details of the fit to theory (dotted lines) and of the limiting molecular masses M_B and M_{BC} for various motional regimes, are given in this reference. Reproduced by permission of Pergamon Press.

its minimum. But it still gives a generally poor fit to more detailed data, e.g. as shown in Figure 4.5.

The second approach was to apply Woessner's calculation for anisotropic diffusion [15, 16], which is formally identical to that for a methyl group rotating fairly rapidly on an isotropically tumbling substrate. This model raises the T_1 minimum in Figure 4.4, but only by a fixed amount, and it also gives rather poor predictions for nuclear Overhauser enhancements. Better results are obtained if one includes multiple but free side-chain rotations [17]. Alternatively, one may consider models [18] with jumps between a limited number of conformations, on top of an underlying isotropic rotation. This class of model is particularly successful when the conformations are rather well defined, as in London's analysis [18] of the ring-puckering motions of proline residues in peptides. The available conformations of a polymer are not as easily defined, although they can be approximated by the rotational isomeric state (RIS) theory described below [19]. Nevertheless, considerable headway can be made even with very crude models. For example, the flexure of a simple linear polymer such as polyethylene can be approximated by considering each carbon atom to be free to jump, subject to its covalent connections, to any neighbouring point on a diamond lattice, i.e. to points arranged in the same way as the carbon atoms in diamond [20]. With this model, if the jump correlation time is τ_D and the isotropic rotational correlation time is τ_0, then the correlation function $G(\tau)$ is given by

$$G(\tau) = \exp(-\tau/\tau_0)\exp(\tau/\tau_D)[1 - \mathrm{erf}(\tau/\tau_D)^{1/2}]. \tag{4.30}$$

This gives a much better fit to experimental data than, say, the log χ^2 distribution [21]. However, it does lead to unreasonably low values of τ_D at low temperatures. They may even become greater than τ_0.

Further improvements to this model are described in chapter 6, along with completely different and rather specific dynamic models which involve the direct simulation of Brownian motions. But at this stage in the analysis of polymer motions, one is beginning to press at the limits of the experimental data. Progress is only possible with extensive data, preferably obtained at several different fields, and even then one must apply further criteria such as the reasonableness of the resulting motional parameters.

4.3.1 *Applications*

4.3.1.1 Dilute, concentrated and plastic polymers. The 'three-τ' model has been applied this way to all the protonated non-methyl carbon resonances of rubber (see above) [9], glutathione dimer and gramicidin-S [7], with T_1 and NOE data obtained at up to four magnetic field strengths. The peptide data provide a severe test, for all data must be consistent with a single overall rotation. In fact, all the data can be fitted within experimental error. An intriguing pattern of motional parameters is thereby revealed, which almost certainly has a broader relevance to the internal motions of polymers. All the *C*HR main-chain resonances correspond to rather rapid motions (correlation times between 20 and 100 ps) with a fairly narrow mean half-angle X. For chains without crosslinks, X is always close to 0.5 rad, implying maximal angle excursions of about 30° away from the mean direction of the CH bond, in any direction, or alternatively, of greater angle in some directions but less in others. Such fluctuations of angle are consistent with dynamic molecular modelling studies. When the group R is crosslinked to another chain, the *C*HR motional rate is not greatly affected, but its extent is reduced. Conversely, it is extended somewhat when steric hindrance is reduced, as with main-chain CH_2 groups, with chain ends [22] and with unrestricted side-chains.

In contrast, the second, slower type of librational motion is much more sensitive to local steric restriction. In conformationally unrestricted side-chains, it can become almost as rapid as the more local motions described above, with semi-angles of up to 1 rad. But where there are serious restrictions of conformation, it can disappear altogether. Thus the extent of the slow libration does seem to correlate sensibly with what is known about the local structure. It may probably be pictured as arising from significant conformational jumps, on the scale of several linked atoms, made possible by an accumulation of smaller-scale distortions. As a final example, the glutathione dimer has a 'rope ladder' structure [7], and so one anticipates, and finds, that the slower librations have much the same angular extent at all positions down the twinned chain. At no point is there local free rotation.

In all the above studies, the bulk chain motions are approximated by those of a rigid rotor. The validity of this approximation is underlined by a study of lithium polystyrene carboxylate in micellar form, undertaken by Raby *et al.* [23]. They showed by electron microscopy that the micelles had a uniform diameter of 6.0 nm, and that near to room temperature their librational (segmental) motions were largely frozen, even though the micelles were still moving freely in the acetone solvent. They then converted the NMR determined value of τ_R to a dynamic micellar diameter using the Stokes–Einstein equation, and thus obtained a value of 5.7 nm. This is satisfyingly close to the microscopy value.

Two broad generalisations may be drawn from these studies and applied to local motions in polymers. The first is that, in the absence of unusual steric restrictions, the atoms of a side-chain will increase in both their extent and rate of motion as the number of bonds from the main-chain increases. This will result in a lengthening of T_1, and in most (but not all) cases an increase of NOE, for both carbons and comparable protons, e.g. in a methylene chain. However, the underlying motions will usually not have the extent anticipated for free rotation at each bond. This generalisation will hold true for both the solution state and the bulk polymer above T_g, where the local motions are similar.

The second generalisation is that the relaxation times of comparable main-chain atoms will increase, at all temperatures, as more conformations become energetically available to the chain. This may lead to a modest dependence of T_1 upon local tacticity. In vinyl polymers, there are indeed several reports of somewhat longer T_1 values for isotactic chain segments relative to syndiotactic segments. These are discussed later in more detail with respect to RIS theory (section 4.5).

4.3.1.2 NMR linewidths in gels and near-solid polymers. One limitation common to all the motional models described so far is that, at least for proton-decoupled carbon spectra, they predict exponential T_2 decay and hence a simple Lorentzian lineshape, even when the correlation function is complex. Furthermore, the predicted linewidths become almost infinite as the assumed underlying isotropic rotation slows to zero. In fact, the linewidths of polymers well below the glass transition temperature reach a maximum, determined by static dipolar interactions. These are discussed further in chapter 7. The lineshape will not normally remain Lorentzian.

If internal librational motions are still present, they will always tend to average the static interactions, and hence reduce the linewidth somewhat. In the case of a lightly crosslinked rubber, described above [7], or gel [10], the local looping motions are sufficient to give a liquid-like spectrum. But the line-narrowing will normally be less than this, so that little resolution is available in more rigid, near-solid polymers. The extent of local motions may also vary in different regions of the solid, e.g. because of partial crystallinity. In this case, the more ordered regions will give broader, and hence less intense resonances, and the lineshape will be highly non-Lorentzian. It will typically have a discernible central peak on top of a broader background. The background may disappear altogether when only solution-state NMR techniques are used, and in this case one must take care not to interpret the residual component as representative of the entire sample (see chapters 6 and 7).

If the local librational motions are of intermediate extent, then the partially averaged dipolar splittings that remain may fall to a few kiloHertz. This is typical for gels and similar phases. In such cases, much of the residual linewidth may also be removed by high-speed magic angle spinning (see chapter 6), to give quite well-resolved spectra [24]. In order to apply this solid-state technique to a non-rigid sample, it is normally necessary to use a spinner with an O-ring seal.

4.3.1.3 Oriented polymers. If a molecule is somewhat oriented with respect to B_0 but still retains liquid-like motions in the other two dimensions, then its spectrum alters significantly. The main reason is that the CH and HH dipolar couplings will not now average to zero. Instead, every resonance becomes sensitive to the orientation and distance of many nearby spins. The result is many line splittings, or at least substantial broadening if the orientation is locally variable. The residual motions do reduce these splittings somewhat, but cannot do so entirely. If they are not too large, as in some liquid crystalline polymer solution phases, then in the case of carbon NMR spectra they can be removed or at least substantially reduced by proton dipolar decoupling.

If the chemical shifts are also anisotropic, as for both carbons and protons in unsaturated moieties, then their resonances will be shifted, because their shift tensors will no longer average to zero. In the case of very slow overall rotation, they will also be anisotropically broadened in proportion to B_0. (This should not be confused with the dynamic contributions that shift anisotropy makes to relaxation. As equation (4.16) shows, these are proportional to the square of the static interaction.) Shift anisotropy is most marked for heavier nuclei, e.g. for attached ^{31}P atoms. The resulting lineshapes can be analysed to reveal the ranges of orientation of the liquid crystal relative to B_0.

Further splittings arise when the nucleus under observation has $I > 1/2$, and is therefore quadrupolar (as discussed in chapter 8). If it lies in an electric field gradient eq, its resonance will be a doublet with splitting

$$\Delta\nu = (3e^2qQ/4)(3\cos^2\theta - 1) \tag{4.31}$$

where Q is the electric quadrupole moment of the nucleus and θ the angle between the gradient direction (typically the bond direction) and B_0. For a typical C–D bond aligned along B_0, the resulting deuterium splitting is close to 180 kHz. This reduces not only as θ alters, but also in proportion to the order parameter S^2 in the case of restricted angular motion. It can therefore be exploited to study partial orientation.

The effects of orientation upon T_1 and NOE are much less dramatic than those upon lineshape. Indeed, T_1 may not jump perceptibly as an isotropic to nematic transition is crossed. The relevant equations are very similar to those derived above for the two-τ or three-τ cases.

4.4 The conformational states of polymers

In principle, the polymer motions described in section 4.3 can be identified not only by broad theories, but also by computer modelling. Polymer modelling is a very active area of research, and has already been successful in explaining a range of bulk and dynamic polymer properties at and above the nanometre scale. However, it is more difficult to model polymers on the atomic scale, because one requires a very precise knowledge of inter- and intramolecular forces. For example, a typical calculation error for two possible conformations involving rotation of a particular bond might easily be 2 kJ mol^{-1}. This corresponds to a factor of two in relative probability at room temperature. Such errors will rapidly propagate down a polymer chain to produce major conformational uncertainties, and will be compounded by small uncertainties in bond angles. Much larger errors are probable in crude calculations which neglect solvent.

Nevertheless, some progress has been made. This section shows how useful conformational calculations have been performed even without the aid of computers, especially on polymers with obvious steric constraints. The next section describes how proton and carbon NMR provide substantial experimental checks on such calculations.

If a 360° rotation takes place about the central bond in n-butane, then the energy of the molecule rises and falls so as to give three minima [19]. Two minima are of equal energy, involving *gauche* arrangements, and the third, which corresponds to an all-*trans* carbon chain, is approximately 3 kJ mol^{-1} lower. The maxima are at least five times higher than this, and more so 180° away from the lowest minimum. Here the two methyl groups have their closest approach. It follows from Boltzmann's law that the central bond in *n*-butane is extremely unlikely to exist more than some 20° either side of these minima, except briefly during a transition between minima. Also, the statistical occupancy of the all-*trans* minimum will be about three times that of either of the other two minima.

Rotational isomeric state (RIS) theory, developed by Flory [19], makes the approximation of describing each of these minima as a single conformational state of the molecule. It is a very general theory, able to cope with any proposed potential and any number of minima, and to translate these into bulk data such as radius of gyration.

For larger molecules, the theory retains the central notion of each bond having a finite and usually small number of isomeric states, but adds factors to account for steric and related longer-range interactions.

We may illustrate RIS analysis by a highly simplified analysis of the conformation of polystyrene (PS), using Flory's later formulation [25] of the theory (his first formulation [19] gives the same results, but partitions the intramolecular interactions less intuitively). In this simple analysis we assume, with some experimental justification, that there is no accessible minimum for any main-chain rotation of PS in which a phenyl group or chain segment lies parallel, or nearly parallel to a phenyl group or chain segment starting at the next chiral centre. Instead, one only considers the four idealised dyad conformations shown in Figure 4.9. We also ignore any consequences of rotating the phenyl group.

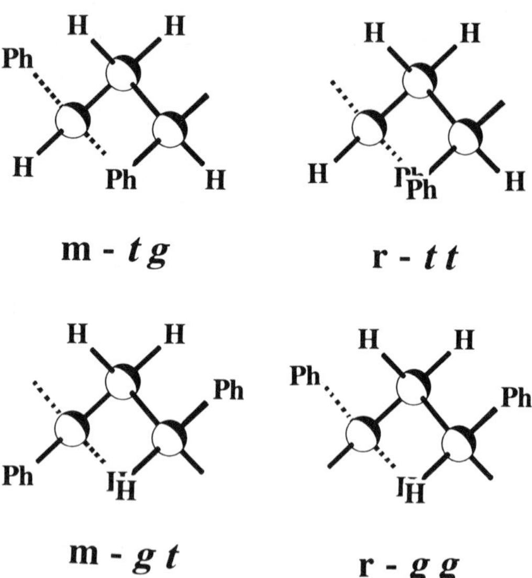

Figure 4.9 The idealised dominant conformations of *m*- and *r*-polystyrene, as described analytically in equation (4.32).

The probabilities of occurrence of each of the above states are now expressed as two by two matrices U_m and U_r (where $g = gauche$, $t = trans$)

$$\overbrace{\quad (t) \qquad (g) \quad}^{\text{(conformation of bond 1)}}$$

$$U_m = \begin{bmatrix} 0 & 1 \\ 1 & 0 \end{bmatrix}\begin{matrix} (t) \\ (g) \end{matrix}\Bigg\} \text{(conformation of bond 2)}, \qquad U_r = \begin{bmatrix} 1 & 0 \\ 0 & x \end{bmatrix} \tag{4.32}$$

Here the probabilities in each matrix are normalised to make the highest probability unity in each. In the more general theory, the factor x is re-expressed in terms of the rotational constraints on individual bonds, and one typically needs five such variables for an acceptable description of an unsymmetrical vinyl polymer in which each bond has three rotational states. In the above case it may simply be defined as the relative probability of the gg state in the racemic dyad.

It is necessary to use matrices, i.e. to consider bonds pairwise, to include the dominant steric interactions between atoms attached to adjacent chiral centres. For the same reason, the conformations of the two bonds on either side of one chiral centre must also be considered pairwise, as a matrix U_I. In the present case, one may reasonably assume that

$$U_I = \begin{bmatrix} 1 & 1 \\ 1 & 0 \end{bmatrix} \tag{4.33}$$

In other words, the local conformations $t|t$, $g|t$ and $t|g$, where the vertical bar indicates the chiral centre, are equally probable but $g|g$ is totally forbidden. Inspection of a simple model shows that the chain interacts strongly with itself if a $g|g$ conformation is forced, but that no serious constraints appear in the other three local conformations.

One now obtains a better estimate of the conformational probability matrix of the original dyad, initially U_m or U_r, by multiplying by U_I, in either direction. For example, U_m would become $U_I U_m U_I$. The effect of this is to disfavour the g contributions to U_m, because unlike the tt component, they are not compatible with all conformations of the neighbouring dyads. One may then proceed to obtain accuracy at any desired level by extending the multiplication sequence, e.g. $U_m U_I U_m U_I U_m U_I U_m U_I U_m$. The key feature of this calculation is that its result will depend significantly not only upon whether the original dyad has m or r stereochemistry, but also on the stereochemistry of the neighbouring units. This results in, for example, chemical shifts which depend upon neighbour tacticity, as described in chapter 1. It can also in principle explain the dependence of bulk solution properties upon tacticity.

The above sketch requires substantial extension for most polymers. As soon as a second non-*trans* conformation is considered, one must take more care with the handedness of the possible *gauche* conformation, and their dependence upon stereochemistry. One may also need to consider repulsive interactions of longer range than above. This may necessitate the use of much larger matrices. However, in many cases this final computational effort is not justified, for the commonest limitation upon RIS calculations is the limited quality and quantity of the experimental data they seek to explain.

One relatively simple extension of RIS theory is to deduce U_m and U_r matrices that are not normalised, from computer-based modelling. From this, one can deduce the relative probabilities of formation of various tacticity sequences during polymer propagation. These may be compared with the experimental probabilities, obtained in polymers where the asymmetric centres can be part-epimerised by chemical means [26].

4.4.1 *Relation of conformation to NMR properties*

Spin–spin couplings. Proton NMR also provides a useful test of the RIS theory, with monosubstituted vinyl polymers, $(CHXCH_2)_n$, for here one may be able to measure all the vicinal interproton coupling constants, ^3J, e.g. by the well-known 2D J-resolved experiment [27] (see chapter 1). More simply, one may choose to study a fully isotactic or syndiotactic chain. The same couplings may be estimated with reasonable accuracy if one calculates the appropriate dyad conformational probabilities, and also the appropriate parameters to enter into the Karplus equation [28], which relates the HCCH dihedral angles to ^3J. In this way, one calculates a weighted average coupling for each proton pair. Some uncertainties will still arise if it is not easy to estimate the bond angles that correspond to the t, g, etc. states, especially as these will in fact also be dynamic averages of highly asymmetric functions. Nevertheless, useful information can be obtained. Experimentally, the couplings for a dyad of given stereochemistry do not seem to depend strongly upon the stereochemistry of its neighbours.

As an example of this, Ferrero and Ragazzi [29] have calculated the average vicinal coupling constants for isotactic and syndiotactic polypropylene, using a 5-state conformational model. They obtain average dihedral angles, for the t bond state, of 173–180° and 65–72°, respectively, for the two vicinal proton pairs, and for the main g state, the exact reverse of these angles. Using their computed weightings at 140°C, which also include more minor contributions from the other three states, they obtain averaged calculated couplings of 6.98–7.76 plus 5.60–6.47 Hz (isotactic) and 8.05–8.82 plus 4.74–5.41 Hz (syndiotactic). The experimental coupling constants at 150°C are 7.0 plus 5.7 Hz (iso) and 8.3 plus 4.8 Hz (syn).

Carbon chemical shifts. The chemical shift patterns in the ^{13}C NMR spectrum of a polymer with chiral centres usually contain much more detailed conformational information than any other comparable measurement. However, these patterns must first be analysed in terms of local tacticity (see chapter 1) and a theory must then be found to relate them to the local conformation. The most successful theory so far is that of the 'γ-*gauche* effect'. This is a broadly accepted experimental conclusion from a wide range of studies [30] of carbon chemical shifts. Essentially, if a unit C*–C–C–X has a *gauche* dihedral angle, where X, the atom γ to C*, is any atom or group other than H, then C* will experience a shift in δ_C of typically -5 ppm, relative to a *trans* dihedral angle. If X is also C, then the effect will be reciprocal, although only qualitatively so if $C^* \neq X$. The theoretical basis for this observation is still not agreed. Experimentally, it seems to be less than -5 ppm for relatively polar groups such as X = OH, and possibly higher for more complex groups, such as $X = COOCH_3$. The dependence of the γ-*gauche* effect upon the precise dihedral angle is also not adequately known.

Nevertheless, Tonelli and co-workers [31, 32] have been remarkably successful in using the γ-*gauche* effect to predict the detailed ^{13}C NMR spectra of several vinyl polymers, such as polypropylene and poly(vinyl chloride). An example of their work is given in Figure 4.10. The fit is almost as good

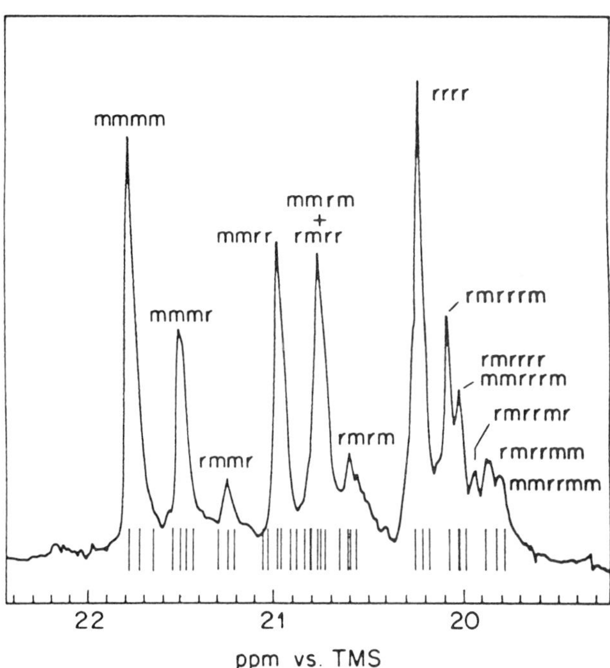

Figure 4.10 Proton-decoupled methyl carbon NMR spectrum of polypropylene, together with shifts calculated for hexads using a 5-state model, from Tonelli [31]. Reproduced by permission of VCH Publishers Ltd.

for the methylene resonances [33], using the same conformational calculations, with five rotational states for each bond. It is also very acceptable for the methine resonances. In addition, it explains the solid-state shifts of polypropylene semi-quantitatively, as discussed in chapter 5.

Interestingly, the carbon shift data in Figure 4.10 may be explained reasonably, albeit less precisely, by a much simpler conformational model for polypropylene, in which each bond is restricted either to a *t* conformation or to that *g* conformation which avoids methyl–methyl, methyl–chain or chain–chain *syn*-axial repulsions. This leads the isotactic chain to be exclusively $(tg)_n$ or its mirror image, and the syndiotactic chain to be either $(tt)_n$ or $(ttgg)_n$, or an admixture of these with appropriate weightings. Table 4.1 shows the fit thus obtained, using the *syn*-axial parameters described below, and a syndiotactic *gg*/*tt* weighting of 0.74. This illustrates how even a very simple, two state model can generate meaningful shift predictions once dyads are linked with the partial or complete exclusion of $g|g$ joins.

The halogenated polymers are slightly less tractable, but they are are in any case known to give shifts that depend somewhat upon solvent, especially in the methylene region. However, the spectrum of poly(methylmethacrylate) (PMMA) cannot be predicted by any reasonable conformational analysis [34]. In this case the shift of the methyl carbon in the all-*trans* chain should be insensitive to the stereochemistry at its neighbouring centres, for the PMMA chain is known from other measurements to have this predominantly all-*trans* stereochemistry. Yet the shifts of its carbons show a very large dependence upon tacticity, in both the solution and crystalline states.

The problem here probably arises from our incomplete understanding of the *γ-gauche* effect. Alternative interpretations exist for explaining the same carbon shift data [35]. One possibility is to perceive the *γ-gauche* effect as arising from the weighted average of three *syn*-axial interactions. Thus the chain–methyl group in isotactic PMMA is *syn*-axial (crudely, parallel) to two neighbouring chain–methyl groups, whereas in syndiotactic PMMA it

Table 4.1 Carbon-13 NMR shift differences for polypropylene calculated using a simplified RIS model

Methyl resonances	Methylene resonances	Methine resonances
mmmm 1.91 [1.99][a]	*mmm* 0.60 [0.79]	(all) 0 [0 to −0.5]
mmmr 1.71 [1.73]	*mmr* 1.35 [1.35]	
rmmr 1.52 [1.46]	*rmr* 2.14 [2.07]	
mmrm 1.04 [1.10]	*mrm* 0 [0]	
rmrm 0.83 [0.81]	*rrm* 0.96 [0.88]	
mmrr 1.18 [1.27]	*rrr* 1.90 [1.81]	
rmrr 0.97 [0.96]		
mrrm 0 [0]		
mrrr 0.18 [0.27]		
rrrr 0.34 [0.49]		

[a] [observed shift differences].

Table 4.2 Carbon-13 NMR shift differences for poly(methylmethacrylate) using a *syn*-axial shift calculation

CH_3		CO	Chain C	CH_2	
mmmm	5.04 [5.30][a]	− 0.08 [0.20]	0.87 [0.98]	*mmm*	0.66 [0.60]
mmmr	4.92 [4.87]	− 0.04 [0.10]	0.89 [0.98]	*mmr*	1.89 [2.10]
rmmr	4.80 [4.60]	0 [0]	0.91 [0.98]	*rmr*	3.13 [3.15]
mmrm	2.49 [2.60]	0.96 [1.00]	0.41 [0.38]	*mrm*	0 [0]
rmrm	2.37 [2.40]	0.99 [0.96]	0.44 [0.38]	*rrrm*	1.44 [1.40]
mmrr	2.64 [2.50]	0.95 [0.80]	0.42 [0.38]	*rrr*	2.85 [2.86]
rmrr	2.52 [2.30]	0.98 [0.80]	0.44 [0.38]		
mrrm	0 [0]	1.83 [2.20]	0 [0]		
mrrr	0.17 [0.00]	1.83 [1.94]	0.00 [0.00]		
rrr	0.33 [0.00]	1.83 [1.64]	0.00 [0.00]		

[a] [observed shift difference].

is *syn*-axial to two ester groups. If the *syn*-axial down-frequency shift from the ester groups was almost twice that from the methyls, then this would fit the experimentally observed solid-state methyl shifts of 23.7 ppm for *i*-PMMA and 15.25 or 17.7 ppm for s-PMMA.

Table 4.2 shows how this approach can also approximately explain the solution-state spectrum of atactic PMMA. In this simple approach, the polymer is modelled via only three conformations per bond, and possible variations in ester conformation are also ignored. The elements of the U_m and U_r matrices are allowed to float freely in the fitting procedure, subject only to dyad symmetry, and to fitting all the available shift data simultaneously. The fitted values confirm the reasonableness of the *tt* conformations in the solid state [34], although other conformations are also important in solution. The *syn*-axial shift from COOMe is not very well defined by this particular data set, but is approximately − 9.5 ppm. That from CH_3 or CH_2R is − 5.0 ppm. The same simple approach gives a slightly better fit for poly(methylacrylonitrile), with a − 2.9 ppm *syn*-axial shift from CN, and considerably better fits for less sterically crowded polymers such as poly(acrylonitrile), poly(vinyl alcohol) and polypropylene. All these calculations use the same *syn*-axial parameter for the same substituent group. One should note that a *syn*-axial H atom contributes a shift of − 4 ppm, whereas H is irrelevant to the simple *γ-gauche* effect. Interestingly, OH contributes only − 3 ppm. This explains some otherwise confusing shift observations in fused-ring systems [35, 36]. It has not so far proved necessary to vary the ability of the different types of carbon atom, under observation, to sense this *syn*-axial interaction.

In principle, it should be possible to use the conformational information that one has obtained from the above analyses in order to also understand the proton NMR shifts. But unfortunately no simple shift theory is available in the general case. It is also difficult to assess the precise orientation, and hence shift effects, of substituent groups such as phenyl.

4.5 Application of conformational theory to relaxation and NOE

It was shown in section 4.3 that T_1 often lengthens significantly for all carbons in isotactic chain segments in dissolved polymers [37]. This may be explained by the RIS analysis in the previous section, given the reasonable assumption that the transitions between conformations is rapid. Furthermore, the analysis shows that the relaxation parameters are insensitive to the rate of the local motions, within this limit. For example, an i-PMMA dyad must have two equally weighted conformations, loosely *tg* and *gt*, on symmetry grounds. The calculation described above also finds that *tt* has a very similar weighting. In contrast, an s-PMMA dyad has only one main conformation, *tt*, with a smaller contribution from *gg*. Thus more extensive local motion is to be anticipated in short i-PMMA sequences than in short s-PMMA sequences, in the same overall chain. Similar observations and calculations have been made with poly(acrylonitrile).

The homonuclear Overhauser enhancements explained in section 4.2.1 may also in principle be exploited for spatial and conformational information. However, this is only possible when the proton shifts are unusually sensitive to tacticity, because otherwise they cannot be resolved. They can be investigated by either 1-D or 2-D methods. The 1-D method involves selective irradiations, but is relatively quick and easy to quantify. The 2-D method, NOESY [38], is more elegant. In both cases one must take precautions to avoid spin-diffusion if quantitative information is required. Spin-diffusion is discussed in chapters 6 and 7 in the context of solid-state NMR, where its effects are more serious. What happens, in brief, is that the W_0 term in T_1 (equation 4.6) can become very large for a proton pair, so that spin energy flows freely between most of the protons in the sample. Thus all selectivity is lost. In practice, spin diffusion is not too serious in a typical semi-mobile polymer solution, so that qualitative distance information may easily be obtained. The method is particularly promising for random copolymers, where more resonances are available for selective irradiation, because of the many variations of local sequence, as well as of tacticity.

If quantitative information on interproton distance is required, then it is necessary to plot the NOESY off-diagonal peak intensity against the mixing time after which mutual proton spin flips are monitored. The slope of this plot, σ, is given by [39]

$$\sigma = 5.7\left(\frac{6\tau_R}{1 + 4\omega^2\tau_R^2}\right) - \tau_R \times 10^{10} r_{HH}^{-6} \tag{4.34}$$

for a rigid molecule. The symbols here have the same meanings as previously. Bovey and co-workers [40] have used such information to evaluate RIS calculations on a strictly alternating poly(styrene-methylmethacrylate) copolymer, and even to estimate the deviations from strict linearity in the all-*trans* state. Figure 4.11 gives an example of their NOESY data. It shows pairwise interactions

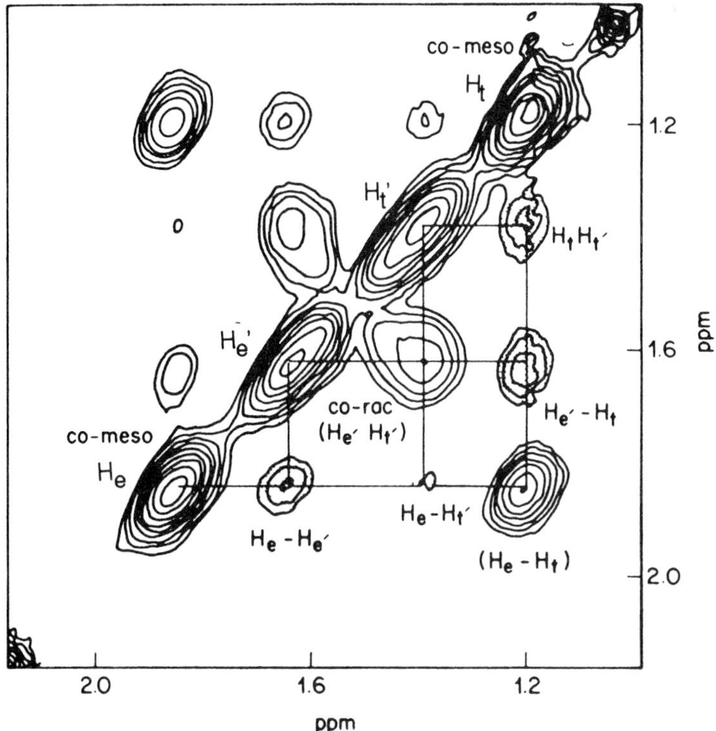

Figure 4.11 500 MHz 2-D ^1H–^1H NOESY spectrum of 1:1 alternating styrene–methylmethacrylate copolymer; methylene resonances only. See section 4.6. From Bovey and Mirau [39]. The subscripts t and e stand respectively for *threo* and *erythro* methylene protons, and the primes refer to the neighbouring methylene unit, which has distinct shifts when an *m* co-*meso* unit is neighbour to a co-*racemic* unit. A detailed analysis of the peak intensities shows that the all-*trans* conformer is the most highly favoured. Reproduced by permission of the American Chemical Society.

between the two types of *erythro* (*e*) and *threo* (*t*) methylene protons, which approach each other closely across $m|r$ dyad joins, as well as the much stronger *e*–*t* pairings within methylene groups in *m* and *r* dyads.

The same homonuclear NOE methods have been used to investigate the interactions between different chains. These typically become obvious in solution at weight concentrations in excess of 40%. Polystyrene (PS) and poly(vinylmethyl ether) have been shown to blend on the molecular scale in concentrated solution in toluene, but not in chloroform [40, 41]. The ether methoxyl protons closely approach all the aromatic protons. With a similar experiment where PS is replaced by polycaprolactone [40], all the latter's chain methylene protons sense the proximity of the ether methoxyl.

4.6 Experimental methods

Almost all the experiments described in this chapter are part of the standard pulsed FT-NMR repertoire, and so their detailed implementation should be obtainable from the manufacturer of the available instrument. However, a few more general comments may be made on each type of measurement.

T_1. This is most commonly measured by the inversion recovery method, for the alternative methods are more relevant to molecules with long relaxation times. The basic pulse sequence is $(D - \pi - t - \pi/2 - \text{FID})_n$, where D is approximately $5T_1$ for full thermal equilibration, and t is typically given 8–12 values ranging from very short to approximately the value of D, and mostly between $0.2T_1$ and $2T_1$. The recovery to thermal equilibrium is then fitted to an exponential curve. If the signal intensity or area is h_0 for a very short interpulse interval t, h_t in general and h_∞ when t is very long, then $h_\infty - h_t = (h_\infty - h_0)\exp(-t/T_1)$. In this calculation, due attention must be paid to the sign of h_t. Care must also be taken when resonances with different T_1 overlap partially. Indeed, this sequence may be used simply to separate, for example, overlapping CH and CH_2 carbon resonances. It is essential to use continuous proton decoupling in order to eliminate the fluctuating passage of spin energy between carbons and protons.

NOE. The measurement of heteronuclear NOE, rather as with T_1, only gives simple results when the protons are decoupled during collection of the carbon FID. Great care is necessary to ensure that this decoupling is switched off absolutely during preparation of the spins in the absence NOE. Even a tiny leakage of power through an electronic switch may give rise to a small NOE, for the power of irradiation that can affect the proton spin population is about one thousandth that for full decoupling. The decoupled carbon spectrum must be obtained following a preparation period of about $10T_1(C)$ and the NOE is the ratio of the peak intensity with and without gentle irradiation during this period.

Interproton NOEs are normally measured in 1-D NMR by difference spectroscopy. Decoupling must not take place, for this affects spin populations substantially. Instead, one irradiates the chosen resonances in cyclic sequence, including one irradiation into a blank region of the spectrum to provide a spectrum from which the other spectra may be subtracted. The irradiations are highly selective, e.g. ± 10 Hz, because of the low irradiation power.

In 2-D NMR, interproton NOEs are measured by the NOESY experiment, which in its simplest form is (D-$\pi/2$-t-$\pi/2$-d-$\pi/2$-FID), where D, t and d are, respectively, the relaxation delay (approx. 1–$2T_1$), the stepped time-variable common to all 2-D experiments, and a fixed, or slightly variable mixing delay of typically 0.2–0.5 s. During this latter delay time d, there is time for, for example, W_0 transitions to take place, and thus to transfer magnetic polarisation from spin to nearby spin. This gives rise to off-diagonal resonances, with the chemical shift coordinates of all the NOE correlated spin pairs.

The NOEs in both 1-D and 2-D experiments will be negative for almost all polymers. Quantitative NOE measurements require several experiments with different mixing or pre-irradiation times, as discussed in the previous section.

T_2. This can be measured directly from the FID length, as in simple miniature spectrometers, provided one does not need to resolve resonances with different shift. Full Fourier transform measurements are normally performed using a Carr–Purcell B sequence. In principle this is $D\text{-}\pi/2(\text{-}t\text{-}\pi\text{-}t)_n$ with symbols as above. The FID refocuses every $2t$ seconds, and decays exponentially with time constant T_2. Thus the decaying echo intensities may be fitted to the same equation as the one above, with T_2 replacing T_1. In practice, phase cycling and/or composite pulses are necessary to reduce pulse errors. Even so, the method is unreliable for proton-decoupled carbon resonances especially when $T_2 > 0.35$ s. The decoupling can upset the carbon resonances via polarisation transfer, in a way that is not readily predictable, and depends upon the precise value of t. Alternatively, one may select the Carr–Purcell A sequence, $(D\text{-}\pi/2\text{-}t\text{-}\pi\text{-FID})_n$, in which the later half of the single echo is transformed, so that the decay of its individual components may be monitored as a function of t.

References

1. O.W. Howarth, *Theory of Spectroscopy*, Nelson, London (1973), ch. 4.
2. A.J. Brown, O.W. Howarth, P. Moore and A.J. Bain, *J. Magn. Reson.* **28** (1977) 317–319.
3. J. Mason, ed., *Multinuclear NMR*, Plenum, New York (1987), ch. 5.
4. C.P. Slichter, *Principles of Magnetic Resonance*, Harper and Row, New York (1963).
5. J. Mason, ed., *Multinuclear NMR*, Plenum New York (1987), ch. 5.
6. O.W. Howarth, *J. Chem. Soc. Faraday Trans. 2* **74** (1978) 1031–1041; **75** (1979) 863–868.
7. O.W. Howarth and L.Y. Lian, *J. Chem. Soc. Perkin Trans. 2* (1982) 263–267
8. G. Lipari and A. Szabo, *J. Am. Chem. Soc.* **104** (1982) 4546–4559.
9. O.W. Howarth, *J. Chem. Soc. Faraday Trans. 2* **76** (1980) 1219–1223.
10. W.T. Ford and T. Balakrishnan, *Adv. Chem. Ser.* (*Polymer Characterisation*) **203** (1983) 475–478.
11. R. Kimmich, G. Schnur and M. Köpf, *Prog. NMR Spectrosc.* **20** (1988) 385–421.
12. M. Doi and S.F. Edwards, *The Theory of Polymer Dynamics*, Clarendon Press, Oxford (1986).
13. N. Bloembergen, *Nuclear Magnetic Relaxation*, Bengamin, New York (1961).
14. J. Schaefer, *Macromolecules* **6** (1973) 882–902.
15. D.E. Woessner, *J. Chem. Phys.* **37** (1962) 647–654.
16. D. Doddrell, V. Glushko and A. Allerhand, *J. Chem. Phys.* **56** (1972) 3683–3689.
17. R.J. Wittebort and A. Szabo, *J. Chem. Phys.* **69** (1978) 1722–1736.
18. R.E. London, *J. Am. Chem. Soc.* **100** (1978) 2678–2685.
19. P.J. Flory, *Statistical Mechanics of Chain Molecules*, Hanser, Munich (1969).
20 B. Valeur, J.-P. Jarry and L. Monnerie, *C. R. Acad. Sci., Ser. C* **278** (1974) 589–592.
21. F. Heatley and M.K. Cox, *Polymer* **18** (1977) 225–232.

22. F.C. Schilling and A.E. Tonelli, *Macromolecules* **19** (1986) 1337–1343.
23. P. Raby, F. Heatley, R.H. Mobbs, C. Price and R.B. Stubbersfield, *Eur. Polymer. J.* **23** (1977) 455–461.
24. J. Forbes, J. Bowers, X. Shan, L. Moran, E. Oldfield and M.A. Moscarello, *J. Chem. Soc. Faraday Trans. 1* **84** (1988) 3821–3849.
25. P.J. Flory, P.R. Sundararajan and L.C. DeBolt, *J. Am. Chem. Soc.* **96** (1974) 5015–5024.
26. U.W. Suter and P. Neuenschwander, *Macromolecules* **14** (1981) 528–532.
27. R.R. Ernst, G. Bodenhausen and A. Wokaun, *Principles of NMR in One and Two Dimensions*, Clarendon, Oxford (1987).
28. C.A.G. Haasnoot, F.A.A.M. de Leeuw and C. Altona, *Tetrahedron* **36** (1980) 2783–2792.
29. D.R. Ferrero and M. Ragazzi, *Macromolecules* **14** (1981) 1830–1831.
30. D.K. Dalling and D.M. Grant, *J. Am. Chem. Soc.* **89** (1967) 6612–6622.
31. A.E. Tonelli, *NMR Spectroscopy and Polymer Microstructure*, VCH, New York (1989).
32. F.C. Schilling and A.E. Tonelli, *Macromolecules* **13** (1980) 270–275.
33. H.N. Cheng and G.H. Lee, *Macromolecules* **20** (1987) 436–438.
34. A.E. Tonelli, *Macromolecules* **24** (1991) 3065–3068.
35. S.Li and D.B. Chestnut, *Magn. Reson. Chem.* **23** (1985) 625–638.
36. S.H. Grover, J.P. Guthrie, J.P. Stothers and C.T. Tan, *J. Magn. Reson.* **10** (1973) 227–230.
37. S.H. Grover and J. B. Stothers, *Can. J. Chem.* **52** (1974) 870–878.
38. J.R. Lyerla Jr., T.T. Horikawa and D.E. Johnson, *J. Am. Chem. Soc.* **99** (1977) 2463–2467.
39. F.A. Bovey and P.A. Mirau, *Acc. Chem. Res.* **21** (1988) 37–43.
40. P.A. Mirau, H. Tanaka and F.A. Bovey, *Macromolecules* **21** (1988) 2929–2933.
41. M.W. Crowther, I. Cabasso and G.C. Levy, *Macromolecules* **21** (1988) 2924–2928.

5 High resolution solid-state NMR studies of polymer chemical and physical structures

A.E. TONELLI

5.1 Introduction

Even though concentrated polymer solutions and melts are macroscopically viscous systems, they exhibit high resolution NMR spectra when recorded under the same conditions employed to obtain high resolution NMR spectra of pure liquids and solutions of low molecular weight molecules (see chapters 1 and 4). Even rubbery, crosslinked polymers, which are completely prevented from flowing by their three-dimensional network structures, produce high-resolution spectra under these same conditions. These observations are the direct consequence of rapid, isotropic motions of the polymer segments.

In polymer solutions, melts and rubbers, the localized motions of polymer segments are generally isotropic and rapid on the NMR timescale of kHz to MHz. The line-broadening effects caused by the direct, through-space, dipolar coupling of neighboring nuclear spins and by the anisotropy of nuclear shielding from the applied magnetic field are therefore largely suppressed. This line-broadening often occurs in glassy and crystalline polymer samples where the chain segments are immobilized.

We are concerned in this chapter with observing and analyzing high resolution NMR spectra of solid polymer samples. The chemical shift or resonance frequency of a spin 1/2 nucleus depends sensitively on the electronic structure of the molecular environment in which it resides. NMR spectra of polymer solutions generally resolve resonances for each type of atomic site in the repeat unit, and can often distinguish between similar atomic sites whose structural environments are distinguishable only by different substituents four, five, and more bonds removed. It is this high sensitivity to the local molecular architecture that makes high-resolution NMR spectroscopy of polymers in solution the method of choice for determining their microstructures [1] as described in chapter 1.

In addition to our desire to gain comparable structural information concerning the molecular architectures of intractable polymer solids, which can neither be dissolved or melted, polymer scientists are also interested in learning about the organization and mobility (see chapter 6) of polymer chains in their various solid phases, because of their intimate connections to the unique and useful physical properties they often manifest. The ability to observe high-resolution NMR spectra for solid polymer samples permits a rational approach to establishing molecular structure (determined from high-resolution, solution or solid-state NMR), physical property relations through determination of the connections between the detailed molecular architectures of constituent polymer chains and their organization and mobility in the solid state. Our purpose here then is to describe how to obtain high-resolution NMR spectra of solid polymers, and to illustrate by means of several applications the utility of the information obtained from their analyses regarding the structural organization, conformation and orientation of polymer chains in the solid phase.

5.2 High-resolution NMR spectroscopy of solids

Generally two types of experiments can produce high-resolution NMR spectra for polymer solids, depending on the type of nucleus under observation. Abundant spin 1/2 nuclei, such as ^{1}H and ^{19}F, may be observed at high resolution in solids by NMR as described in chapter 7. Here, we deal exclusively with spin 1/2 nuclei, such as ^{13}C, ^{15}N, ^{29}Si, and ^{31}P, which are dilute in the solid sample either because of their low natural isotopic abundance (^{13}C, ^{15}N, ^{29}Si) or because of their relatively uncommon occurrence in solid polymer samples. Both for its frequent occurrence in synthetic polymers and for the relative ease with which it can be observed, the ^{13}C nucleus at natural abundance (1%) is the specific focus of our discussion.

There exist two interactions between nuclear spins and their neighbors or with the applied magnetic field that result in severe broadening of their solid-state NMR spectra when recorded under conditions (see chapter 1) that produce high-resolution NMR spectra for their solutions. Both of these nuclear interactions, the direct dipolar coupling of nuclei through-space (the spin–dipolar interaction) and the anisotropic electronic shielding of the nucleus from the applied magnetic field, are present in the liquid, but do not lead to resonance line-broadening because they are averaged to zero by the rapid and essentially isotropic motions occurring there. Motional averaging of these nuclear interactions is usually incomplete in rigid, solid samples like glassy and crystalline polymers and produce spectra like the one shown in Figure 5.1(a).

5.2.1 *Dipolar coupling*

Two neighboring nuclear spins (see Figure 5.2) will respond to, in addition to the applied magnetic field B_0, the local magnetic field B_{loc} produced by each other, which is given by

$$B_{loc} = \pm \mu r^{-3} (3\cos^2\Theta - 1) \qquad (5.1)$$

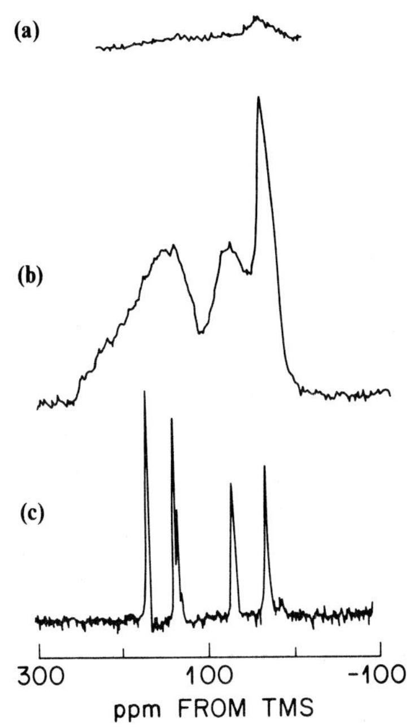

Figure 5.1 ^{13}C NMR spectra of bulk poly(butylene terephthalate) obtained using lower-power decoupling (a), high-power (dipolar) decoupling (b), and dipolar decoupling with rapid magic-angle sample spinning (c). (Reprinted with permission from Jelinski [2].)

where r is the distance between the nuclear spins of magnetic moment μ and Θ is the angle between their internuclear vector and the applied magnetic field. The $\pm$ sign reflects whether or not the neighboring magnetic dipole is aligned with or against the applied field B_0. This spin–spin interaction is called dipolar coupling.

^{13}C nuclei observed at natural abundance are dipolar-coupled to the abundant and nearby ^1H nuclei (see Figure 5.3(a)) resulting in the splitting D of the ^{13}C resonance given (in Hertz) by

$$D = [\hbar\gamma_c\gamma_H/2\pi r^3]\,(3\cos^2\Theta - 1) \tag{5.2}$$

This splitting is illustrated in Figure 5.3(b) and corresponds to the dipolar coupling of a ^{13}C nucleus with the two spin states (up and down) of a ^1H nucleus located at a distance r and orientation Θ (to B_0). The magnitude of

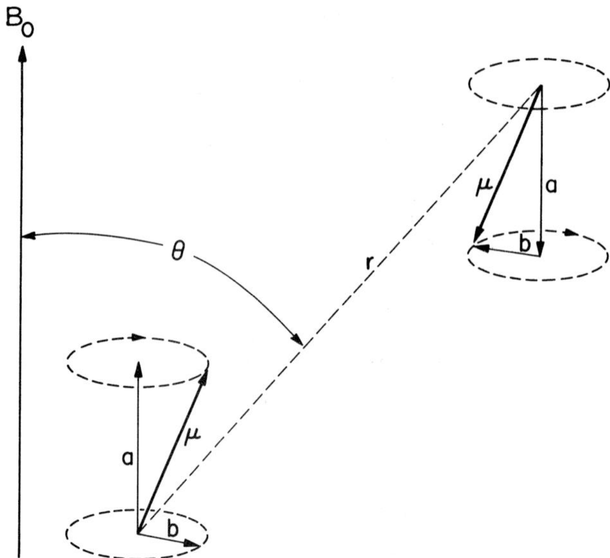

Figure 5.2 A pair of precessing nuclear moments (μ) with static (a) and rotating (b) components indicated. The second moment is at a distance r from the first, at an angle θ with respect to the magnetic field B_0.

this splitting is approximately 10 kHz [3]. In a rigid, glassy or crystalline powder polymer solid the ^{13}C nuclei and their nearby protons are randomly arranged and their C–H vectors assume all possible angles with respect to the external, applied magnetic field. This results in a Pake pattern [4] of ^{13}C resonances, as shown in Figure 5.3(c), if all C–H vectors are of the same distance r. In addition, the ^{13}C nuclei in a rigid, solid polymer sample are dipolar-coupled to protons located at more than a single inter-nuclear distance r. When the dipolar interaction between ^{13}C and ^{1}H nuclei is averaged over both the distances and orientations of all the C–H vectors present in the sample, the broad Gaussian lineshape presented in Figure 5.3(d) is obtained.

As a result of the dipolar interactions with nearby abundant protons, the ^{13}C NMR linewidths observed in rigid, organic polymers are typically tens of kiloHertz. Because the range of ^{13}C NMR resonance frequencies, or chemical shifts, observed [5] in a given polymer is usually less than 200 ppm, which at an applied field of 4.7 T (50 MHz for ^{13}C) corresponds to a frequency range of 10 kHz, ^{13}C NMR spectra of solids whose lines are broadened by ^{1}H dipolar coupling (approx. 20 kHz) cannot resolve their chemically shifted, resonance frequencies. Without removing the ^{13}C–^{1}H dipolar coupling, ^{13}C NMR spectra of solid polymers, like poly(butylene terephthalate)(PBT) in Figure 5.1(a), cannot provide useful structural information.

In a polymer solution, where the rates of segmental motion are generally faster than the strength of the dipolar ^{13}C–^{1}H interactions, i.e. approx. 20 kHz,

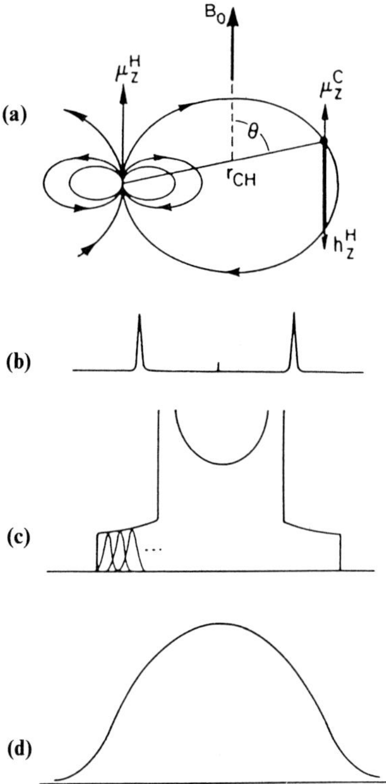

Figure 5.3 (a) Dipolar interaction between a ^{13}C and proton spin. The μ are the z components of the magnetic moments, and h_z^H is the z component of the proton dipolar field at the ^{13}C nucleus. (b) Dipolar splitting of isolated CH pairs at one angle relative to the magnetic field. (c) Pake pattern expected for isolated C–H pairs distributed at all angles as in polycrystalline or glassy materials. Several components from specific angles are illustrated schematically. (d) Approximate Gaussian lineshape observed for non-isolated C–H pairs, where all dipolar interactions are operative.

the resonance lines are not broadened by dipolar coupling, because the time-average of $(3\cos^2\Theta - 1)$ (see equation (5.2)) can be replaced by its space-average, which vanishes. If, however, the proton spins could be driven to flip at a rate that is rapid compared to the static ^{13}C–^{1}H dipolar interaction, then the resonance lines observed in solid state ^{13}C NMR spectra would likewise no longer be broadened by these heteronuclear, spin–dipolar interactions.

In Figure 5.1(b) an RF field B_1 at the resonant frequency of protons was applied to the PBT sample in a direction perpendicular to the applied field B_0 (analogous to the broadbond ^{1}H scalar decoupling of ^{13}C NMR solution spectra described in chapter 1). When the strength of B_1 is approximately 50 kHz, we note that a substantial increase in the spectral resolution results (compare (a) and (b) in Figure 5.1). Although the high-power ^{1}H dipolar decoupling (DD) of the ^{13}C NMR spectrum has improved resolution, the spectrum in Figure 5.1(b) falls far short of the high resolution ^{13}C NMR spectra recorded from polymer solutions (see chapters 1, 2 and 3). The remaining line-broadening is due primarily to chemical shift anisotropy (CSA).

5.2.2 *Chemical shift anisotropy*

The first four chapters of this book make abundantly clear the power of high-resolution, liquid-state NMR. Resonance frequencies or chemical shifts are highly sensitive to the local microstructures of polymers, and their analyses lead to a detailed structural characterization. We seek comparable information concerning solid polymers, but to achieve this we must be able to observe high-resolution NMR spectra of solid samples to fully exploit the sensitivity of the magnetic resonance phenomenon to the local structural environment of the observed nucleus.

The external applied magnetic field B_0 produces electronic currents in a molecule, and these currents in turn produce local magnetic fields at the various constituent nuclei. Because the electronic environment about a nucleus is not uniform, i.e. the distribution of electrons is directionally sensitive or anisotropic, the local magnetic field experienced by a nucleus is three-dimensional, and its magnitude and molecular orientation may be described [5] by the chemical shift tensor σ, given by

$$\sigma = \sigma_{11}\lambda_{11}^2 + \sigma_{22}\lambda_{22}^2 + \sigma_{33}\lambda_{33}^2 \tag{5.3}$$

The principal values of the chemical shift tensor (σ_{11}, σ_{22}, σ_{33}) give the magnitude of the tensor σ in three mutually perpendicular directions (Cartesian coordinates), and the λs are direction cosines specifying the orientation of the molecular principal coordinate system with respect to the applied field B_0. The rapid molecular motion experienced by polymer segments in solution results in the observation of their isotropic chemical shifts, σ_i, obtained by averaging the chemical shift tensor σ over all orientations:

$$\sigma_1 = 1/3(\sigma_{11} + \sigma_{22} + \sigma_{33}) = 1/3 \text{ trace } \sigma \tag{5.4}$$

It is apparent from equation (5.3) that in a rigid, solid sample the chemical shift of a particular nucleus will depend on its orientation with respect to the applied field. A sample having all carbon nuclei with the same orientation, as in a single crystal, will exhibit a chemical shift that varies as the crystal is rotated in the applied magnetic field. In a powdered sample, all possible crystalline orientations are present, and the NMR spectrum will consist of the chemical shift tensor powder pattern.

Two theoretical [6] chemical shift tensor powder patterns are illustrated in Figure 5.4(a),(b). Principal values σ_{11}, σ_{22} and σ_{33} are indicated, and their isotropic averages, σ_i, are given as dotted lines. In the axially symmetric case (b), $\sigma_{\parallel}$ and $\sigma_{\perp}$ are the resonant frequencies observed when the principal-axis system is aligned $\parallel$ and $\perp$ to the applied field. Molecular motion will narrow

(a)

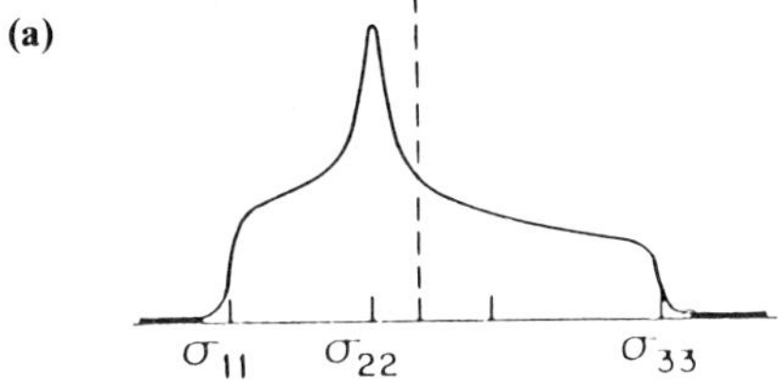

(b)

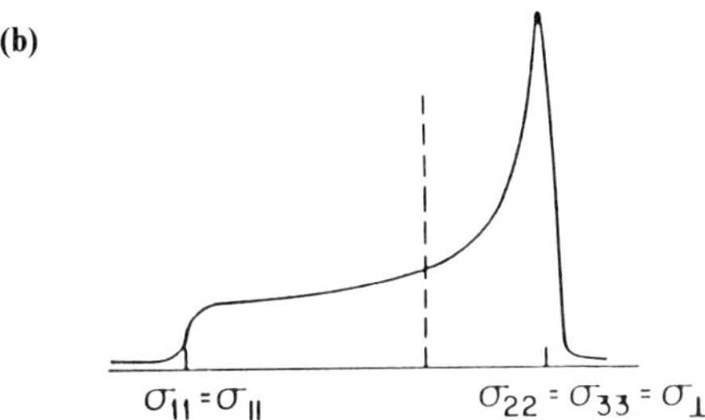

(c)

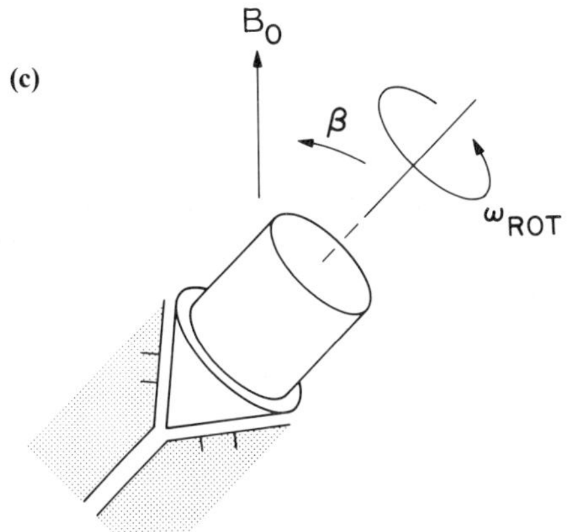

Figure 5.4 Schematic chemical shift tensor powder pattern for an axially asymmetric (a) and axially symmetric (b) tensor. The isotropic chemical shift values σ_i are indicated as dashed lines. (c) Typical Andrews [7] design sample holder (rotor) rotating on air bearings within a stator (shaded).

the chemical shift tensor, by partial averaging, and the resulting powder pattern will then contain information concerning the axis and angular range of the motion (see below and chapters 6 and 8).

Although potentially capable of providing motional and orientational information, the chemical shift tensor powder pattern contributes significantly to the broadening of solid state NMR spectra (see Figure 5.1(b)) and often obscures the structural information available from the isotropic chemical shifts. The chemical shift tensor broadening of resonances in the solid state may be removed by high-speed spinning of the sample at the magic angle.

If the solid sample is rotated rapidly about an axis making an angle β with respect to the applied field B_0 (see Figure 5.4(c)), the direction cosines (λ_{11}, λ_{22}, λ_{33}) in equation (5.3) vary during each rotation period. For rapid sample rotation the time average of equation (5.3) becomes

$$\sigma = 1/2 \sin^2 \beta(\sigma_{11} + \sigma_{22} + \sigma_{33}) + 1/2(3\cos^2\beta - 1)$$
$$\times \text{(functions of the direction cosines)} \qquad (5.5)$$

When the angle β between the axis of sample rotation and the applied magnetic field is 54.7° (the magic angle), $\sin^2\beta = 2/3$, $3\cos^2\beta - 1 = 0$, and thus $\sigma = 1/3$ $(\sigma_{11} + \sigma_{22} + \sigma_{22}) = \sigma_i$, the isotropic chemical shift. Magic-angle spinning (MAS) reduces the anisotropic chemical shift tensor powder pattern (see Figure 5.4(a),(b)) to the isotropic average. The broad overlapping carbonyl and aromatic carbon chemical shift anisotropies observed for PBT in Figure 5.1(b) are reduced to their isotropic averages, leading to the high-resolution spectrum shown in Figure 5.1(c).

The chemical shift anisotropies (CSA) of ^{13}C nuclei in different structural environments [8] vary from about 30 ppm for CH_2 carbons to about 200 ppm for aromatic carbons. Unlike the dipolar interactions between ^{13}C and ^{1}H spins, the strength of the CSA depends on the strength of the applied magnetic field B_0, because the strength of the local magnetic field B_0 experienced by a ^{13}C nucleus depends on the strength of the electronic currents in the vicinity of that nucleus induced by B_0, i.e.

$$B_{\text{loc}} = (1 - \sigma)B_0 \qquad (5.6)$$

Consequently, the speed at which a solid sample must be spun at the magic angle in order to collapse CSA tensors to their isotropic averages, σ_i, increases with the applied field. For example, at a field strength of 4.7 T (50 MHz for ^{13}C), the CSA of aromatic carbons requires MAS at (200 ppm) (50 MHz) = 10 000 Hz = 10 kHz for complete collapse to σ_i. In practice MAS at speeds lower than the full strength of the CSA yield a relatively narrow centre band and a spectrum of side bands [9], whose separations from the center band and intensities increase and decrease, respectively, with the speed of MAS.

When selecting a NMR spectrometer to record high-resolution solid-state spectra, it must be borne in mind that the increased sensitivity gained by operating at higher static field strengths must be balanced against the concomitant increase in the MAS speed required to collapse CSAs to their isotropic values. Increased rates of MAS on the order of 10 kHz generally require the preparation of homogeneous, well-packed rotor samples, which may preclude observation of heterogeneous samples such as some polymer composite materials.

Any solid-state interactions displaying a $3\cos^2\beta - 1$ dependence can be removed by MAS, including the $^{13}C-^{1}H$ dipolar interactions (see equations (5.1) and (5.2)). We have seen that the strength of this interaction is approximately 20 kHz. Consequently, MAS of solid samples alone (without high power ^{1}H DD) should yield high resolution ^{13}C NMR spectra unbroadened by dipolar and/or CSA interactions provided they are spun at speeds greater than 20 kHz. Improvements in rotor, spinning assembly, and probe designs have recently [10] permitted these rapid sample spinning rates, but their achievement is not yet routine, so most high resolution ^{13}C NMR spectra reported for solids are recorded with both MAS and DD.

5.2.3 *Cross-polarization*

In addition to $^{13}C-^{1}H$ dipolar coupling and chemical shift anisotropy, one other obstacle must be overcome before high-resolution solid-state ^{13}C NMR spectra can be practically obtained. As mentioned in the discussion of high-resolution solution ^{13}C NMR performed by pulsed FT techniques (see chapter 1), the rate at which signal averaging can be repeated, or the pulse repetition rate, is dictated by the spin–lattice relaxation T_1 values of the ^{13}C nuclei. Because most solids exhibit little motion in the megaHertz frequency range, which is required for coupling of the spins to their surrounding nuclei or to the lattice, ^{13}C T_1 values are long for solids. Rare nuclei, such as ^{13}C (1.1% natural abundance), require signal averaging, and the repetition rate of RF pulses becomes an important consideration in their observation by NMR.

How can we circumvent the long signal accumulation times required by the low repetition rate for ^{13}C nuclei with long T_1 values in solid samples? The answer lies in the ability to transfer the polarization of abundant ^{1}H nuclear spins with short T_1 values to the rare ^{13}C nuclei. The repetition rate for signal averaging is now determined by the short ^{1}H T_1 values, because energy is being transferred from the protons to the carbons. This process of polarization transfer from abundant to rare spins is termed cross-polarization (CP) and was introduced by Pines *et al.* [11].

Although ^{1}H and ^{13}C nuclei have Larmor frequencies different by a factor of four, Hartmann and Hahn [12] showed that energy may be transferred between them in the rotating reference frame when

$$\gamma_C B_{1C} = \gamma_H B_{1H}. \tag{5.7}$$

Equation (5.7) results in a match of the rotating-frame energies for ^{1}H and ^{13}C and is called the Hartmann–Hahn condition. The match is produced when the applied carbon RF field (B_{1C}) is four times the strength of the applied proton RF field (B_{1H}), because $\gamma_H/\gamma_C = 4$. In Figure 5.5 the vector diagrams and pulse sequence are presented for this double rotating-frame experiment which results in CP.

The proton and carbon spin systems are equilibrated in the magnetic field

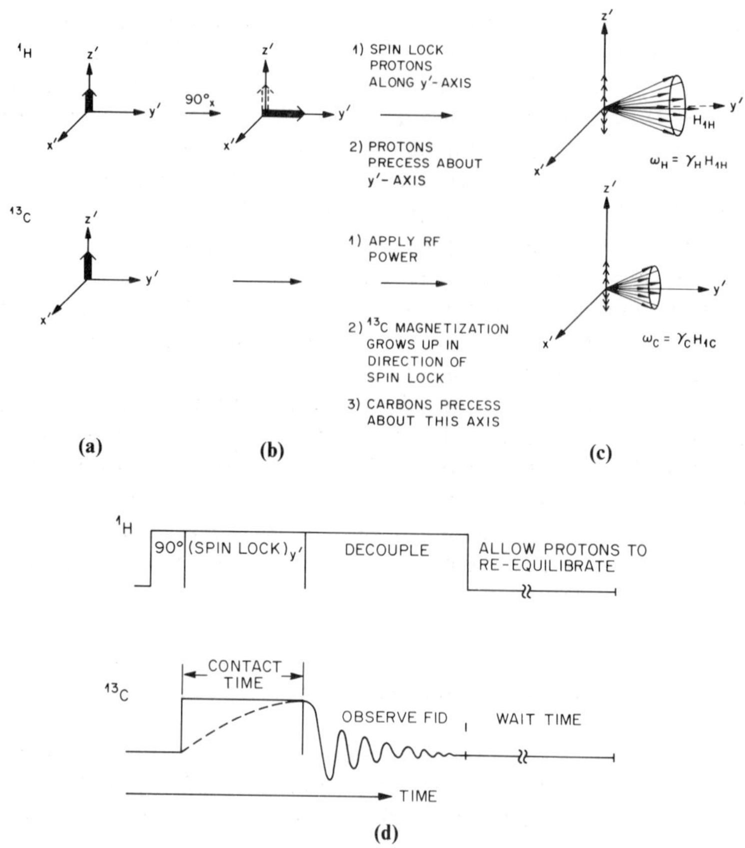

Figure 5.5 (a), (b), (c) Vector diagrams for a ^{1}H and ^{13}C double rotating-frame CP experiment. (d) CP pulse sequence (Reprinted with permission from Jelinski [2].)

in Figure 5.5(a). An RF pulse $B_{1\text{H}}$ at the proton Larmor frequency is applied along the x'-axis in (b) for a duration sufficient to tip the proton magnetization 90° along the y'-axis. The proton spins are forced to precess about the y'-axis of their rotating frame with frequency $\omega_\text{H} = \gamma_\text{H} B_{1\text{H}}$ for the duration of the strong $B_{1\text{H}}$ pulse (c), a process called spin-locking. During the time of spin-locking of the proton nuclei, the carbon RF field $B_{1\text{C}}$ is applied, thereby establishing contact between the two types of nuclei. ^{13}C magnetization grows in the direction of the spin-lock field (y'-axis) as the carbons precess about this axis with frequency $\omega_\text{C} = \gamma_\text{C} B_{1\text{C}}$.

Polarization is transferred between the proton and carbon nuclei as they both precess about the y'-axis by adjusting the power levels of the applied fields $B_{1\text{H}}$ and $B_{1\text{C}}$ until the Hartmann–Hahn condition is matched ($\gamma_\text{H} B_{1\text{H}} = \gamma_\text{C} B_{1\text{C}}$). The transfer of polarization is made possible because the z'-components of both ^{1}H and ^{13}C magnetizations have the same time dependence

(Figure 5.5(c)), resulting in mutual spin flips. CP can be thought of as a 'flow' of polarization from the abundant ^{1}H spins to the rare ^{13}C spins.

Since the ^{13}C nuclei obtain their polarization from the ^{1}H spins, it is the proton spin–lattice relaxation time (T_1) which determines the repetition rate of the CP experiment. This circumvents the problem of the long ^{13}C T_1 values normally found in solids. In addition, the ^{13}C signal shows an enhancement in its intensity, which can be as large as $\gamma_H/\gamma_C = 4$. The CP experiment results in both a time-saving and an improvement in the signal-to-noise ratio in the ^{13}C NMR spectrscopy of solid samples.

The first truly high resolution NMR spectra of solid polymers were reported by Schaefer and Stejskal [13]. They combined the three previously developed techniques of high-power proton decoupling (DD), cross-polarization (CP) and magic-angle sample spinning (MAS) to obtain these spectra. Since their pioneering work, much progress has been made in the field of high-resolution solid-state NMR, including the availability of commercial spectrometers that perform a wide variety of solid-state NMR experiments. These developments permit the study of the structures, conformations and mobilities of solid polymer samples by high-resolution NMR. Several examples are presented in the remainder of this chapter and in chapter 6.

5.3 Acquisition and analyses of solid-state ^{13}C NMR spectra

Commercial spectrometers currently available permit the routine recording of high-resolution NMR spectra via the combination of CP, MAS and DD, CPMAS/DD, for solid polymer samples. These are available in a variety of field strengths (B_0), have the ability to perform variable-temperature observations, and can be used to observe ^{15}N, ^{19}F, ^{29}Si, and ^{31}P nuclei as well as ^{13}C. In addition, specially designed and commercially available solid-state probes and high-power RF pulse generators can be added to existing spectrometers to provide almost any NMR spectroscopist with the capability of recording high-resolution spectra for solid polymer samples using the CPMAS/DD scheme.

For soluble polymers, comparison with their solution spectra offers the simplest means to assign the resonances in their CPMAS/DD solid-state spectra. On the other hand, those polymers that cannot be dissolved, melted or sufficiently swollen to provide high-resolution liquid-state spectra for comparison, can be observed in the solid with methods comparable to those developed for editing solution-state spectra. It is easier to describe these methods with the aid of a qualitative understanding of how the dilute ^{13}C and abundant ^{1}H spins interact in a typical solid, organic sample.

Figure 5.6 presents a schematic diagram of an organic solid where all ^{13}C and ^{1}H nuclear spins have been isolated from each other and from the lattice. Here, the lattice can be simply viewed as being all other species whose motions

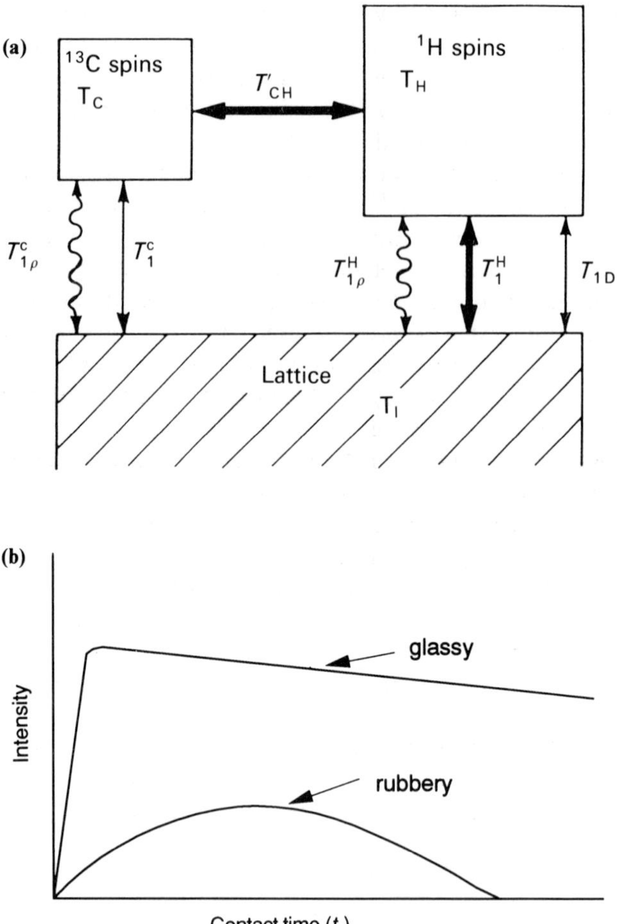

Figure 5.6 (a) Schematic diagram of the carbon and proton spin systems in a typical organic solid. T_C, T_H and T_l are the spin temperatures of the carbons, protons and lattice. $T_{1\rho}$, T_1 and T_{1D} are spin-lattice relaxation times in the rotating frame, laboratory frame and local dipolar field. T_{CH} is the cross-polarization time. Efficient cross-polarization of dilute ^{13}C spins by the abundant ^1H spins requires $T_1^C > T_1^H \geqslant T_{1\rho}^H > t_C > T_{CH}$. (b) Schematic diagram of the behaviour of carbon resonance intensity with varying contact time (t_c) in a CP experiment. Glassy materials result in rapid growth (short T_{CH}) and slow decay (long $T_{1\rho}^H$) Rubbery materials result in slower growth and more rapid decay, meaning that full equilibrium may not be reached.

provide a mechanism for transferring thermal energy between nuclei and their surroundings. Because we wish to accumulate free induction decays (FIDs) and later transform them to obtain FT-NMR spectra with sufficient ^{13}C signal intensity, it would be desirable to apply pulses at the carbon frequency with minimal repetition delay. However, the repetition rate for spectral accumulation is ordinarily limited by T_1^C, the spin–lattice relaxation time for ^{13}C nuclei. This is the time constant for the loss of spin thermal energy (temperature T_C) to the lattice, as shown in the diagram. T_1 values are long for rigid solids (10s or 100s of seconds) and require corresponding long delays between pulse accumulations. If instead we permit the ^1H spins, whose T_1^H values are much shorter, to interact via energy conserving, mutual spins flips, then the polarization [11] of ^{13}C nuclei occurs via this pathway and the delay between pulse accumulations can be shortened to the order of T_1^H. Communication or contact between ^{13}C and ^1H spins is established by application of two transverse RF fields B_{1C} and B_{1H} precessing at $\omega_C = \gamma_C B_{1C}$ and $\omega_H = \gamma_H B_{1H}$, where $\omega_C = \omega_H$ ensures the Hartmann–Hahn [12] condition and transfer of their spin polarization by mutual spin flips. In addition, an enhancement of ^{13}C signals ($\gamma_H/\gamma_C = 4$) also accrues from the experiment [11] (see previous section and Figure 5.5).

By bringing the ^{13}C and ^1H nuclear spins in contact for a time t_C under the Hartmann–Hahn condition, where both spin systems are in reference frames rotating at the same rate, we achieve cross-polarization of the carbon spins. The cross-polarization (CP) transfer is characterized by a growth rate of carbon intensity with a corresponding cross-polarization time (T_{CH}), which must be carefully considered when optimizing the experiment. It is important to set t_C (see Figure 5.5(d)) at a greater time than T_{CH} (see Figure 5.6(a)) to ensure that all ^{13}C nuclei are in equilibrium with the abundant ^1H thermal reservoir (at temperature T_H) during CP contact, thereby causing t_C to act somewhat like the pulse repetition time in the one-pulse experiment used for liquid systems (see chapter 1). Contact times of the order of 1–2 ms are commonly used for the acquisition of spectra of glassy or crystalline polymers (see Figure 5.6(b)).

For fully quantitative ^{13}C NMR spectra of solid samples recorded with CPMAS/DD, a contact time study (see Figure 5.7) [14] needs to be conducted; to determine a contact time at which all ^{13}C nuclei are fully at equilibrium with the ^1H reservoir. (This is particularly critical because carbons with different distances to neighboring protons and attendant mobilities may exhibit different CP rates). To compound the difficulty, relaxation of the proton and carbon spins will occur in the rotating frame ($T_{1\rho}^H$ and $T_{1\rho}^C$, respectively) causing the observed signal intensities to fall at longer contact times. $T_{1\rho}^H$ in particular is fairly short in polymer solids (of the order of milliseconds) so the contact time study will usually have to take it into account. For example, the top spectrum in Figure 5.7, obtained using the longest contact time, reveals that the protonated aromatic carbon intensity is slightly lower than in the spectrum just below, obtained using a shorter contact time. A simple theoretical model for the behavior of carbon intensity with contact time, which ignores the normally small contribution from $T_{1\rho}^C$, gives the following equation (see chapter 6):

$$M(t_C) = M_0[\exp(-t_C/T_{1\rho}^H) - \exp(-t_C/T_{CH})]/(1 - T_{CH}/T_{1\rho}^H) \quad (5.8)$$

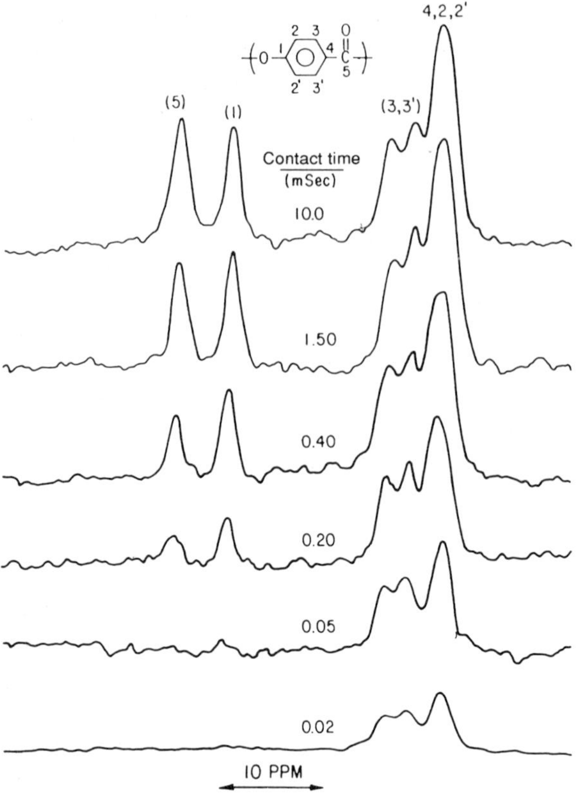

Figure 5.7 Cross-polarization MAS ^{13}C NMR spectrum of poly(hydroxybenzoic acid) as a function of contact time (in ms). All spectra were obtained from 2 K FTs of 1000 FID accumulations at a 3.5 s repetition time. (Reprinted with permission from C.A. Fyfe *et al.* [14].)

where $M(t_C)$ is the ^{13}C intensity at contact time t_C, and M_0 is the full equilibrium intensity. Typically, the fitting of data to equation (5.8) leads to values of T_{CH} of 20–100 μs for rigid protonated carbons. Only by a full analysis of this kind can we directly connect the observed resonance intensities (peak areas) with the numbers of contributing nuclei. Note from Figure 5.7 that the non-protonated carbons cross-polarize most slowly as a consequence of their weaker ^{13}C–^{1}H dipolar interactions. In general, mobile rubbery polymer samples also give rise to longer values of T_{CH}, due to the motional attenuation of the dipolar interaction. They also exhibit shorter $T^{H}_{1\rho}$ and T^{H}_{1} relaxation times than rigid, glassy or crystalline polymers (see Figure 5.6(b)). In fact, T_{CH}, $T^{H}_{1\rho}$, T^{H}_{1} and T^{C}_{1} are all diagnostic of local and/or regional polymer motions; different aspects of such studies are discussed in detail in chapters 6 and 7.

Differences in the ^{13}C–^{1}H dipolar interaction among ^{13}C nuclei may be used to routinely differentiate between protonated and non-protonated carbons by employing a simple technique [15] termed 'dipolar dephasing'. If in the usual CP pulse sequence (see Figure 5.5) a short delay (T_{dd}) is inserted after t_C, when the ^{13}C signal has been created by CP, but before signal acquisition, and during this delay the ^1H spins are not decoupled, then ^{13}C nuclei strongly coupled to protons by virtue of C–H chemical bonds are totally dephased while the non-protonated carbon nuclei are not if T_{dd} is appropriately selected.

Two-dimensional NMR techniques [16] (see chapter 1) may also be utilized to assign the ^{13}C resonances observed in high-resolution solid-state spectra. In normal one-dimensional FT-NMR experiments, an FID results from the application of an RF pulse and is sampled as a function of time. Standard frequency spectra are then obtained by FT of the time domain signal. A second time variable or dimension is introduced in the two-dimensional NMR experiment (see Figure 5.8), and during this time (evolution), the spin system is allowed to evolve under the influence of some NMR interaction or state selected by the observer. A non-equilibrium spin system is created by RF pulses during the preparation period and is allowed to evolve during t_1, the evolution period. A mixing period is sometimes employed to transfer spin information among different sets of spins. Signal detection occurs during t_2.

The experiment is repeated for different values of t_1 to fill the second dimension and yield a series of spectra after FT during just t_2. Spectra recorded with different values of t_1 reflect the spin interactions operating during the evolution period. Among these spin interactions are the chemical shift, scalar or J-coupling, dipolar coupling and the CSA. An example [17] of recovery of CSA information is presented in Figure 5.9, where both the pulse sequence and results obtained for *p*-dimethoxybenzene are presented. Note that the sample spinning axis is flipped between the evolution and detection periods. CSA information is encoded during t_1, when the sample is spun at 90°, and then stored along the z-axis by the 90° (y) RF pulse until the spinning axis is flipped to the magic angle (54.7°) for high resolution acquisition.

At the bottom of Figure 5.9 the isotropic (σ_1) ^{13}C spectrum of solid *p*-dimethoxybenzene is displayed along with four cross-sections from the 2-D

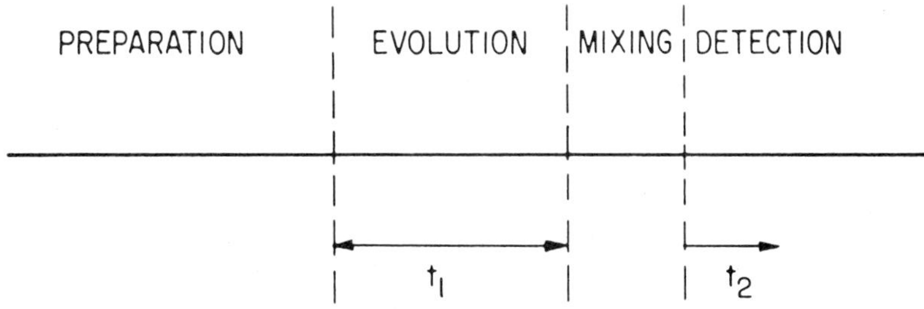

Figure 5.8 Generalized diagram of the 2-D NMR experiment.

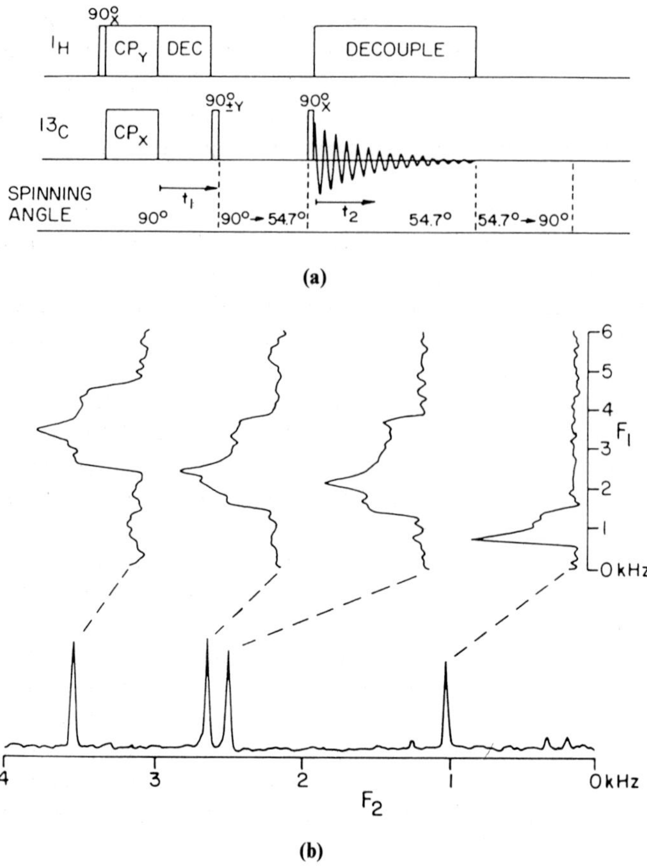

Figure 5.9 (a) 2-D NMR experiment for mapping out the CSA pattern in the F_1 dimension. In the present case a sample spinning at 2.5 kHz undergoes the angle change between t_1 and t_2 in 0.5 s. (b) Projection of the 2-D spectrum of *p*-dimethoxybenzene with cross-sections showing the CSA patterns. Because of the way the experiment was performed, the CSA patterns are half as wide as in the static case. (Adapted from Figures 1 and 2 of [17]. Reprinted with permission.)

spectrum of the CSA patterns. This two-dimensional NMR experiment [17] permits the association of σ_1 and CSA for solid state resonances and can assist spectral assignments. More important, however, is the fact that this type of experiment, although not routine, provides both a high resolution spectrum and the CSA information, so we are able to learn about sample mobility and orientation, as well as determine its microstructural characteristics.

As first suggested and demonstrated by Schaefer and Stejskel [13], combination of the techniques of high-power ^{1}H dipolar decoupling (DD), rapid magic-angle spinning (MAS) [7], and cross-polarization (CP) [11] of ^{1}H and ^{13}C nuclear spins permits observation of high-resolution ^{13}C NMR spectra for solid polymers. We have just summarized some of the most commonly employed means to assign the CPMAS/DD observed ^{13}C resonances to the structurally unique carbon atoms in the sample. Aside from establishing the important identification of structurally different carbon atoms in the sample, what else can we learn about solid polymer samples from their observed ^{13}C chemical shifts?

It was demonstrated in the preceding chapter that the ^{13}C chemical shifts ($\delta^{13}C$) observed for dissolved polymers depend sensitively on their local microstructures, and that the connection [1] between $\delta^{13}C$ and microstructure is the local conformation about the observed carbon nucleus. This connection is provided by the conformationally sensitive γ-*gauche* substituent effect [18] on ^{13}C chemical shifts.

A portion of an *n*-alkane is illustrated in Figure 5.10, where the (a) *trans*(t) and (b) *gauche*(g) conformations about the central C–C bond are presented in Newman projections. Note that the observed (C^{o}) and γ-substituent (C^{γ}) carbons are proximal in the *gauche* conformation (3 Å) and distant (4 Å) in the *trans* conformation. It has been demonstrated that when a carbon nucleus is *gauche* to a substituent carbon in the γ position it is shielded by approximately -5 ppm and therefore resonates 5 ppm upfield from a similar carbon nucleus that is *trans* to its γ-substituent carbon.

In Figure 5.10(c) three γ-related examples of the conformationally sensitive γ-*gauche* shielding of carbon nuclei are presented. Division of the difference between methyl carbon chemical shifts observed [19] in *n*-butane, 1-propanol and 1-chloropropane, and that measured for propane, whose methyl carbons are without a γ-substituent, by the fraction of *gauche* character possessed by their central bonds, results in the determination of γ-*gauche* shieldings of $\gamma_{c,c} = -5.2$, $\gamma_{c,OH} = -7.2$, and $\gamma_{c,cl} = -6.8$ ppm. Consequently, any variation in the microstructure of a molecule or polymer chain that affects its local conformation can be expected to be reflected in its ^{13}C NMR spectrum. This expectation has been abundantly confirmed in the ^{13}C NMR solution spectra of polymers (see chapter 4), where the observed ^{13}C resonances are reflecting the average local conformations of their mobile ^{13}C nuclei.

In solid samples, especially crystalline and glassy samples, polymer chains are generally not able to rapidly sample all of the conformations accessible in the solution or molten state. As a consequence, for solid polymers, the chemical shifts of ^{13}C nuclei observed by the CPMAS/DD techniques reflect their rigid conformation(s) and not the average conformational environments experienced in solution. It is therefore possible to derive the solid-state conformations of solid, rigid polymers by observation of their ^{13}C CPMAS/DD NMR spectra.

An example of the effect polymer chain conformation can have on high-resolution solid-state ^{13}C CPMAS/DD NMR spectra is presented for the semi-crystalline polymer polyethylene (PE) in Figure 5.11. These spectra [20] were recorded with Torchia's CP–T_1 pulse sequence [21], as illustrated, which

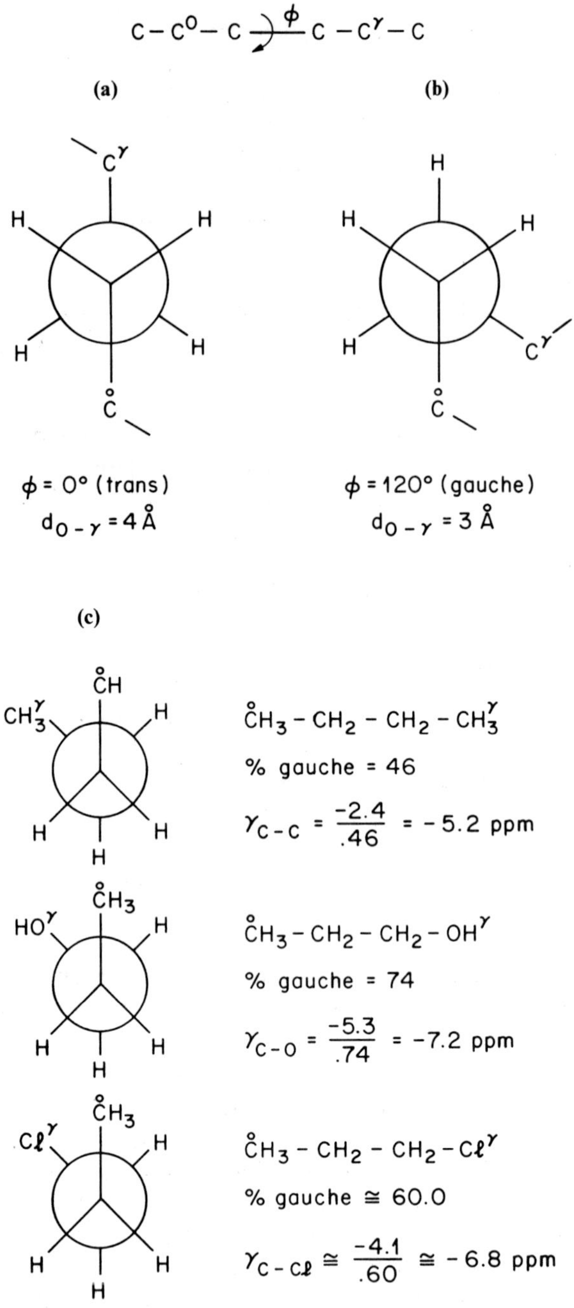

$C - C^{o} - C \overset{\phi}{-} C - C^{\gamma} - C$

(a)

$\phi = 0°$ (trans)
$d_{o-\gamma} = 4$ Å

(b)

$\phi = 120°$ (gauche)
$d_{o-\gamma} = 3$ Å

(c)

$\overset{o}{C}H_3 - CH_2 - CH_2 - CH_3^{\gamma}$

% gauche = 46

$\gamma_{C-C} = \frac{-2.4}{.46} = -5.2$ ppm

$\overset{o}{C}H_3 - CH_2 - CH_2 - OH^{\gamma}$

% gauche = 74

$\gamma_{C-O} = \frac{-5.3}{.74} = -7.2$ ppm

$\overset{o}{C}H_3 - CH_2 - CH_2 - Cl^{\gamma}$

% gauche ≅ 60.0

$\gamma_{C-Cl} \cong \frac{-4.1}{.60} \cong -6.8$ ppm

Figure 5.10 Newman projections of a *n*-alkane chain in the (a) *trans* ($\phi = 0°$) and (b) *gauche* ($\phi = 120°$) conformations. (c) Derivation of the γ-*gauche* shielding produced by the γ-substituents C, OH, and Cl.

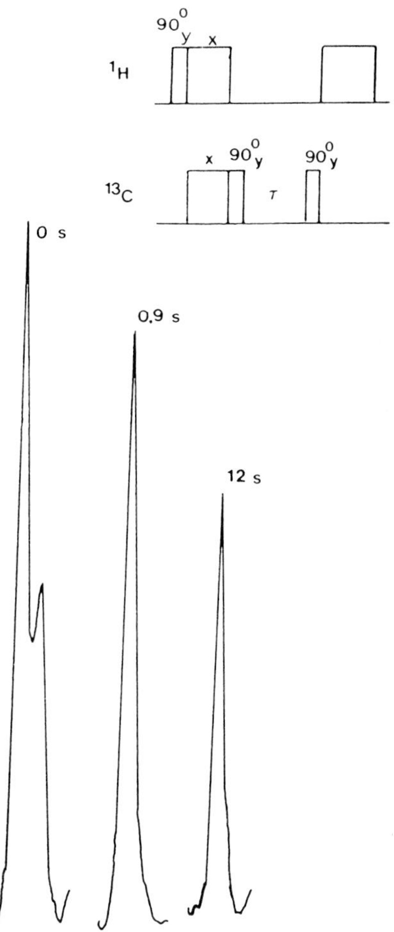

Figure 5.11 Effect of increasing delay time on resolution in a CP–T_1 pulse sequence of semi-crystalline polyethylene [20] ($\tau = 0.096$ s, $\tau = 0.9$ s, $\tau = 12$ s.) The pulse sequence is shown at the top.

permits measurement of ^{13}C spin–lattice relaxation times, T_1^C, during CPMAS/DD spectral acquisition. Note that the first spectrum on the left recorded with a 96 ms T_1 relaxation delay (τ) between spin-locking and decoupling of the proton spins shows two resonances separated by about 2.5 ppm. The downfield resonance has a $T_1 = 100$ s and the upfield shoulder a $T_1 = 170$ ms. When the delay between ^1H spin-locking and decoupling exceeds $5T_1 = 5$ (0.17 s) = 0.85 s, as in the other two spectra, only the downfield resonance remains, and is attributed to the slower moving all-*trans*, planar zigzag crystalline PE chains. Amorphous PE chains are responsible for the upfield shoulder observed with short or no delays, because both *trans* and shielding, *γ-gauche* effect conformations are permitted for the PE chains in the disordered faster moving portions of the sample.

The nearly 1000-fold difference in the values of T_1 of the crystalline and amorphous carbons in PE permit their separate observation. However, for semi-crystalline polymers with rigid amorphous phases characterized by high glass-transition temperatures, T_g, such a clean separation may not be possible. Although the chains in a glassy polymer sample are disordered, they may not be sufficiently mobile to rapidly sample all potentially accessible conformations. As a consequence, each amorphous conformational environment will contribute a different solid state chemical shift, producing a broad envelope of resonances for the disordered chains. Warming the semi-crystalline sample above T_g will

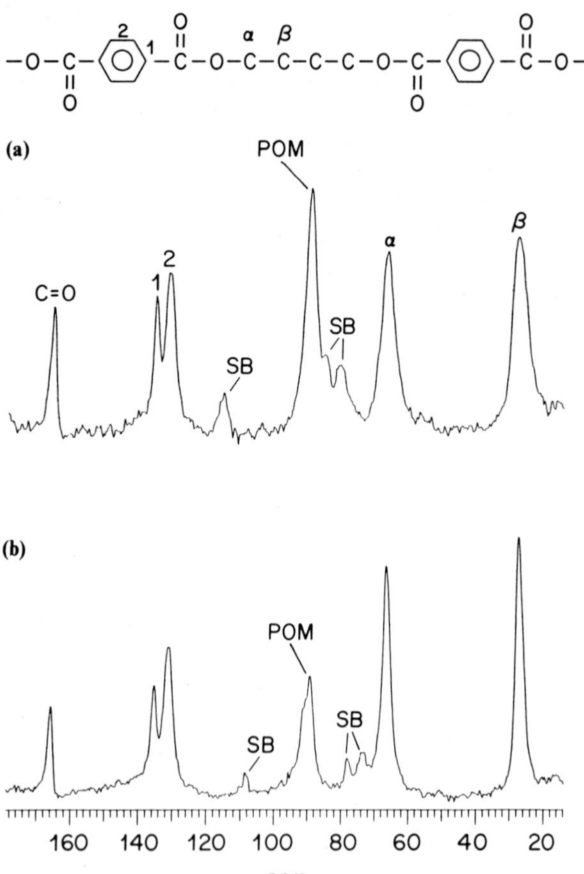

Figure 5.12 CPMAS/DD spectra of α-PBT at ambient temperature (a) and 105°C (b). (Reprinted with permission from Gomez *et al.* [22].) POM, poly(oxymethylene) signal from rotor; SB, spinning side-band.

permit the amorphous chains to rapidly average among their conformational environments, dramatically reducing their T_1, and allowing their observation separate from the conformationally ordered, immobile, carbon nuclei in crystalline environments.

Figure 5.12 shows the CPMAS/DD ^{13}C NMR spectra of semi-crystalline α-PBT [poly(butylene terephthalate)] recorded [22] below (a) and well above (b) its T_g of 55°C. At ambient temperatures, both crystalline and amorphous carbons are sufficiently rigid to cross-polarize efficiently, but at 105°C the amorphous carbons are mobile and conformationally averaging. The ambient spectrum is broadened by the rigid and conformationally diverse amorphous carbon nuclei which resonate at a variety of frequencies. At 105°C the amorphous carbons, which are mobile with small T_1, are conformationally averaging to σ_1, and no longer cross-polarize efficiently. Thus at elevated temperatures, only the crystalline carbons in the sample are contributing to the CPMAS/DD spectrum.

This example illustrates just one advantage of being able to perform variable-temperature (V-T) CPMAS/DD measurements. Aside from enabling the discrimination between different phases in a polymer solid, solid polymers are sometimes capable of undergoing motions in the kiloHertz regime which nullify their observation under CP, because $T_{1\rho}^{H}$ decreases significantly below t_C and T_{CH} (see Figure 5.6 and the related discussion). Changing the temperature of observation can shift the relaxation behavior of solid ^{13}C and ^{1}H nuclei and permit or prevent their observation under CPMAS/DD. We will shortly present several examples of this phenomenon.

5.4 Applications of high-resolution CPMAS/DD NMR to polymer solids

5.4.1 NMR determination of crystalline polymer conformation

Having illustrated the possibility of separately observing the fixed conformations of rigid crystalline chain segments and the motionally averaged conformations of mobile amorphous chain segments in semi-crystalline PE, we turn our attention to the conformations adopted by polymer segments in the crystals formed by different stereoregular polymers. Figure 5.13 presents the CPMAS/DD ^{13}C spectra of crystalline isotactic (i) [23] and syndiotactic (s) [24] polypropylenes (PP). The only major difference between their spectra is the single CH_2 resonance observed for i-PP and the CH_2 doublet seen for s-PP, a difference attributable to the different crystalline conformations [25] adopted by the two stereoregular PPs.

The 3_1-helical, ...*tgtgtgtg*... conformation is adopted by i-PP chains in all four of its crystalline polymorphs, while s-PP chains adopt the 4_1-helical, ...*ttggttgg*... conformation in its prevalent, most stable polymorph. Each crystalline CH_2 carbon in i-PP is *trans* in one direction along the backbone

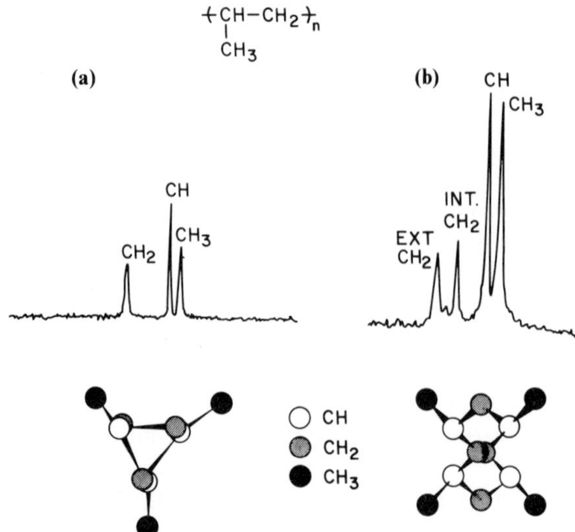

Figure 5.13 Solid-state ^{13}C NMR spectra of polypropylene. (a) Solid-state ^{13}C NMR spectrum of isotatic polypropylene (reproduced with modification from Fleming *et al.* [23] by permission of the American Chemical Society) plus representation of the helical conformation of isotactic polypropylene. (b) Solid-state ^{13}C NMR spectrum of syndiotactic polypropylene, plus representation of the conformation of syndiotactic polypropylene. (Reproduced from Bunn *et al.* [24] by permission of the Royal Chemical Society.)

and *gauche* in the other direction to its methine carbon γ substituents, because of its ... *tgtgtgtg*... conformation, and therefore experiences a single $\gamma_{CH_2;CH}$ *gauche* shielding. Half the crystalline CH_2 carbons in s-PP experience no *gauche* shielding from their γ-CH substituents. These CH_2s lie on the periphery of the 4_1-helix and are denoted EXT according to Figure 5.7. The remaining CH_2 carbons (INT) line the core of the 4_1-helix and are in a shielding, *gauche* arrangement with both of their γ-CH substituents.

It is not surprising that crystalline s-PP exhibits two CH_2 carbon resonances separated by $2\gamma_{CH_2,CH}$ shieldings, or approximately $2(-5\,\text{ppm}) = -10\,\text{ppm}$ (-8.7 ppm is actually observed). The fact that the single CH_2 resonance observed for crystalline i-PP lies midway between the s-PP CH_2 doublet is a consequence of the fact each CH_2 carbon in crystalline i-PP is shielded by a single $\gamma_{CH_2;CH}$ interaction, while the CH_2 carbons in s-PP are either shielded by 0 or 2 identical $\gamma_{CH_2;CH}$ interactions.

Syndiotactic polystyrene (s-PS) can be crystallized into two polymorphs [26], forms I and II. The all-*trans*, planar zigzag conformation is adopted by the chains in form I crystals, while in form II crystals the s-PS chains adopt the same 4_1-helical, ...*ttggttgg*... conformation found for the prevalent form of crystalline s-PP. Notice the CH_2 carbon resonances [26] in form II

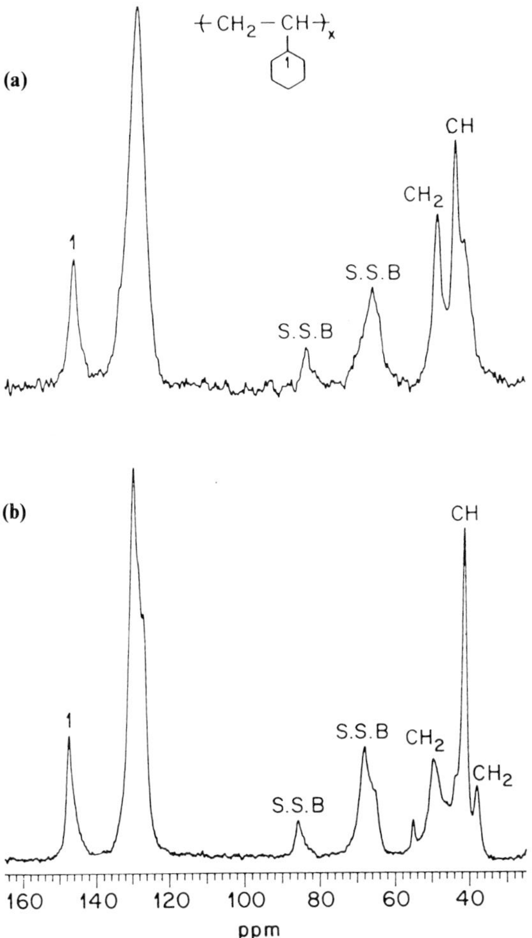

Figure 5.14 CPMAS/DD ^{13}C NMR spectra [26] of (a) form I and (b) form II s-PS. The form II sample of (b) was obtained by absorption of dichloromethane into an amorphous, melt-quenched film of s-PS. SSB, spinning side-band.

s-PS (see Figure 5.14(b)) are split into a doublet, as was observed for s-PP in Figure 5.13(b), and are separated by 11 ppm, or approximately (2–0) $\gamma_{CH_2;CH}$. The single CH_2 resonance observed [26] in Figure 5.14(a) for the all-*trans*, planar zigzag PS chains in form I crystals is virtually superimposeable with the form II CH_2 resonance observed at the highest chemical shift in Figure 5.14(b). This near coincidence in CH_2 resonance frequencies originates from the absence of *gauche* shielding interactions with γ-CH substituents for all of the CH_2 carbons in form I s-PS and for half of the CH_2 carbons in the form II polymorph.

5.4.2 *Solid-state NMR observations of copolymer sequences and their distribution between crystalline and amorphous phases*

A series of ethylene-vinyl chloride (E-V) copolymers was generated [27] by the reductive dechlorination of poly(vinyl chloride) (PVC) [28]. Each E-V copolymer has the same average chain length (approx. 1000 repeat units) and polydispersity. High-resolution solution ^{13}C NMR revealed [27] their 'random-like' comonomer sequence distributions, which were quantitatively determined to the comonomer sequence triad-level (see chapter 2). E-V copolymers with less than 40% V units were observed [29] to crystallize, and the stabilities, structures and morphologies of their crystals were observed [29, 30] and analyzed by DSC, X-ray and electron diffraction, and electron microscopy.

X-Ray diffraction observations of E-V powders and fibers showed an expansion of the unit-cell basal plane, mainly in the *a*-direction, and a constant fiber period, respectively, as their chlorine content increased. Together these observations indicate that at least some V units, with their attendant Cl atoms, are incorporated into the crystals, resulting in an increase in interchain separation, but no alteration in the all-*trans*, planar zigzag conformation found for crystalline PE.

Figure 5.15 presents a series of CP spectra obtained [3, 0] at 62°C for E-V 13.6 (13.6 mol% V) by varying the contact times. This E-V copolymer melts at 78°C, so at 62°C it is still largely crystalline, but its amorphous phase is much more mobile than at room temperature. The rapidly cross-polarized peak at 33.4 ppm was believed to be due to inner CH_2 carbons (Cn in Figure 5.16) in the rigid crystalline phase, and the slowly cross-polarized peak at 31.0 ppm was believed to be due to the equivalent carbon types in the mobile amorphous phase. At room temperature, measurements at different contact times did indicate two phases for the CHCl carbons, although the results are most apparent at high temperature where the amorphous carbons have the greatest mobility. In Figure 5.15(b) a sharp resonance at 64.8 ppm and a small shoulder at 66 ppm are observed for the CHCl carbons. These are assigned to the amorphous and crystalline phases of the sample, respectively, by comparison with the melt spectrum, where only a single CHCl resonance at 64.7 ppm is observed (see Figure 5.16(c)). This assignment is confirmed by spin–lattice relaxation time measurements which show that the downfield resonance at 66 ppm has a T_1 three times longer than the more shielded resonance.

The β-CH_2 resonance at 40 ppm (assigned in Figure 5.16) exhibits a small downfield shoulder in the CP spectra in Figure 5.15, which is also ascribed to the crystalline phase. All of these results constitute clear experimental evidence for the inclusion of Cl inside the crystals of these semi-crystalline E-V copolymers.

However, it is difficult to obtain quantitative information in a CP NMR experiment because of differences in the efficiency of CP for carbon nuclei in different phases, as demonstrated by the differing contact time behavior

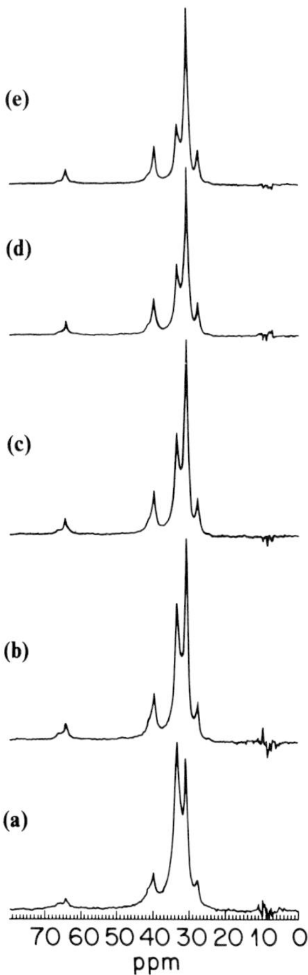

Figure 5.15 CPMAS/DD ^{13}C NMR spectra [30] of E-V 13.6 recorded at 62°C with different contact times: (a) 1.0, (b) 2.0, (c) 2.5, (d) 3.0 and (e) 3.5 ms.

of the CH_2 carbons in Figure 5.15. Consequently, we resorted to an alternative set of experiments to obtain more detailed information concerning the distribution of Cl substituents between the crystalline and amorphous phases of E-V copolymers. Figure 5.16 presents the ^{13}C NMR spectra of E-V 13.6 obtained at different temperatures under DD and MAS, but without CP. Spectra were recorded with different delays in order to establish conditions under which only the amorphous carbons were observed. Previously the T_1 of each of the amorphous resonances was measured by the inversion-recovery method [31]. The spectrum at 86°C (Figure 5.16(c)), where E-V 13.6 is in the melt, includes all of the carbons in the sample. The spectrum recorded at 62°C,

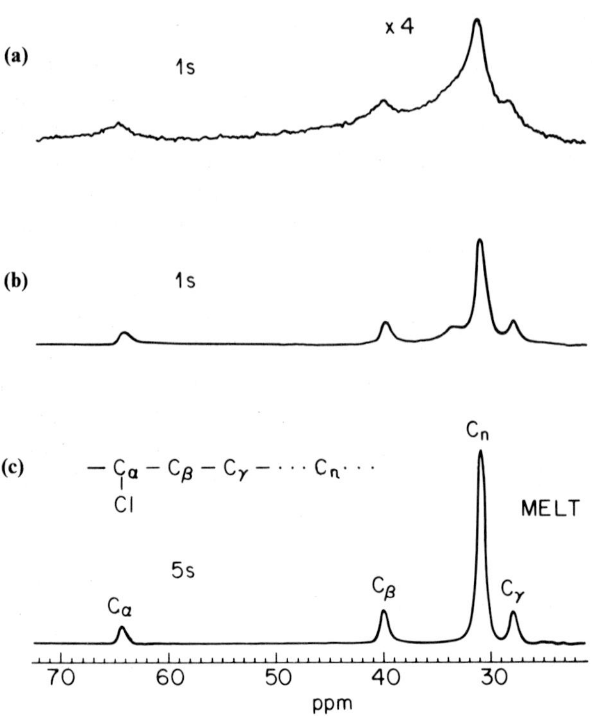

Figure 5.16 MAS/DD ^{13}C NMR spectra [30] of E-V 13.6 recorded at (a) 25, (b) 62 and (c) 86°C.

where the sample is still substantially crystalline, was obtained with a 1 s delay in order to observe only the amorphous carbons. The sample was quenched from the melt, and its spectrum was recorded at 62°C under quantitative conditions, immediately following the observations made at 86°C.

Intensity differences between both spectra give a measure of the amount of Cl included inside the crystals. The inner-CH_2 peak (Cn) shows a 35% loss of intensity compared with the total inner-CH_2 resonance observed at 86°C in the melt. This is in close agreement with the crystallinity measured for this sample by X-ray and DSC methods. The CHCl resonance shows a 20% loss of intensity at 62°C indicating at least 20% of the Cls are included in the crystals. At 62°C some partial melting occurs resulting in an increase in CHCl intensity and a decrease in our estimate of Cl incorporation. We have to point out that, although the spectrum recorded at 62°C was acquired with a very short delay, a small portion of the inner-CH_2 crystalline resonance at 33.4 ppm is still observed. Even though a similar observation is not detectable for the CHCl resonance, we nevertheless consider the 20% figure as a lower limit for the degree of Cl incorporation in the crystals.

5.4.3 NMR observations of solid-state polymer reactions

When organic polymers are exposed to γ-radiation they are observed to crosslink and/or degrade via chain scission reactions. O'Donnell and Whittaker [32] were able to monitor the crosslinking of bulk *cis*-1,4-polybutadiene (*cis*-PBD) upon exposure to γ-radiation by observing the CPMAS/DD ^{13}C NMR spectra of the irradiated *cis*-PBD samples. Because peak intensities in a CP-NMR spectrum depend on: (1) the rate of cross-polarization from the ^{1}H nuclei to near-neighbor ^{13}C nuclei (proportional to $1/T_{CH}$); and (2) the relaxation of proton magnetization during CP, described by the relaxation time $T^{H}_{1\rho}$, quantitative equilibrium peak intensities were obtained by fitting contact time plots (T_{CH} versus peak intensities) to the two time constants T_{CH} and $T^{H}_{1\rho}$ in equation (5.8), using a Simplex minimization procedure.

The results of their quantitative analyses are displayed in Figure 5.17, where the contents of double bonds (*cis* and *trans*) and crosslinks found for γ-irradiated *cis*-PBDs are presented. Clearly, some isomerization of the double bonds is occurring upon γ-irradiation, and their decrease with radiation dose closely parallels the corresponding increase in crosslink content. It is apparent that crosslinking has occurred primarily through the double bonds. The quantities of crosslinks produced/100 eV of energy absorbed ($G_{\times\text{-link}}$) was much higher than the values derived from physical observations (elastic modulus, swelling ratio, etc.) [33–39] performed on the insoluble gel fractions.

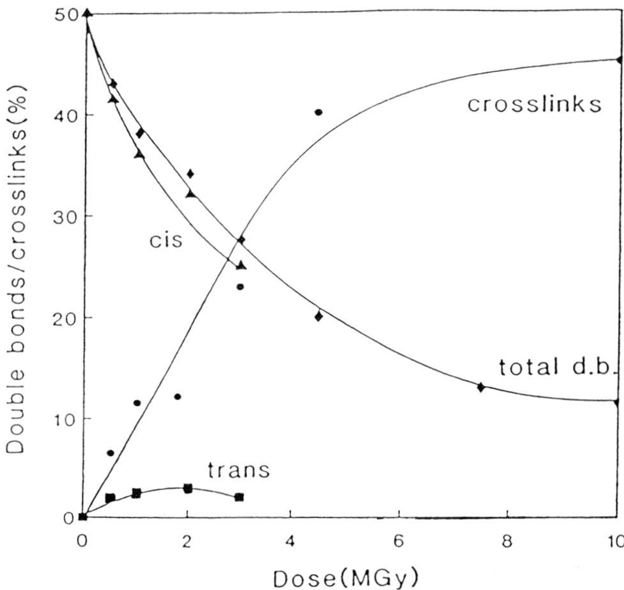

Figure 5.17 Changes in double bond (db) and crosslink content in irradiated *cis*-PBD [32].

This implies that solid-state NMR measures the total crosslink content, both elastically effective interchain and wasted intrachain crosslinks [40–42].

When poly(ethylene oxide) (PEO) powders are γ-irradiated under vacuum they are observed [43] to crosslink. The CPMAS/DD and MAS/DD ^{13}C NMR spectra of virgin and vacuum γ-irradiated POE (see Figure 5.18) show [44] marked differences. The MAS/DD spectrum of unirradiated PEO shows a well-resolved resonance at 72.36 ppm due to carbons in mobile amorphous regions, which have relatively short T_1^C values, whilst at 73.43 ppm the CPMAS spectrum shows a weak, broad envelope of resonances due to rigid crystalline carbon environments. The corresponding irradiated PEO sample, on the other hand, presents well-resolved resonances for both crystalline (under CPMAS/DD) and amorphous (under MAS/DD) carbons. Figure 5.18(b) shows that under CP conditions, the crystalline resonances are efficiently cross-

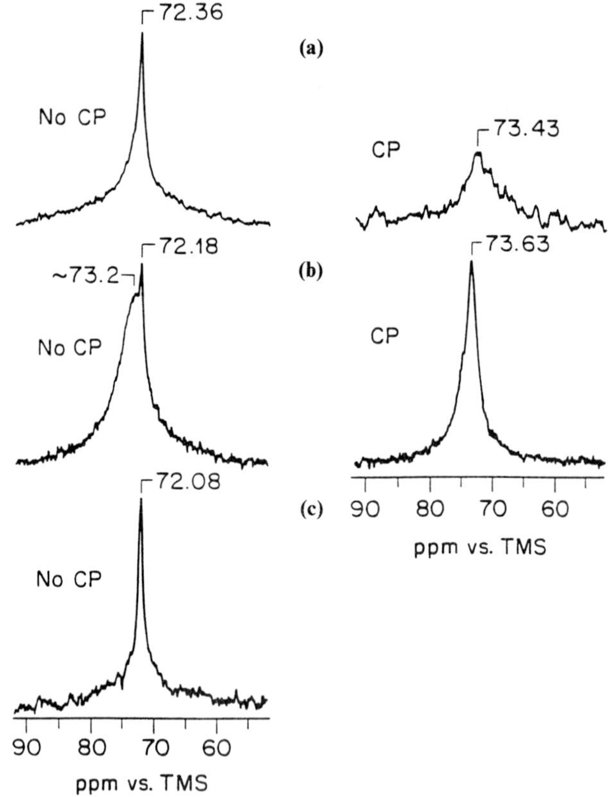

Figure 5.18 50.31 MHz ^{13}C NMR MAS spectra [44] of (a) PEO control sample, (b, c) PEO sample irradiated with 29 Mrad. Spectra (a) and (b) were recorded at + 23°C with and without cross-polarization (CP, No CP). Spectrum (c) was reporded at + 23°C without cross-polarization. The pulse delays used without CP were (a) 5 s, (b) 60 s and (c) 0.5 s.

polarized to give a peak at 73.63 ppm, whilst with a very long relaxation delay both crystalline resonances, at 73.2 ppm, and amorphous resonances, at 72.18 ppm, can be observed without cross-polarization. Only the amorphous carbons are seen without CP when a shorter relaxation delay is used as in Figure 5.18(c). Upon recrystallization from their melts both PEO samples yield solid state ^{13}C NMR spectra that are closely similar (Figure 5.19) to that of the virgin, unheated sample (Figure 5.18(a)). Observation [44] of both melt-recrystallized samples at $-60°C$ (Figure 5.20) yields similar spectra under CP conditions with a set of well-resolved crystalline resonances. The major peak at 73.94 ppm agreed with that from the crystalline phase at room temperature. The two smaller peaks were presumed to be due to different conformational structures.

Crosslinking is the predominant chemical change occurring during the γ-irradiation of PEO [43] under vacuum and produces a change in the motional character of the crystalline phase. This change is not the result of a reduction in crystallinity as evidenced [44] by DSC observations. The most probable explanation is that crosslinks are concentrated at the surface of the crystalline PEO lamellae and produce a change in the low frequency molecular motions of the crystalline chains. This motional change lengthens $T^{H}_{1\rho}$ such that the

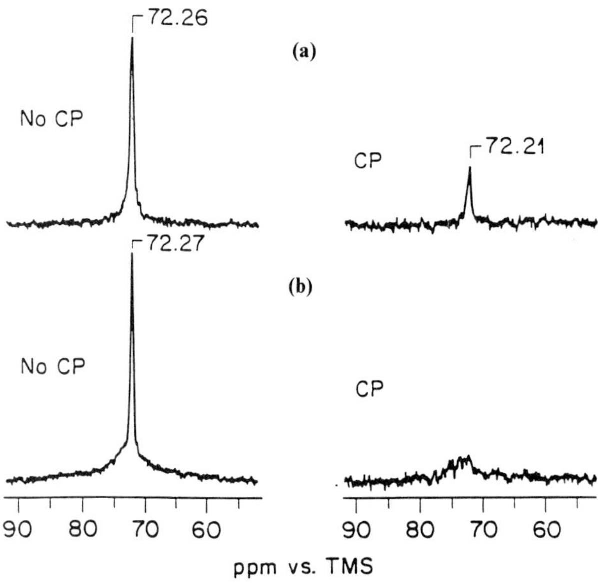

Figure 5.19 50.31 MHz ^{13}C NMR MAS spectra [44] of (a) PEO control sample and (b) PEO sample irradiated with 29 Mrad; recorded at $+23°C$ with and without cross-polarization (CP, No CP) after both samples were heated to melt and recrystallized by cooling to $+23°C$. The pulse delay used without CP was 3 s.

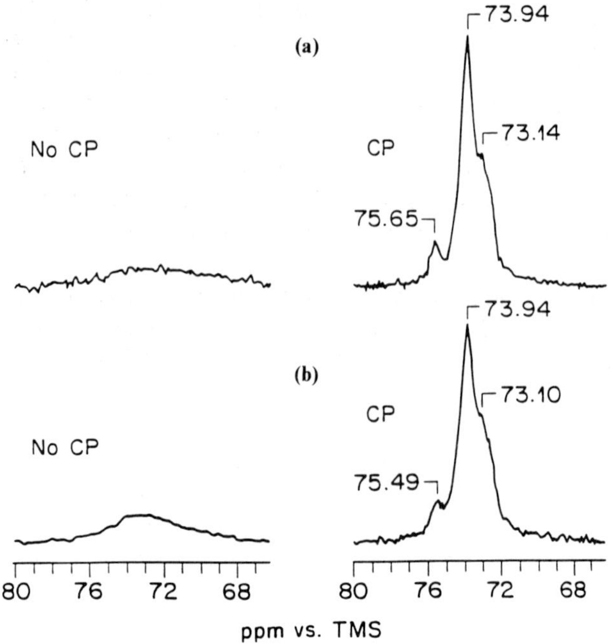

Figure 5.20 50.31 MHz ^{13}C NMR MAS spectra [44] of (a) PEO control sample and (b) PEO sample irradiated with 29 Mrad; recorded at -60°C with and without cross-polarization (CP, No CP) after samples were heated to melt and recrystallized by cooling to $+23$°C. The pulse delay used without CP was 1 s.

crystalline carbon nuclei can now be cross-polarized at room temperature and the resonance linewidth is reduced as seen in Figure 5.18(b) with CP. A comparable shift in $T^{\mathrm{H}}_{1\rho}$ can also be achieved by lowering the sample temperature (see Figure 5.20(a) with CP). Following melting and recrystallization the motional characteristics of the irradiated PEO are nearly identical to those of the unirradiated sample, as shown in Figure 5.19, probably as a result of a redistribution of the crosslinks throughout the amorphous phase of the sample during recrystallization.

5.4.4 *Solid-state NMR observation of polymer blends and interphases*

As the use of polymers continues to increase in multi-component and multiphase applications, such as in polymer blends and composites, the behavior of polymers at interphases and the degree of mixing polymer segments in their blends become critical issues. Understanding how the conformations and mobilities of polymer chains in these heterogeneous systems compare with those observed in their homogeneous bulk systems may provide a point of departure in the discussions of their physical characteristics.

Blum *et al.* [45] have observed the CPMAS/DD ^{13}C NMR spectra of poly(isopropyl acrylate) (PIPA) adsorbed on fine silica powders. Note in Figure 5.21 the differences in the observed spectra that result from various coverages of the silica powder surface by PIPA. At PIPA monolayer coverage of silica and below, a marked improvement in the resolution of all resonances except the methyl carbons is observed compared with the bulk PIPA spectrum. They attribute the enhanced resolution observed in the silica surface specimens to the decrease in motion occurring when PIPA chains are bound to the silica surface.

The PIPA chains in a bulk sample observed at ambient temperature, which is approximately 30°C above the T_g of PIPA, are probably sufficiently mobile to make the cross-polarization process inefficient. PIPA chains attached to

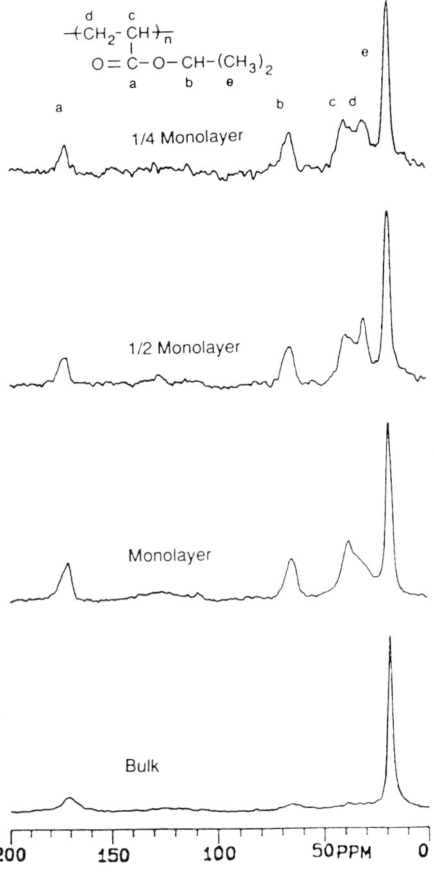

Figure 5.21 Spectra of PIPA at varying surface concentrations on silica surfaces and in bulk [45].

the surface of silica powder, on the other hand, probably cross-polarize more efficiently because of their enhanced rigidity. Clearly the mobilities of bulk and interfacial PIPA chains are different.

Most homopolymer blends and block copolymers are phase separated systems which permit the segregation of incompatible polymer segments into separate domains. ^{13}C-detected $T_{1\rho}^{\mathrm{H}}$ measurements can be utilized to monitor the phase separation that can occur in homopolymer blends and in copolymers. In a homogeneous sample, all protons relax at about the same rate via spin diffusion [46, 47] as discussed in chapters 6 and 7, but if the phase separated domains in a heterogeneous system are large enough (> 10–20 Å), it may be possible to observe different proton relaxation rates in the different sample domains through observation of ^{13}C-detected $T_{1\rho}^{\mathrm{H}}$. If carbon resonances associated with one homopolymer, but not the other, or with one block, but not the other, can be uniquely observed, then whether or not the homopolymer blend or block copolymer are homogeneous may be determined by observation of a single or two different $T_{1\rho}^{\mathrm{H}}$, respectively, via these carbon signals. Insertion of a spin-locking period in the proton RF channel between its 90° excitation pulse and its cross-polarization with ^{13}C nuclei (see Figure 5.5) permits the measurement of $T_{1\rho}^{\mathrm{H}}$ by detection in the ^{13}C channel. Further examples of this technique are described in chapter 6.

Deuteration of one polymer component, either one homopolymer in a blend, or one block of a copolymer, can also be utilized to study phase separation. In cross-polarized experiments, such as CPMAS/DD or $T_{1\rho}^{\mathrm{H}}$, the deuterated species will not appear in the spectrum for heterogeneous samples, or will be observed but with the relaxation time $T_{1\rho}^{\mathrm{H}}$ of the protonated component for homogeneous samples.

Sankar *et al.* [48] have studied A/B block copolymers of polystyrene (PS) and poly(methylmethacrylate) (PMMA) by observation of ^{13}C-detected $T_{1\rho}^{\mathrm{H}}$, where neither, one or the other, and both blocks were perdeuterated. They found that the carbons in each block of the copolymers were experiencing primarily their own protons, indicating some degree of phase separation, but the $T_{1\rho}^{\mathrm{H}}$ for each block differed somewhat from the $T_{1\rho}^{\mathrm{H}}$ measured for the PS and PMMA homopolymers. Annealing quenched samples apparently increased the degree of phase separation, because their $T_{1\rho}^{\mathrm{H}}$ values more closely approached the homopolymer values.

Note in Figure 5.22 that the block copolymer with perdeuterated PS (D) still shows, albeit at much reduced intensities, the same PS resonances observed in the fully protonated block copolymer. Because ^{13}C nuclei in PS (D) can only cross-polarize with protons from the PMMA blocks, it is apparent that these freeze-dried, quenched block copolymer samples are not completely phase separated. There is mixing between the PS (D) and PMMA block copolymer segments sufficient to produce a modest degree of cross-polarization between the protons of PMMA and the deuterated carbon nuclei of PS (D).

(a)

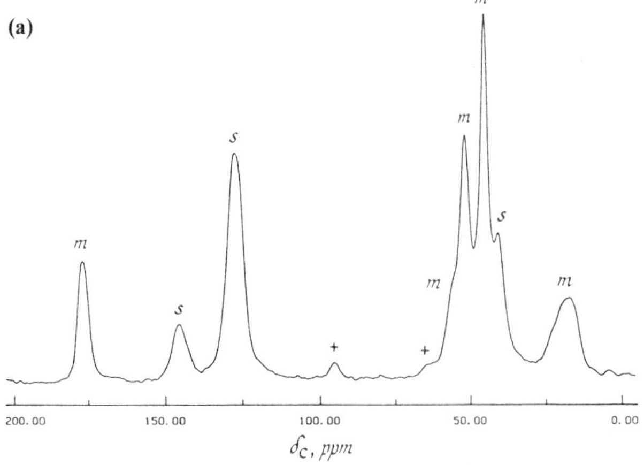

(b)

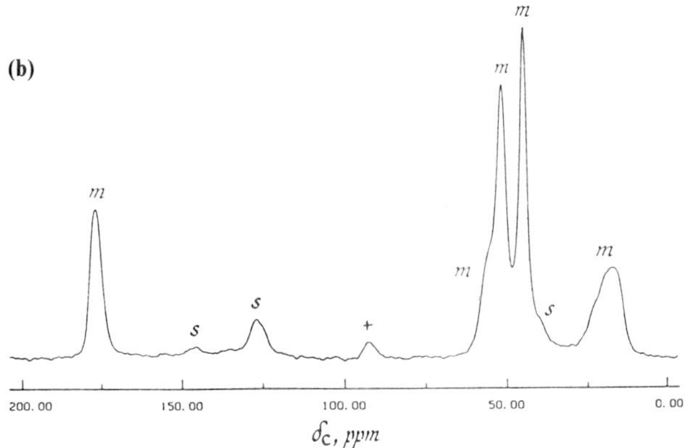

Figure 5.22 50 MHz CP/MAS ^{13}C NMR spectra of (a) polystyrene/poly(methyl methacrylate) block copolymer (PS/PMMA) and (b) perdeuteropolystyrene/poly(methyl methacrylate) block copolymer (PSD/PMMA) [48]. *m*, PMMA; *s*, PS; and +, spinning sideband.

5.4.5 NMR observations of orientation in solid polymers

Although until now we have discussed only high resolution NMR spectra recorded on solid polymer samples via the CPMAS/DD technique, useful information regarding the degree of orientation of rigid polymer chains in their solid samples can be obtained by observing spectra recorded without MAS (non-spinning). Such a spectrum [49] is presented in Figure 5.23(b) for a sample of powdered crystalline PE. The chemical shift anisotropy (CSA) powder pattern for the $-CH_2-$ carbons in PE is collapsed to its isotropic value (σ_i) in the spectrum in part (a), where the solid PE powder has been

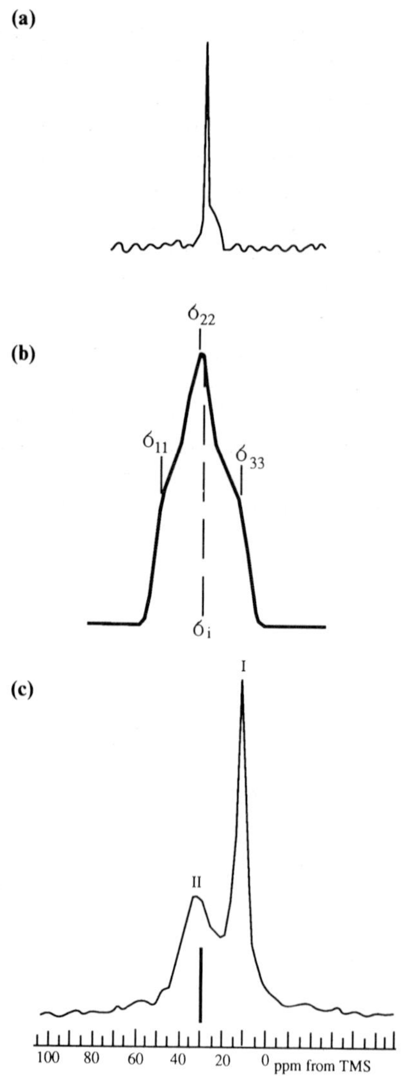

Figure 5.23 (a) CPMAS/DD spectrum of PE powder. (b) CP/DD spectrum of PE powder with principal shift tensor values shown. (c) DD spectrum of uniaxially oriented PE film with draw direction parallel to B_0. I is assigned to oriented crystallites and II to unoriented amorphous material [49]. The stick at 30 ppm in (c) corresponds to the resonance frequency of dissolved PE.

rapidly spun at the magic angle, and when the spectrum of dissolved PE is recorded, as indicated in part (c) by the stick at 30 ppm.

The full spectrum in part (c) was recorded on uniaxially oriented PE film by Nakagawa *et al.* [49], who also did not employ cross-polarization in an attempt to obtain a quantitative spectrum. The narrow resonance labelled I corresponds to the σ_{33} component of the methylene carbon CSA pattern shown in part (b) for unoriented PE powder. The broader resonance labeled II in part (c) is assigned to the carbons in the amorphous portions of the drawn PE film, which unlike the crystalline PE chains are not oriented by the drawing process, because they are also observed in the upfield (lower δ) shoulder seen in part (a) for the unoriented PE powder by CPMAS/DD.

By means of varying the alignment of a drawn, oriented polymer sample with respect to the applied magnetic field B_0, and comparing the resulting non-spinning spectra to the non-spinning and CPMAS/DD spectra recorded for the same sample in powder form, it is possible to determine the microscopic extent and direction of polymer chain orientation in the drawn film sample. Similar information can also be obtained from a single two-dimensional NMR spectrum (see Figure 5.9) recorded [17] while flipping the MAS axis. Most recently, a rotor-synchronized three-dimensional CPMAS/DD technique has been developed and described [50] which permits observation of the correlation between a polymer's microstructure, its orientations, and its mobility in uniaxially drawn samples.

5.5 Summary

By way of examples, we have attempted to illustrate the utility of high resolution NMR spectroscopy to the study of solid polymers. Although we have only presented applications of ^{13}C NMR, other spin 1/2 nuclei (^{15}N, ^{19}F, ^{29}Si, ^{31}P) have also been used to probe the structures, conformations, phase behavior, and orientations of solid polymers. In this regard, the ^{15}N NMR observations of crystalline nylons by Murthy and co-workers [51, 52] and the ^{29}Si NMR observations of the thermo- and piezochromic polysilanes by Schilling and co-workers [53–58] are particularly noteworthy. In chapter 6, many of the same high resolution techniques are used to probe the mobilities of polymer chains in solid samples.

References

1. A.E. Tonelli, *NMR Spectroscopy and Polymer Microstructure: The Conformational Connection*, VCH, New York (1989).
2. L.W. Jelinski, in *Chain Structure and Conformation of Macromolecules*, ed. F.A. Bovey, Academic Press, New York, (1982), Chap. 8.
3. T.M. Duncan and C.R. Dybowski, *Surf. Sci. Rep.* **1** (1981) 157.
4. G.E. Pake, *J. Chem. Phys.* **16** (1948) 327.
5. F.A. Bovey, *Nuclear Magnetic Resonance Spectroscopy*, Academic Press, New York (1988), Chap. 3, p. 95.
6. M. Mehring, *High Resolution NMR in Solids*, 2nd edition, Springer-Verlag, Berlin (1983).
7. E.R. Andrews, A. Bradbury and R.G. Eades, *Nature (London)* **183** (1959) 1802.
8. T.M. Duncan, *J. Phys. Chem. Ref. Data* **16** (1987) 125.
9. M.M. Maricq and J.S. Waugh, *J. Chem. Phys.* **70** (1979) 3300.

10. S.F. Dec, R.A. Wind and G.E. Maciel, *Macromolecules* **20** (1987) 2754.
11. A. Pines, M.G. Gibby and J.S. Waugh, *J. Chem. Phys.* **56** (1972) 1776; *Chem. Phys. Lett.* **15** (1972) 373.
12. S.R. Hartmann and E. Hahn, *Phys. Rev.* **128** (1962) 2042.
13. J. Schaefer and E.O. Stejskal, *J. Am. Chem. Soc.* **98** (1976) 1031.
14. C.A. Fyfe, J.R. Lyerla, W. Volksen and C.S. Yannoni, *Macromolecules* **12** (1979) 757.
15. S.J. Opella and M.H. Frey, *J. Am. Chem. Soc.* **101** (1979) 5854.
16. R.R. Ernst, G. Bodenhausen and A. Wokaun, *Principles of Nuclear Magnetic Resonance in One and Two Dimensions*, Oxford University Press, New York (1987).
17. A. Bax, N.M. Szeverenzi and G.C. Maciel, *J. Magn. Reson.* **55** (1983) 494.
18. A.E. Tonelli and F.C. Schilling, *Acc. Chem. Res.* **14** (1981) 233.
19. J.B. Stothers, *Carbon-13 NMR Spectroscopy*, Academic Press, New York (1972), Chap. 3.
20. D.E. Axelson, in *High Resolution NMR Spectroscopy of Synthetic Polymers in Bulk*, ed. R.A. Komoroski, VCH, New York (1986), Chap. 5.
21. D.A. Torchia, *J. Magn. Reson.* **30** (1978) 613.
22. M.A. Gomez, M.H. Cozine and A.E. Tonelli, *Macromolecules* **21** (1988) 388.
23. W.W. Fleming, C.A. Fyfe, R.D. Kendrick, J.R. Lyerla, H. Vanni and C.S. Yannoni, in *Polymer Characterization by ESR and NMR*, eds. A.E. Woodward and F.A. Bovey, ACS Symp. Ser No. 142, ACS, Washington, DC (1980).
24. A. Bunn, E.A. Cudby, R.K. Harris, K.J. Eacker and B.J. Say, *Chem. Commun.* (1981) 15.
25. H. Tadokoro, *Structure of Crystalline Polymers*, Wiley–Interscience, New York (1979).
26. M.A. Gomez and A.E. Tonelli, *Macromolecules* **23** (1990) 3385; **24** (1991) 3533; *J. Am. Chem. Soc.* **112** (1990) 5881.
27. F.C. Schilling, A.E. Tonelli and M. Valenciano, *Macromolecules* **18** (1985) 356.
28. W.H. Starnes, Jr., F.C. Schilling, I.C. Plitz, R.E. Cais, D.J. Freed, R.L. Hartless and F.F. Bovey, *Macromolecules* **16** (1983) 790.
29. T.N. Bowmer and A.E. Tonelli, *Polymer (British)* **26** (1985) 1195.
30. M.A. Gomez, M.H. Cozine, A.E. Tonelli, F.C. Schilling, A.J. Lovinger and D.D. Davis, *Macromolecules* **22** (1989) 4441.
31. T.C. Farrar and E.D. Becker, *Pulse and Fourier Transform NMR*, Academic Press, New York (1971).
32. J.H. O'Donnell and A.K. Whittaker, *J. Polym. Sci. Polym. Chem. Ed.* **30** (1992) 185.
33. A.S. Kuzminsky, T.S. Nikitina, E.V. Zhuravskaya, L.A. Oksentievich, L.L. Sunitsa and N.I. Vitushkin, *Proc. 2nd Int. Conf. on Peaceful Uses of Atomic Energy*, Geneva (1959), Vol. 29, p. 258.
34. V.T. Koslov, A.G. Yevseyev and P.I. Zubov, *Polym. Sci. U.S.S.R.* **11** (1969) 2539.
35. V.T. Koslov, M.V. Gurev, A.G. Yevseyev, N.G. Kasheveskaya and P.I. Zubov, *Polym. Sci. U.S.S.R.* **12** (1970) 665.
36. E. Witt, *J. Polym. Sci.* **41** (1959) 507.
37. B. Jankowski and J. Kroh, *J. Appl. Polym. Sci.* **9** (1965) 1363.
38. B. Jankowski and J. Kroh, *J. Appl. Polym. Sci.* **13** (1969) 1795.
39. S.M. Miller, M.W. Spindler and R.L. Vale, *J. Polym. Sci.* **A-1** (1963) 2537.
40. A.E. Tonelli, *Polymer (British)* **15** (1974) 194.
41. A.E. Tonelli and E. Helfand, *Macromolecules* **7** (1974) 59.
42. E. Helfand and A.E. Tonelli, *Macromolecules* **7** (1974) 832.
43. A.L. Cholli, F.C. Schilling and A.E. Tonelli, in *Solid State NMR of Polymers*, ed. L.J. Mathias, Plenum, New York (1991), p. 117.
44. F.C. Schilling, A.E. Tonelli and A.L. Cholli, *J. Polym. Sci. Polym. Phys. Ed.* **30** (1992) 91.
45. F.D. Blum, R.B. Funchess and N. Meesiri, in *Solid State NMR of Polymers*, ed. L.J. Mathias, Plenum, New York (1991), p. 271.
46. N.M. Szeverenyi, M.J. Sullivan and G.E. Maciel, *J. Magn. Reson.* **47** (1982) 462.
47. P. Caravatti, P. Neuenschwander and R.R. Ernst, *Macromolecules* **18** (1985) 119.
48. S.S. Sankar, E.O. Stejskal, R.E. Fornes, W.W. Fleming, T.P. Russell and G.C. Wade, in *Polymer and Fiber Science: Recent Advances*, eds. R.E. Fornes and R.D. Gilbert, VCH, New York, (1992), Chap. 11.
49. M. Nakagawa, F. Horii and R. Kitamaru, *Polymer (British)* **31** (1990) 323.
50. Y. Yang, A. Hagemeyer, K. Zemke and H.W. Spiess, *J. Chem. Phys.* **11** (1990) 7740.
51. N.S. Murthy, G.R. Hatfield and J.H. Glans, *Macromolecules* **23** (1990) 1342.

52. G.R. Hatfield, J.H. Glans and N.S. Murthy, *Macromolecules* **23** (1990) 1654.
53. F.C. Schilling, F.A. Bovey, A.J. Lovinger and J.M. Zeigler, *Macromolecules* **19** (1986) 2660.
54. F.C. Schilling, A.J. Lovinger, J.M. Zeigler, D.D. Davis and F.A. Bovey, *Macromolecules* **22** (1989) 3055.
55. F.C. Schilling, F.A. Bovey, D.D. Davis, A.J. Lovinger, R.B. Macgregor, Jr., C.A. Walsh and J.M. Zeigler, *Macromolecules* **22** (1989) 4645.
56. A.J. Lovinger, D.D. Davis, F.C. Schilling, F.A. Bovey and J.M. Zeigler, *Polymer* (*British*) **30** (1989) 356.
57. F.C. Schilling, F.A. Bovey, A.J. Lovinger and J.M. Zeigler, *Silicon-Based Polymer Science*, eds. J.M. Zeigler and F.W.G. Fearon, ACS, Adv. Chem. Ser. No. 224, ACS (1990), p. 341.
58. A.J. Lovinger, D.D. Davis, F.C. Schilling, F.J. Padden, Jr., F.A. Bovey and J.M. Zeigler, *Macromolecules* **24** (1991) 132.

6 High-resolution solid-state ^{13}C NMR studies of local motions and spin dynamics in bulk polymers

F. LAUPRÊTRE

6.1 Introduction

As shown by the numerous studies described in the literature, high-resolution ^{13}C NMR has proven to be a most powerful tool for investigating local dynamics in polymers. Unlike fluorescence anisotropy or electron spin resonance techniques, it does not require any labeling of the molecule under study, and yields direct information on the compound under study. As a selective technique, it allows the observation of one signal per magnetically inequivalent carbon, and therefore the dynamic behaviour of each part of a molecule can be followed independently. Moreover, many NMR parameters are sensitive to molecular motions. They include the different relaxation times as well as the spectrum lineshape, the strength of the dipolar coupling and the chemical shift anisotropy. The available spectral windows depend on the type of measurement that is performed. They range from about 10^{-1} Hz for slow processes to several hundreds of MHz for very fast modes. In the case of bulk polymers at temperatures well above the glass-transition temperature, the fast processes of the local dynamics can be investigated by determining the spin–lattice relaxation time, T_1, and the nuclear Overhauser enhancement, that probe modes in the Larmor frequency region, whereas measurements probing slower motions are more appropriate for glassy state investigations.

Whereas modern NMR techniques permit high-resolution ^{13}C NMR spectra in bulk polymers to be obtained at temperatures either above or below their glass-transition temperature, T_g, the strengths of the interactions that govern the spectral parameters are very different in both cases. At temperatures well above the glass-transition temperature, the local motions of bulk polymers are fast processes. As a consequence, the main tensorial interactions (chemical shift anisotropy, homonuclear and heteronuclear dipolar couplings, and quadrupolar couplings in the case of spins higher than 1/2) are averaged to a large extent by fast local motions. For *cis*-1,4-polybutadiene at room temperature, for example, Cohen-Addad *et al.* [1] and English *et al.* [2, 3] have shown that most, i.e. 90% or more, of the dipolar interaction is averaged by rapid motions. Therefore, for bulk polymers at temperatures well above T_g, high-resolution ^{13}C NMR spectra can be obtained by using the conventional spectrometers for solution investigations. From an NMR point of view, bulk polymers in this temperature range and polymers in solution share a number of common features.

In contrast, motions that may occur in a glassy polymer are much slower than modes observed in the melt. In addition, they are very localized and involve only side-groups or short sequences of the main chain. Therefore, the tensorial interactions listed above are only partly averaged, or even not averaged at all when no local modes exist. Therefore, in order to obtain high-resolution ^{13}C NMR spectra, one has to use the specific line-narrowing techniques that have been described in chapter 5. These techniques, based on proton dipolar decoupling and magic-angle sample spinning, perform an efficient averaging of the carbon–proton dipolar interactions and chemical shift anisotropy and allow the recovery of high resolution.

It is important to mention at this point that, in the solid state, some of the above spectral parameters are not only determined by molecular motions, but also by the dynamics of the spin system. As a consequence, in bulk polymers at temperatures below the glass-transition temperature, the competition between molecular dynamics and spin dynamics may introduce some difficulties in the data interpretation. On the other hand, the spin dynamics phenomenon itself can be a most useful tool for studying the solid-state organization of heterogeneous materials.

In this chapter, we consider the sensitivity of the different NMR parameters to both molecular motions and spin dynamics. The first two parts illustrate the capability of NMR to study the local dynamics of bulk polymers at temperatures above and below the glass-transition temperature, T_g, respectively. The third part is devoted to the NMR investigation of the solid-state organization of heterogeneous polymer systems. In this last section, examples are taken mainly from the field of polymer blends.

6.2 ^{13}C NMR investigation of local dynamics in bulk polymers at temperatures well above the glass-transition temperature

6.2.1 *Models for local dynamics*

In polymers, due to the constraint resulting from the connectivity of the chain, the local motions are usually too complicated to be described by a single isotropic correlation time τ_c, as discussed in chapter 4. Indeed, fluorescence anisotropy decay experiments, which directly yield the orientation autocorrelation function, have shown that the experimental data obtained on anthracene-labelled polybutadiene and polyisoprene in solution or in the melt cannot be represented by simple motional models. To account for the connectivity of the polymer backbone, specific autocorrelation functions, based on models in which conformational changes propagate along the chain according to a damped diffusional process, have been derived for local chain motions in solution [4–7]. The concept of the autocorrelation function, and the use of multiple correlation time models for polymer local dynamics in solution have been discussed in chapter 4, as has the use of such models for predicting NMR relaxation behaviour. As shown by fluorescence anisotropy decay studies carried out on polybutadiene or polyisoprene, the autocorrelation functions associated with these models can also represent experimental data recorded in the bulk state [8, 9].

Among the various expressions that are based on a conformational jump model and have been proposed for the orientation autocorrelation function of a polymer chain, $G(t)$, the formula derived by Hall and Helfand (HH) [4] leads to a very good agreement with fluorescence anisotropy decay data. It is written as

$$G(t) = \exp(-t/\tau_2)\exp(-t/\tau_1)I_0(t/\tau_1) \tag{6.1}$$

the Fourier transform of which is

$$J(\omega) = \mathrm{Re}\,[1/(\alpha + i\beta)^{1/2}] \tag{6.2}$$

with $\alpha = \tau_2^{-2} + 2\tau_1^{-1}\tau_2^{-1} - \omega^2$ and $\beta = -2\omega(\tau_1^{-1} + \tau_2^{-1})$.

I_0 is the modified Bessel function of order 0, τ_1 is the correlation time associated with correlated jumps responsible for orientation diffusion along the chain, and τ_2 corresponds to damping which consists either of non-propagative specific motions or of distortions of the chain with respect to its most stable conformations. However, it must be pointed out that the molecular origin of the damping is still not well identified. Molecular processes such as specific isolated jumps, fluctuations of internal rotation angles, and chemical defects (head-to-head linking, stereochemical sequences...) are expected to affect the propagation of conformational changes along the chain sequence leading to a damping effect.

The expression for the autocorrelation function derived by Hall and Helfand can be identified, in a generalized diffusion and loss equation, with the orientation cross-correlation function of two neighbouring bonds inside the polymer chain. To account for motional coupling of non-neighbouring bonds, resulting for example from the presence of side-chains, Viovy *et al.* (VMB) [5] have introduced cross-correlation functions of a pair of bonds separated by j bonds into the orientation autocorrelation function. These functions are written

$$C_j(t) = \exp(-t/\tau_2)\exp(-t/\tau_1)I_{|j|}(t/\tau_1) \tag{6.3}$$

where $I_{|j|}$ is a modified Bessel function of order $|j|$. The expression for $G(t)$ is then

$$G(t) = \exp(-t/\tau_2)\exp(-t/\tau_1)(I_0(t/\tau_1) + \textstyle\sum_i g_i I_i(t/\tau_1)) \tag{6.4}$$

where g_i is a rapidly decreasing function of i. The Hall–Helfand autocorrelation function is a particular case of equation (6.4), corresponding to $g_i = 0$, whatever the value of i.

Another expression for the orientation autocorrelation function of chains undergoing three-bond jumps on a tetrahedral lattice has been developed by Jones and Stockmayer [7]. Analysis [10] of the derived orientation autocorrelation function has shown that this function can be considered as a particular case of expression (6.4).

From a practical point of view, all the above expressions for the orientation autocorrelation function lead to very similar numerical results for NMR spin–lattice relaxation time (T_1) calculations.

6.2.2 Experimental studies

Figure 6.1 shows the behaviour of carbon-13 NMR nT_1 as a function of the reciprocal of temperature T for the CH and CH_2 carbons of bulk poly(vinylmethyl ether) at temperatures well above the glass-transition temperature [11]. The nT_1 minimum is observed at 90°C at 62.5 MHz and 70°C at 25.15 MHz. nT_1 values at the minimum are 0.177 and 0.070 s at 62.5 and 25.15 MHz, respectively. These experimental values of the nT_1 minima are listed in Table 6.1. They are compared with those calculated from the different models

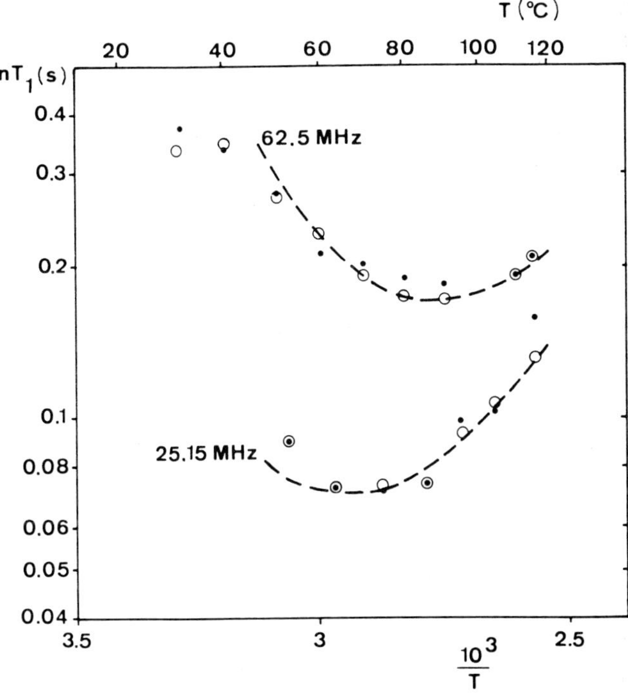

Figure 6.1 62.5 and 25.15 MHz ^{13}C spin–lattice relaxation times nT_1 in bulk poly(vinylmethyl ether): (○) CH carbon: (●) CH_2 carbon; (---) best fit calculated from DLM autocorrelation function: $a = 0.40$, $\tau_1/\tau_0 = 200$, $\tau_2/\tau_1 = 2$. After [11].

Table 6.1 Comparison of experimental and calculated nT_1 (second) values at the minimum in bulk poly(vinylmethyl ether). After [11]

ω_C	62.5 MHz	25.15 MHz
Experimental	0.177	0.070
Isotropic model.	0.100	0.040
HH	0.128	0.050
VMB	0.135	0.054

by using the nT_1 expressions given in chapter 4 under the assumption of a purely $^{13}C-^{1}H$ dipolar relaxation mechanism and taking r_{CH} as 1.08 Å for the unsaturated CH group and 1.09 Å for the saturated CH and CH_2 groups [12]. It can be seen that the experimental values are always higher than the calculated values, in proportions ranging from about 35% in the case of the HH and VMB models to 75% in the case of isotropic reorientation. It is important to note that this large discrepancy cannot be accounted for by internuclear distance imprecisions.

For the CH and CH_2 carbons of poly(vinylmethyl ether) in 15% $CDCl_3$ solution, the minimum of nT_1 is observed at 25.15 MHz at a temperature about 110°C lower than that of the bulk poly(vinylmethyl ether) T_1 minimum. At the minimum, nT_1 is 0.070 s, as observed in bulk, and is much higher than the values that can be derived from specific polymer chain motional models.

The nT_1 values at the nT_1 minimum are listed in Table 6.2 for a number of amorphous polymers in bulk at temperatures well above the glass-transition temperature. For all the polymers considered, the same result is observed; at the minimum, nT_1 is always higher than the values that are derived from specific motional chain models [13–15].

To account for the large deviations between calculated and experimental nT_1 values at the minimum, two different assumptions can be made. The first assumption is the existence, inside the polymer chain, of an additional motion

Table 6.2 Experimental nT_1 (second) values at the minimum in bulk polymers. After [11, 13–15]

	Saturated CH		CH_2	
	25.15 MHz	62.5 MHz	25.15 MHz	62.5 MHz
Poly(vinylmethyl ether)	0.070	0.177	0.070	0.177
Poly(propylene oxide)	0.068	0.168	0.096	0.250
Polyisobutylene			0.060	0.18
	Unsaturated CH		**CH_2**	
	25.15 MHz	62.5 MHz	25.15 MHz	62.5 MHz
cis-1,4-Polyisoprene	0.064	0.154	0.080	0.21
			0.095	0.22
cis-1,4-Polybutadiene	0.072	0.166	0.095	0.24

that is not considered in the previous models and that contributes to a partial reorientation of the CH vectors with a characteristic time that differs from the correlation time for orientational diffusion along the chain. The alternative explanation is that the chain segmental motions have to be described by a distribution of correlation times. As shown in Table 6.2, for all polymers except poly(vinylmethyl ether), the nT_1 values at a given experimental frequency are different for adjacent CH and CH_2 carbons of the polymer main chain. Therefore, a different distribution of correlation times would be required for each of these adjacent sites, which is not consistent with the concept of cooperative motions involving a few monomer bonds. If one now considers the existence of an additional motion, the assignment of this additional motion to a process slower than the orientation diffusion along the chain associated with τ_1 must be precluded since results derived from fluorescence anisotropy decay [8] and NMR [1, 3] on polybutadiene have shown that such slow modes contribute only to a very small extent to the decay of the orientation autocorrelation function in this temperature range. Therefore, this additional motion has to be a faster mode and thus more local than the orientation diffusion process along the chain. A similar fast process has recently been observed by neutron scattering experiments [16]. It can be assigned to molecular librations of limited extent of the CH vector about their equilibrium conformation and corresponds to oscillations inside a potential well.

The Dejean–Lauprêtre–Monnerie (DLM) orientation autocorrelation function is based on the above description. It takes into account independent damped conformational jumps, described by the Hall–Helfand autocorrelation function, and librations of the internuclear vectors represented, as proposed by Howarth [17] (see chapter 4) by a random anisotropic fast reorientation of the CH vector inside a cone of half-angle Θ and axis the rest position of the internuclear vector. The resulting orientation autocorrelation function can be written as

$$\begin{aligned} G(t) = {} & (1-a)\exp(-t/\tau_2)\exp(-t/\tau_1)I_0(t/\tau_1) \\ & + a\exp(-t/\tau_0)\exp(-t/\tau_2)\exp(-t/\tau_1)I_0(t/\tau_1) \end{aligned} \qquad (6.5)$$

where the correlation time τ_1 characterizes the conformational jumps, the correlation time τ_2 is associated with the damping and τ_0 is the characteristic time of the libration, a is related to the half-angle Θ of the libration cone through the relation

$$(1-a) = ((\cos\Theta - \cos^3\Theta)/2(1-\cos\Theta))^2. \qquad (6.6)$$

Assuming that τ_0 is much shorter than τ_1 and τ_2 (a situation that is often encountered), the second term in expression (6.5) can be simplified and $G(t)$ written as

$$G(t) = (1-a)\exp(-t/\tau_2)\exp(-t/\tau_1)I_0(t/\tau_1) + a\exp(-t/\tau_0) \qquad (6.7)$$

Under this condition, the spin–lattice relaxation time T_1 is written as [11]

$$(nT_1)^{-1} = ((1-a)\hbar^2\gamma_C^2\gamma_H^2(J_{HH}(\omega_H-\omega_C)+3J_{HH}(\omega_C)+6J_{HH}(\omega_H+\omega_C)) + a\hbar^2\gamma_C^2\gamma_H^2(J_0(\omega_H-\omega_C)+3J_0(\omega_C)+6J_0(\omega_H+\omega_C)))/10r_{CH}^6 \quad (6.8)$$

$J_{HH}(\omega)$ and $J_0(\omega)$ are defined as

$$J_{HH}(\omega) = (\alpha + i\beta)^{-1/2}$$

$$J_0(\omega) = \tau_0/(1+\omega^2\tau_0^2)$$

where $\alpha = \tau_2^{-2} + 2\tau_1^{-1}\tau_2^{-1} - \omega^2$, $\beta = -2\omega(\tau_1^{-1} + \tau_2^{-1})$ and ω is the frequency in radians (H = proton, C = carbon).

The first term, which is proportional to $1-a$, contains only factors originating from the HH function and thus depends only on τ_1 and τ_2. The second term, which is proportional to a, corresponds to the libration of limited angular extent and depends on τ_0. Under the assumption of fast librations and segmental motions, $\tau_0 \ll \tau_1, \tau_2$, $(\omega_H+\omega_C)\tau_1 < 1$, the second term in expression (6.8) can be neglected, giving

$$(nT_1)^{-1} = (1-a)\hbar^2\gamma_C^2\gamma_H^2(J_{HH}(\omega_H-\omega_C)+3J_{HH}(\omega_C) + 6J_{HH}(\omega_H+\omega_C))/10r_{CH}^6 \quad (6.9)$$

For bulk poly(vinylmethyl ether), nT_1 values of the minimum correspond to $a = 0.40$ and $\Theta = 33°$. With this parameter a thus determined, the best fit obtained from equation (6.8) is shown in Figure 6.1 for the same polymer. As can be concluded from this figure, the overall agreement between theoretical prediction and the experimental data is very good. Only in the temperature range of 50–60°C, where the measurements are less precise due to the broadening of the lines, do deviations occur.

It is of interest to note that, as shown by results reported in [11], the relaxation data obtained from poly(vinylmethyl ether) in $CDCl_3$ solution are described by exactly the same values of the ratios of the correlation times τ_1/τ_0 and τ_2/τ_1 and the angle of the libration cone Θ as those listed in Table 6.3 for the bulk local dynamics. Only the temperature dependence of the correlation times is different. Such a similarity in behaviour shows that the very nature of the motions involved in the magnetic relaxation at a given frequency

Table 6.3 $T_{ref/NMR}$ reference temperatures (K) and half-angle Θ_{CH} and Θ_{CH_2} (°) of the libration cone for the CH and CH_2 carbons. After [18]

	Θ_{CH}	Θ_{CH_2}	$T_{ref/NMR}$
Poly(vinylmethyl ether)	33	33	344
Poly(propylene oxide)	25	37	270
Polyisobutylene		23	333
cis-1,4-Polyisoprene	20	33, 37	297
cis-1,4-Polybutadiene	26	36	234

is identical in bulk at temperatures well above T_g and in solution, as discussed also in chapter 4. For example, in poly(vinylmethyl ether) the libration that is revealed by the high value of the T_1 minimum has the same amplitude in both cases.

For all the polymers listed in Table 6.3 that have been investigated in bulk at temperatures well above the glass-transition temperature, the autocorrelation function defined by relation (6.5) gave good fits of the experimental data [13–15]. The values of the half-angle of the libration cone Θ determined from NMR experiments are listed in Table 6.3. They are smaller for a methine carbon than for a methylene carbon except for poly(vinylmethyl ether). Θ is large for polypropylene oxide whereas it takes the smallest value for polyisobutylene. These results indicate that Θ is strongly related to the steric hindrance at the considered site; the larger the steric hindrance, the smaller the half-angle of the cone, as can be expected.

Although the τ_1/τ_0 ratio is not determined to a good accuracy when it reaches large values, interpretation of T_1 data shows that τ_0 is at least 150 or 200 times shorter than τ_1 for each polymer under investigation. This implies that the anisotropic mode, which is observed in addition to the segmental reorientation in the NMR relaxation experiments, is indeed a very fast process. The high frequency of this motion is one of the main reasons for its assignment to a libration of the internuclear vector about its rest position.

6.2.3 *Temperature dependence*

In the above description of local motions, τ_1 characterizes the segmental modes. In order to know whether these segmental motions observed by NMR in bulk at temperatures well above the glass-transition temperature belong to the glass-transition processes, it is of interest to compare the variations of τ_1 as a function of temperature with the predictions of the Williams–Landel–Ferry (WLF) equation [19]. The WLF equation describes the frequency dependence of the motional processes associated with the glass-transition phenomena. It can be written as [20]

$$\log(a_{T/T_g}) = -C_1^g(T - T_g)/(C_2^g + T - T_g) \tag{6.10}$$

where $a_{T/T_g} = \tau_C(T)/\tau_C(T_g)$, $\tau_C(T)$ is the viscoelastic relaxation time at temperature T, and $\tau_C(T_g)$ is the viscoelastic relaxation time at T_g, which serves here as a reference. C_1^g and C_2^g are two parameters that depend on the reference temperature. Using the temperature $T_\infty = T_g - C_2^g$ for which a_{T/T_g} tends to infinity, the WLF equation can be written

$$\log(a_{T/T_g}) = -C_1^g + C_1^g C_2^g/(T - T_\infty) \tag{6.11}$$

T_∞ and the product $C_1^g C_2^g$ are constants characteristic of a given polymer. They do not depend on the reference temperature. Variation of $\log(a_{T/T_g})$ as a

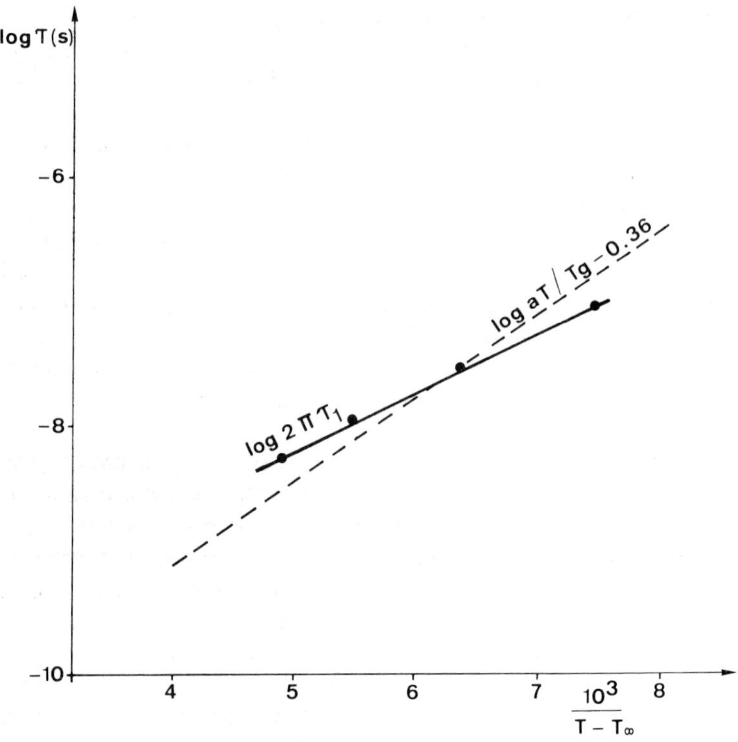

Figure 6.2 Comparison of $\log(2\pi\tau_1)$ (———) and $\log(a_{T/T_g} - 0.36)$(----) dependences on $10^3/(T - T_\infty)$ for bulk poly(vinylmethyl ether), see text. After [11].

function of $1/(T - T_\infty)$ is linear, and a change in the reference temperature only induces a translation of the line without any modification of its slope.

WLF coefficients for bulk poly(vinylmethyl ether) are $T_\infty = 188$ K, $C_1^g = 11.46$, and $C_1^g C_2^g = 671.74$ K [21]. In Figure 6.2, the variations of log (a_{T/T_g}) and of $\log(2\pi\tau_1)$ are plotted as a function of $1/(T - T_\infty)$ for bulk poly(vinylmethyl ether) (PVME). The slopes of the two lines are quite similar, which shows that the segmental motions associated with the τ_1 process have the same temperature dependence as the viscoelastic relaxation times. Therefore, the segmental motions described by τ_1 belong to the processes that are involved in the glass-transition phenomena. Identical results have been obtained for poly(propylene oxide) (PPO) [13], *cis*-1,4-polybutadiene (PB) and *cis*-1,4-polyisoprene (PI) [14]. In the case of polyisobutylene (PIB), the agreement between the variations of $\log(a_{T/T_g})$ and of $\log(2\pi\tau_1)$ as a function of $(T - T_\infty)$ is not so good. In this case, the temperature dependence of the segmental motions as observed by NMR can be understood by considering both glass-transition and secondary relaxation processes [15].

6.2.4 *Factors controlling the local dynamics*

Data plotted in Figure 6.3 show that each polymer has its own dependence of $\log(2\pi\tau_1)$ as a function of $(T - T_\infty)$. Moreover, at a constant $(T - T_\infty)$ difference, the segmental mobility depends on the polymer considered. These results indicate that the differences in segmental mobility at a given temperature observed for the above polymers cannot be interpreted in terms of the differences in the T_∞ temperatures. Similar results are observed when comparing the $\log(2\pi\tau_1)$ variations as a function of $(T - T_g)$ for the different polymers; at a given $(T - T_g)$ difference, each polymer has its own dependence of $\log(2\pi\tau_1)$ as a function of $(T - T_g)$. Although the τ_1 processes are controlled by the segmental motions of the polymer chains involved in the glass-transition phenomena, the polymer matrices do not share the same local dynamics at a given $(T - T_\infty)$ or $(T - T_g)$ difference. Therefore, neither T_∞ nor the glass-transition temperature T_g can be considered as good descriptors for rescaling the segmental motions in bulk polymers at temperatures well above T_g.

It is of interest to define a reference state in which all the polymers are in equivalent states from the point of view of their local mobility. In order to define such equivalent states, one has to look for a property in relation to the frequency window of each experiment. For the ^{13}C NMR relaxation experiments, the temperature, $T_{ref/NMR}$, at which the spin–lattice relaxation

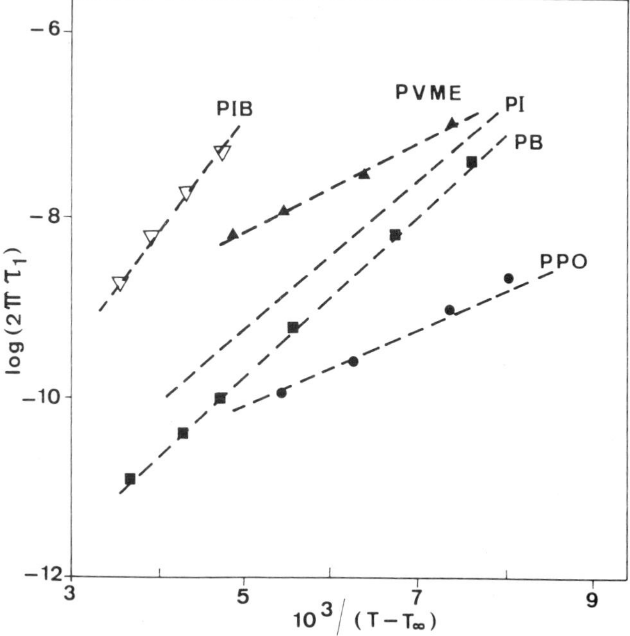

Figure 6.3 Dependences of log $(2\pi\tau_1)$ (– – – –) derived from ^{13}C T_1 NMR experiments, as a function of $10^3/(T - T_\infty)$, see text. After [18].

time T_1 reaches its minimum may constitute such an appropriate reference state. The T_1 minimum is directly related to the local mobility in the frequency range defined by $(\omega_H - \omega_C)$, (ω_C) and $(\omega_H + \omega_C)$ and can be easily determined experimentally. It is independent of the model used to describe the local dynamics. However, due to the flat character of the T_1 minimum, the uncertainty in determining the experimental values of $T_{ref/NMR}$ is quite large. Moreover, the precise value of τ_1 at the T_1 minimum slightly depends on the τ_2/τ_1 ratio. An alternative definition of the NMR reference temperature is the temperature $T_{ref/NMR\,10^{-9}}$ at which τ_1 is equal to 10^{-9} s, in the centre of the ^{13}C T_1 frequency window. In the latter case, $T_{ref/NMR\,10^{-9}}$ is model dependent. $T_{ref/NMR}$ reference temperatures have been obtained at 25.15 MHz for several polymers [18]. They are listed in Table 6.3 and show that $(T_{ref/NMR} - T_g)$ strongly varies from one polymer to an other. For example $(T_{ref/NMR} - T_g)$ is 36° higher in PIB than in PI which implies that the same mobility in terms of correlation time τ_1 is obtained at 36° higher in PIB than in PI.

The next step of this approach is to relate the reference temperatures to data obtained from viscoelastic experiments. Fractional free volumes f_T are expressed as

$$f_T = \alpha_f(T - T_\infty) \tag{6.12}$$

where α_f is the thermal expansion coefficient of the free volume. Fractional free volumes f_{ref} at the $T_{ref/NMR}$ reference temperature defined above have been calculated for the different polymers by using α_f values reported in [20]. f_{ref} values are strongly dependent on the polymer considered [18], which clearly shows that the same fractional free volume is not required for all polymers to perform conformational jumps characterized from NMR by the same value of the correlation time τ_1.

The free volumes $v_f(T_{ref})$ at the reference temperatures can be derived from f_{ref} and Van Krevelen's data on amorphous polymers [22]. They represent the free volume per mole of repeat unit of the polymer divided by the number of main-chain atoms per repeat unit. Once more, the values differ from one polymer to another, which implies that other molecular parameters have to be considered.

Another quantity of interest for the local dynamics is the monomeric friction coefficient ζ_0 which characterizes the resistance encountered by a monomer unit moving through its surroundings [23]. It has been shown to follow the WLF law. The variations of the monomeric friction coefficient ζ_0 as a function of temperature are plotted in Figure 6.4 for several polymers. PI and PIB data have been taken from Ferry's book [20]. For polybutadiene (PB), the approximation $\zeta_0 = \zeta_1$, where ζ_1 is the friction coefficient of a foreign molecule of like size, has been used. For a given $(T - T_\infty)$ difference, the ζ_0 value strongly depends on the polymer considered. However, $\log \zeta_0$ at $T_{ref/NMR}$ and $T_{ref/NMR\,10^{-9}}$ have the same order of magnitude for the polymers considered. Such a result implies that, at these reference temperatures, the polymers share

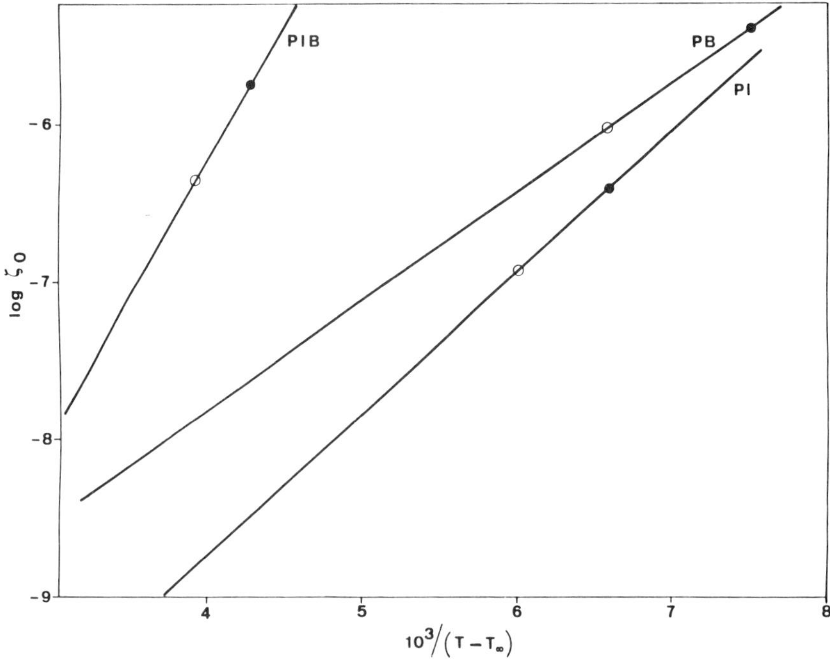

Figure 6.4 Logarithm plots of the monomeric friction coefficient, ζ_0, as a function of $10^3/(T - T_\infty)$ and $\log \zeta_0$ values at the $T_{ref/NMR}$ (●) and $T_{ref/NMR\,10^{-9}}$(○) reference temperatures of the polymer matrices. After [18].

a nearly common value of ζ_0. Therefore, the monomeric friction coefficient of the bulk polymer appears as one of the main factors that control the local dynamics in bulk polymers at temperatures well above the glass-transition temperature. Extensive discussion concerning the role of ζ_0 can be found in [18] where comparison between NMR and excimer fluorescence studies of local dynamics in bulk polymers at temperatures well above T_g shows that the amplitude of the local jumps observed by NMR depends on the precise chemical nature of the polymer chain.

6.2.5 *Conclusion*

The detailed analysis of carbon-13 spin–lattice relaxation times of a number of polymers either in solution or in bulk at temperatures well above the glass-transition temperature T_g has led to a general picture involving several types of motions. The segmental reorientation can be interpreted in terms of correlated conformational jumps which induce a damped orientation diffusion along the chain. It is satisfactorily described by the well-known autocorrelation functions derived from models of conformational jumps in polymer chains [4, 5] which have proven to be very powerful in representing fluorescence anisotropy data [8]. Besides these conformational jumps, there exists an additional fast process that is not observed in the fluorescence depolarization experiments but which has been revealed by the NMR technique in all the polymers under investigation, independently of their chemical structure. For example, it has been observed in polymers with very simple chemical structure such as poly(vinylmethyl ether), polyisobutylene or poly(propylene oxide), and in polymers containing both single and double bonds such as polybutadiene or polyisoprene. The generality of this process and its high frequency have led us to assign it to a libration of limited but significant extent of the internuclear vector about its rest position. It is a general feature of flexible molecules in bulk and in solution. Very recent neutron scattering investigations have led to identical conclusions [24].

Another important result is the similarity of the temperature variation of the correlation time τ_1, associated with conformational jumps, and observed for all the polymers considered except polyisobutylene, to the predictions of the Williams–Landel–Ferry equation for viscoelastic relaxation, which indicates that the segmental motions observed by NMR belong to the glass-transition phenomenon. Moreover, the frequency of these intramolecular motions is mainly controlled by the monomeric friction coefficient of the polymer matrix.

6.3 ^{13}C NMR investigation of local dynamics in bulk polymers at temperatures below the glass-transition temperature

As recalled in the introduction to this chapter, the frequencies and amplitudes of the localized motions that may occur in a bulk polymer below its glass-transition temperature are not sufficient to achieve a complete averaging of the magnetic tensorial interactions. Therefore, the line-narrowing techniques described in chapter 5 have to be used. In the case of ^{13}C nuclei, the combination of proton dipolar decoupling (DD) and magic-angle sample spinning (MAS), which realize a coherent averaging of the ^{13}C–^{1}H dipolar coupling and chemical shift anisotropy, respectively, allows the recovery of high resolution [25]. Moreover, these interactions, which are orientation- and distance-dependent, may be partly averaged by molecular motions. Therefore, measurements of the strengths of the tensorial interactions provide information that can be used in the investigation of local dynamics in the solid state. In addition, as for polymers above their glass-transition temperature, information about local motions can also be derived from spectrum lineshapes and relaxation phenomena.

In the following section, we consider the dependence of the different NMR parameters on molecular motions and we illustrate the capability of NMR by a few examples. Unless otherwise specified, the experiments discussed below were performed by using the classical cross-polarization (CP) pulse sequence, discussed in chapter 5, which leads to an improvement in the sensitivity of the ^{13}C nuclei. Basically, CP is achieved by spin-locking the protons with a RF field, B_{1H}, parallel to the proton rotating frame magnetization, while applying a second RF field, B_{1C}, to the carbons at the carbon resonance frequency under the Hartmann–Hahn condition: $\gamma_H B_{1H} = \gamma_C B_{1C}$, or $\omega_{1H} = \omega_{1C}$, in angular frequency units.

6.3.1 *Chemical shift anisotropy*

The chemical shift of a nuclear spin is a tensorial quantity. Its value depends on the orientation of the electronic distribution about the nucleus with respect to the external magnetic field. In a liquid, due to the rapid molecular motions, this interaction is averaged to zero and the observed chemical shift is the trace of the tensor. In contrast, in a powder, in the absence of motions, all the orientations have the same probability and the signal obtained for each carbon is the sum of the elementary chemical shifts corresponding to the different orientations. When local motions occur in the bulk below T_g, they usually induce a partial averaging of the chemical shift anisotropy.

As described in chapter 5, when a solid sample is spun at an angular frequency, ω_r, about an axis making an angle, ψ, with the external field B_0, equal to the magic angle, $\psi = 54.7°$, the resulting ^{13}C NMR spectrum depends on the extent of chemical shift anisotropy modulation by magic-angle sample spinning. If the spinning speed $\omega_r/2\pi$ is large relative to the chemical shift anisotropy $\Delta\sigma$ expressed in Hertz, the spectrum consists of one peak per magnetically inequivalent carbon, the location of which is given by the trace of the shielding tensor. Conversely, if the spinning speed is smaller than the chemical shift anisotropy, there appear, together with the main resonance, spinning side-bands at a distance $\omega_r/2\pi$ from each other. The envelope of the spinning side-bands is related to the shape of the chemical shift tensor [26].

The chemical shift tensor parameters can be obtained from the proton-decoupled ^{13}C NMR spectra recorded without magic-angle sample spinning. The extent and type of motional averaging can then be interpreted. As the selectivity of this technique is poor, which may justify the use of ^{13}C-enriched samples, other methods have been proposed, based on either spinning out of the magic angle [27], slow magic-angle spinning [27–29] or two-dimensional spectroscopy, placing the anisotropic information into a second frequency dimension, while maintaining the isotropic spectrum in the first [30–40]. In the case of slow magic-angle sample spinning, by referring to the theoretical calculations performed by Herzfeld and Berger [26], it is possible to determine the values of the principal elements of the shielding tensor from the relative intensities of the spinning side-bands of different orders.

A typical example of a partial motional averaging of the chemical shift anisotropy is that of an unprotonated aromatic carbon belonging to a *para*-substituted phenyl ring rotating about its local symmetry axis. For such a carbon, the principal elements of the tensor are parallel and perpendicular to the frame defined by the C_1–C_4 axis within the plane of the phenyl ring, and perpendicular to the plane of the phenyl ring. In the rotation about the *para* axis, the principal element parallel to this axis is not modified by the motion. On the contrary, the two components that are perpendicular to this axis are averaged by the motion. The resulting lineshape corresponds to an axially symmetrical tensor as described in chapter 5 and is markedly different from the rigid-lattice pattern. Such a behaviour has been observed in thermotropic polymers having a mesogen unit inside the main chain [41]. Below the crystal-smectic C transition temperature, the observed chemical shift anisotropy (CSA) is characteristic of a rigid-lattice. Above the crystal-smectic C transition temperature, it has the typical lineshape of the symmetrical tensor described above, indicating the internal rotation of the aromatic rings about their local symmetry axis.

Variable-temperature ^{13}C CSA lineshapes for a solid polycarbonate (PC) sample, with single-site ^{13}C enrichment at one of the two phenylene carbons *ortho* to the carbonate, are shown in Figure 6.5 as a function of temperature

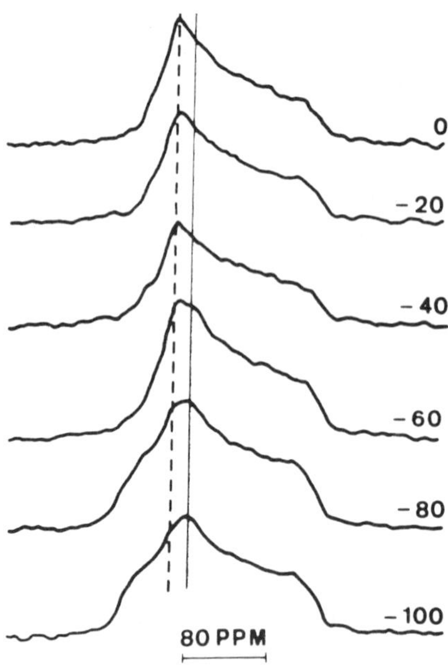

Figure 6.5 Variable-temperature 62.9 MHz ^{13}C CSA lineshapes of one of the two phenylene carbons *ortho* to the carbonate in solid polycarbonate. The solid vertical line indicates the position of the maximum in the rigid lineshape corresponding to σ_{22}. The dashed vertical line indicates the position of the maximum in the high-temperature, motionally averaged spectra. At the intermediate temperatures of $-60°$ and $-80°C$, two maxima corresponding to the two lines are observed, which is indicative of an inhomogeneous distribution of dynamic processes. After [43].

[42, 43]. Their detailed analysis shows that the phenylene group in this polymer undergoes two types of motion simultaneously, both about the C_1–C_4 axis. The primary motion consists of large-angle jumps between two sites whereas the secondary motion involves restricted rotational diffusion over limited angular amplitude. These motions have been described by an inhomogeneous distribution of correlation times [43].

Slow (1 kHz to 0.1 Hz) molecular motions that induce a change in the orientation of the chemical shielding tensor of a particular spin can be detected by a 2-D exchange experiment [44–46]. The pulse scheme and the principle of the experiment are given in Figure 6.6. If during the mixing time, τ_m, a motion of a spin packet (isochromat) has been accompanied by a change in its resonance frequency, this will manifest itself as off-diagonal intensity in the resulting 2-D spectrum. When experiments are carried out on samples rotating about the magic angle with a frequency smaller than the chemical shift anisotropy, molecular motions occurring during the mixing time induce off-diagonal cross-peaks between the different spinning side-bands [45, 46]. Figure 6.7 shows the results of the 2-D exchange experiment applied to poly(oxymethylene) at 50°C. Although the effect is more pronounced for $\tau_m = 2$ s, even at a mixing time of 0.5 s off-diagonal peaks are observed, indicating the presence of motional processes in the crystalline phase of the material [45].

These slow motions which induce a change in the orientation of the chemical shielding tensor of a particular spin can also be studied by ^{13}C spin echo spectroscopy [47]. The experiment consists of producing and digitizing a whole ^{13}C spin-echo train while simultaneously decoupling the protons in a static sample. The Fourier transform of the train gives a spectrum in which the powder pattern of the chemical shift anisotropy is split into a number

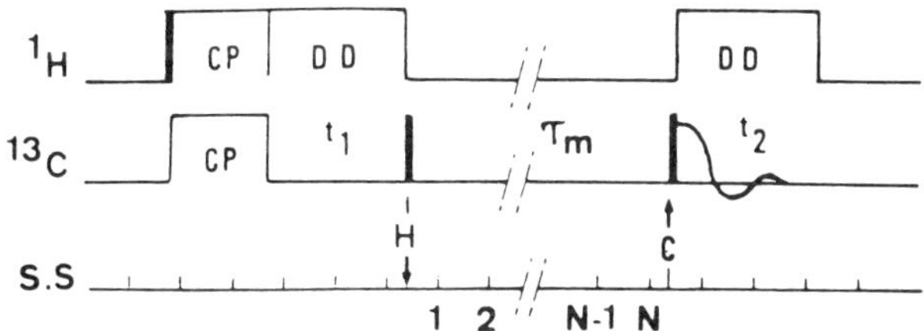

Figure 6.6 Pulse sequence for the 2-D exchange experiment with cross-polarization (contact time = 500 μs) and dipolar decoupling. 90° pulses (6 μs) are in black. In the case of slow magic-angle spinning, a synchronous mixing time is obtained with the Synchro-Spin (SS). With an optical signal from the spinner, the Synchro-Spin holds the pulse programmer at point H and lets it continue after N spinner rotations. ^{13}C magnetization is created via cross-polarization. Then the spins are allowed to precess freely during the evolution time. At time t_1, a 90° pulse is applied to the ^{13}C system, creating for each spin isochromat a magnetization vector along the z axis whose magnitude depends on the precession frequency of the isochromat during evolution. The remaining transverse magnetization dephases quickly because proton decoupling is turned off during the mixing time τ_m. At the end of the mixing period, a second 90° pulse is applied and the ^{13}C free induction decay is acquired during t_2. After [45].

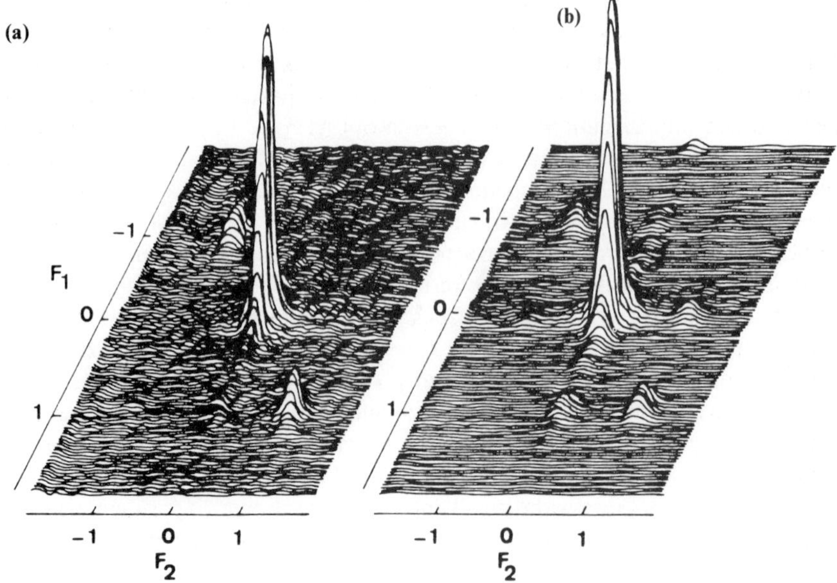

Figure 6.7 ^{13}C exchange NMR spectra of the single peak in poly(oxymethylene) at 50°C. Numbers on the two axes indicate the order and position of the spinning side bands. (a) Mixing time, τ_m, ~0.5 s; (b) mixing time, τ_m, ~2 s. After [45].

of peaks separated by the reciprocal of the pulse spacing in the echo train. The intensities of the peaks provide a map of the powder pattern to a first approximation, and the width of each peak is a measure of the echo decay time or the transverse relaxation rate. The formation of the echo depends on the Larmor frequency of the spin remaining essentially unchanged during the time period between the refocusing pulses. If the Larmor frequency of a spin is changed, as a consequence, for example, of a local motion, then that spin contributes much less to the subsequent echo. On increasing mobility, spin-echo peaks are progressively collapsed to the residual powder lineshape that results from the Fourier transform of the free induction decay at the beginning of the echo train. Such a behaviour has been observed for polyvinyl acetate above its glass-transition temperature [47].

6.3.2 Spectrum lineshape

In the same way as observed in solution NMR, chemical exchange can modify the lineshape of high-resolution solid-state ^{13}C NMR spectra. Carbons that are magnetically inequivalent in the absence of motion and yield two distinct peaks on the NMR spectrum can be rendered equivalent by specific motions, leading to a single NMR line. Between these two extreme situations, the slow exchange and the rapid exchange, the spectrum lineshape is strongly dependent on the rate of the motion in the range of 10^{-1}–10^{6} Hz. One of the first examples evidenced by high-resolution solid-state ^{13}C NMR is that of poly(2,6-dimethyl phenylene oxide)(PMPO) whose protonated aromatic carbons present only one NMR line in solution and two distinct lines in bulk at room temperature [25]. In solution, the phenyl rings of PMPO are freely rotating about their symmetry axis and the two *ortho* carbons are rendered equivalent by the rotation. In the solid state, this motion is frozen as demonstrated by the presence of two lines in the ^{13}C NMR spectrum. Figure 6.8 shows the high-resolution solid-state ^{13}C NMR spectra of an epoxy resin studied by Garroway *et al.* [48]. The spectrum lineshape in the aromatic region is affected by the same type of exchange phenomena as those observed in PMPO. At 151 K, the resonances of the carbons *ortho* to the oxygen are split into two resolved peaks. The resonance of the *meta* carbons is just barely split at 151 K. At higher temperatures, each set of peaks merges into a single line and that line continues to narrow even at the highest temperature of 352 K. The observed coalescence of these spectral lines is an indication of 180° rotations of the phenylene rings. The full temperature dependence has been described in terms of a distribution of correlation times or, equivalently, a non-exponential

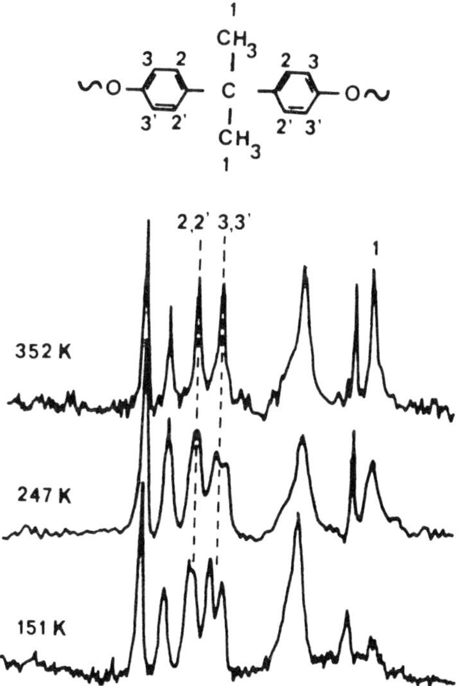

Figure 6.8 Variable-temperature ^{13}C MAS/CP/DD NMR spectra of an epoxy resin diglycidyl ether of bisphenol A (DGEBA) cured with piperidine. After [48].

autocorrelation function. Such an approach has led to a precise characterization of the ring motion in terms of correlation time distribution, whose central values depend on the exact model considered, and the determination of an activation energy of about 63 kJ/mol. [48].

6.3.3 Relaxation times and linewidths

Relaxation times and linewidths can also provide a variety of information on local dynamics below the glass-transition temperature. The main relaxation times that are of interest in high-resolution solid-state ^{13}C NMR for the carbon-13 nucleus are the spin–lattice relaxation time $T_1(^{13}C)$, the spin–lattice relaxation time in the rotating frame $T_{1\rho}(^{13}C)$ and the spin–spin relaxation time T_2. Some of these relaxation times are affected by modes with a frequency range of the order of the intensity, in frequency units, of the RF fields. This is the case for the spin–lattice relaxation time in the rotating frame $T_{1\rho}(^{13}C)$ and the linewidth $(1/\pi T_{2m})$ when determined by motional modulation of the ^{13}C–^{1}H dipolar coupling, whose frequency windows are of the order of a few tens of kiloHertz. In contrast, the spin–lattice relaxation time $T_1(^{13}C)$ and the nuclear Overhauser enhancement are determined by motional components close to the Larmor frequencies, i.e. of the order of a few hundreds of megaHertz.

As already mentioned in the Introduction, magnetic relaxation in the solid state may be governed not only by dynamic phenomena but there may also exist a contribution from static phenomena such as spin diffusion, which consists of mutual exchanges of spin state, or 'flip-flops', between strongly coupled nuclei that have the same precession frequency but antiparallel spins. This mechanism is explained in more detail in chapter 7. It does not involve any variation in the energy of the system. When it occurs, magnetic relaxation times cannot be interpreted in terms of local dynamics only, but the two contributions have to be separated.

6.3.3.1 Spin–lattice relaxation times $T_1(^{13}C)$ and $T_{1\rho}(^{13}C)$. Due to the low gyromagnetic ratio and natural abundance of the ^{13}C spins, the spin diffusion between carbon-13 nuclei is slow and usually negligible in polymers as compared with the dynamic contribution to spin–lattice relaxation. Therefore $T_1(^{13}C)$ can be interpreted in terms of local motions having frequencies of the order of a few hundreds of megaHertz. An interesting example of the usefulness of $T_1(^{13}C)$ is observed in semi-crystalline polymers studied at temperatures above the glass-transition of the amorphous phase: due to the rapid motions occurring in the amorphous phase, the $T_1(^{13}C)$ relaxation times of the amorphous carbons are markedly shorter than the $T_1(^{13}C)$ values of the crystalline carbons. As discussed in section 6.4, this property has been used to study the partitioning of semi-crystalline polymers in their different phases. On the other hand, in the solid state, at temperatures below T_g the

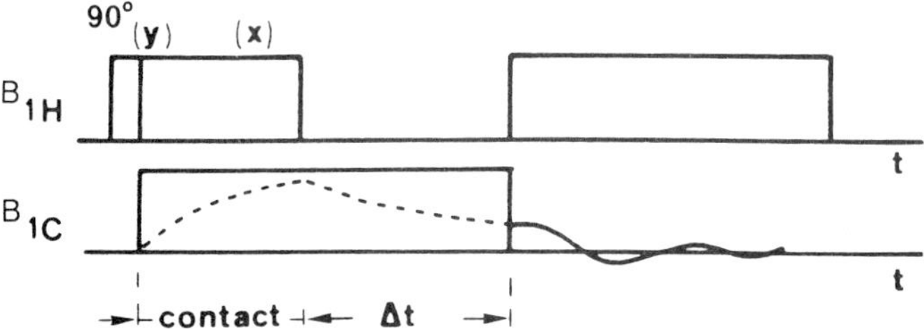

Figure 6.9 Pulse sequence used for $T_{1\rho}(^{13}C)$ determination [25]. After the contact, the ^{13}C magnetization evolves in the presence of the RF field, B_{1C}, applied to the carbon spins during a time Δt. The decrease of the magnetization as a function of Δt is characterized by the $T_{1\rho}(^{13}C)$ relaxation time.

interest in $T_1(^{13}C)$ measurements is limited by the near-absence of high-frequency range motional modes other than methyl rotations and librations.

The situation is quite different in the rotating frame. As for $T_1(^{13}C)$, $T_{1\rho}(^{13}C)$ is usually not determined by ^{13}C–^{13}C spin diffusion. However, as a consequence of the MAS based pulse sequences used to measure $T_{1\rho}(^{13}C)$ among the mechanisms involved in the relaxation, there may be a contribution from the spin diffusion between carbon-13 and proton spins [49–53], as discussed below.

One of the cross-polarization pulse sequences used to measure $T_{1\rho}(^{13}C)$ is shown in Figure 6.9. During the evolution time, Δt, of this pulse sequence, the ^{13}C magnetization is spin-locked along B_{1C}. Any reorientation of the ^{13}C–^1H internuclear vectors induces fluctuations of the dipolar local fields, and the ω_{1C} component of these fluctuating fields participates in the relaxation of the ^{13}C spins. This is a 'spin–lattice' relaxation mechanism which is related to molecular motions and is characterized by the relaxation time [49–53]:

$$(T_{1\rho}(^{13}C))^{-1}_{\text{spin-lattice}} = \tfrac{1}{2}\langle \Delta M^2_{CH}\rangle_m J(\omega_{1C}) \tag{6.13}$$

where $\langle \Delta M^2_{CH}\rangle_m$ is the part of the ^{13}C–^1H second moment, M^2_{CH}, which is averaged by the motions and $J(\omega_{1C})$ is the spectral density associated with the fluctuating fields at the frequency ω_{1C}.

During the same Δt period, there is no RF field applied to the protons. Therefore, the protons are in their dipolar local field. Their magnetization decreases towards thermal equilibrium with a relaxation time T_{1D} which is strongly dependent on the speed of the rotor. For rapid magic-angle spinning, the relaxation time T_{1D} is very short ($< 100\,\mu s$) and the proton magnetization decreases very quickly. Under these conditions, there is magnetization transfer between the ^{13}C spins that have a relatively low spin temperature, and the proton nuclei that have lost their magnetization. This magnetization transfer is characterized by the spin–spin cross-polarization time for adiabatic demagnetization in the rotating frame, T_{CH}(ADRF) [49–53]. The complete expression for $T_{1\rho}(^{13}C)$, as measured by the pulse sequence represented in Figure 6.9 is

$$(T_{1\rho}(^{13}C))^{-1} = (T_{1\rho}(^{13}C))^{-1}_{spin-lattice} + T_{CH}(ADRF)^{-1} \quad (6.14)$$

where T_{CH}(ADRF) essentially represents the dipolar fluctuations and not the local dynamics. It can be determined by a nine-step procedure [54] including a measurement of the proton local dipolar fields [55].

In rigid materials such as crystalline polymers where there is almost no motion,

$$T_{1\rho}(^{13}C) = T_{CH}(ADRF) \quad (6.15)$$

In contrast, when the spectral density at ω_{1C} is important, the $T_{1\rho}(^{13}C)$ relaxation time essentially reflects the local dynamics.

$T_{1\rho}(^{13}C)$ relaxation times have been extensively used to study the local dynamics in bulk polymers below T_g [54, 56–65]. A significant example is that of polycarbonate for which $T_{1\rho}$ (^{13}C) determinations show that there is a large distribution of motions with frequencies of a few tens of kiloHertz [57, 62]. The addition of diluents having a relatively low glass-transition temperature is accompanied by an increase in the spectral density of micro-Brownian motions in the blends in this frequency range, which is well reflected by the $T_{1\rho}(^{13}C)$ variation [65]. $T_{1\rho}(^{13}C)$ determinations have also demonstrated that small amounts of CO_2 in PVC induce an increase of the main-chain motions of the polymer, which shows the effect of the penetrant gas on the interchain packing [66]. In the case of poly(vinyl chloride) (PVC), $T_{1\rho}(^{13}C)$ measurements have demonstrated that antiplasticization–plasticization by tricresyl phosphate affects the cooperative main-chain motions of the polymer; the main-chain motions are reduced when the additive acts as an antiplasticizer whereas they are increased when the polymer is plasticized [67].

6.3.3.2 Linewidths and T_{2m} and $T_{2\sigma}$ spin–spin relaxation times. Line-broadening mechanisms in amorphous polymers examined under MAS and proton dipolar decoupling conditions have been reviewed by VanderHart *et al.* [68]. The different sources of line-broadening are listed in Table 6.4 together with their order of magnitude and their static magnetic field dependence. They include both static and dynamic contributions. Static contributions do not exceed 6 ppm in linear polymers and are due to conformational inequivalence (see chapter 5), bond distortions or variation of the local susceptibility. In contrast, broadenings of dynamic origin may be larger. They originate from motions that occur at frequencies close to the spinning speed of the rotor or close to the intensity of the proton RF field, expressed in frequency units, and reduce the effectiveness of the line-narrowing techniques, i.e. magic-angle sample spinning and proton dipolar decoupling, respectively. In the first case, the relaxation mechanism is the motional modulation of the chemical shift anisotropy [69]. It affects unsaturated carbons which present a large chemical shift anisotropy and may induce noticeable line-broadenings

Table 6.4 Relative contribution of different mechanisms to ^{13}C line-broadening in MAS/DD NMR spectra of glassy organic solids. After [68]

Source of broadening	Extent of broadening[a]	Dependences[b]				
		B_0	ω_{1H}	ω_r	T	Orientation
Conformational non-equivalence	2–6 ppm	1	0	0	m	0
Packing effects		1	0	0	m	?
Bond distortion	?					
Variation in local susceptibility	0.5–2 ppm					
Motional modulation of the CH coupling	4 kHz max	0	−2/0	0	s	w
Motional modulation of the chemical shift anisotropy	200 Hz max at 15 MHz 1 kHz max at 50 MHz	2/0	0	−2/0	s	w

[a] Broadening is either in Hertz or parts per million; a broadening linear in B_0 (first power dependence) leaves the resolution (in ppm) independent of B_0.
[b] For power law dependence, the exponent or range of exponents is given. ω_{1H} is the proton RF field intensity expressed in frequency units, ω_r is the MAS rotor spin rate. Temperature (T) or orientation dependence is weak (w), modest (m), or strong (s). Orientation refers to the question of whether linewidths are influenced by the use of a highly ordered sample (e.g. a single crystal) versus a powder sample.

$(\pi T_{2\sigma})^{-1}$ (see Table 6.4). By using this property, the nature and correlation times of the pentadienyl ring motion in decamethylferrocene have been determined [69]. In the latter case, the relaxation mechanism is the motional modulation of the ^{13}C–^{1}H dipolar interaction which is of particular interest for protonated carbons having a strong ^{13}C–^{1}H dipolar coupling [70].

Assuming that the protons are irradiated at their resonance frequency and that the spinning speed is much smaller than ω_{1H}, the expression for the spin–spin relaxation time T_{2m} arising from the motional modulation of the ^{13}C–^{1}H dipolar interaction and yielding a $(\pi T_{2m})^{-1}$ contribution to the linewidth is written [70] as

$$(T_{2m})^{-1} = \tfrac{1}{2}\langle \Delta M^2_{CH} \rangle_m J(\omega_{1H}) \tag{6.16}$$

where $J(\omega_{1H})$ is the spectral density at the angular frequency corresponding to the intensity of the ^1H RF field. Under the Hartmann–Hahn condition for cross-polarized spectra, $\omega_{1H} = \omega_{1C}$ and from equation (6.13),

$$(T_{1\rho}(^{13}C))_{spin-lattice} = T_{2m} \tag{6.17}$$

Therefore the dynamic contributions to the relaxation times $T_{1\rho}(^{13}C)$ and T_{2m}, which are due to the motional modulation of the dipolar coupling, are equal. For a carbon strongly coupled with a proton and placed in a not too intense magnetic field B_0 so that the motional broadening may be not negligible as compared to the static linewidth, the range of motional frequencies that is accessible from the two relaxation parameters is quite similar (10^{-4}–10^{-7} s). A motion in this frequency range, which is able to dominate the spin-diffusion magnetization transfer, will also induce a broadening of the lines corresponding to the carbons that are involved in this process.

For example, in solid poly(cyclohexyl methacrylate) [64, 71], the broadening of the 12 MHz NMR lines assigned to the carbons of the cyclohexyl ring as well as their low $T_{1\rho}(^{13}C)$ is due to the existence of a slow chair–chair ring inversion at room temperature, in very good agreement with results obtained from mechanical measurements.

6.3.4 ^{13}C–^{1}H dipolar interaction

The ^{13}C–^{1}H dipolar interaction, which is one of the factors determining the values of relaxation times, can also be measured independently. For a powder, its expression is given by

$$\langle b^2 \rangle = \tfrac{4}{5}(\gamma_C \gamma_H \hbar^2 / r_{CH}^3)^2 \tag{6.18}$$

where r_{CH} is the carbon–proton distance.

The reduction in the strength of the dipolar interaction by molecular motions of frequencies comparable to or greater than the dipolar interaction itself is a measure of the amplitude of the motion. Essentially two techniques have been proposed for this purpose [64, 72]. The first method consists of measuring the intensity of the dipolar side-band patterns, obtained from dipolar rotational spin-echo ^{13}C NMR and arising from the heteronuclear dipolar interaction, while the homonuclear ^{1}H–^{1}H dipolar interactions are suppressed by multiple-pulse ^{1}H–^{1}H (WAHUHA) decoupling. (Multiple-pulse sequences are described in more detail in chapter 7). The second method is based on the behaviour of the rises of ^{13}C magnetization in cross-polarization experiments at very short contact times [64].

In the case of plasticized poly(butyral-co-vinyl alcohol) [73], use of dipolar rotational spin-echo ^{13}C NMR in conjunction with $T_{1\rho}(^{13}C)$ determinations, has shown that the frequencies but not the amplitudes of cooperative main-chain motions of the polymer in the hard regions, corresponding to solid polymer associated with partially immobilized plasticizer, are influenced by interactions with the soft regions attributed to liquid plasticizer containing mobile polymer. From this result, a schematic representation of the partitioning of the polymer and plasticizer in terms of a two-phase domain model has been proposed.

The molecular mechanism of the ring-flip process in polycarbonate has also been investigated by this technique [74]. To match the observed dipolar side-band intensities requires all the rings to undergo 180° flips with an average flipping rate greater than 100 kHz at room temperature. This means that there is only one dynamic ring population, not one undergoing flips and another static. The best fit of the data is obtained by superimposing some independent random wiggles to the ring flips. The short methyl-carbon $T_{1\rho}(^{13}C)$ and the methyl-carbon dipolar side-band intensities are consistent with the existence of main-chain wiggles which indicates that the polycarbonate chain is flexible in the glass. Such a behaviour is in contrast with results observed in poly(2,6-dimethyl phenylene oxide) where the protonated aromatic-carbon dipolar pattern is indistinguishable from that of a rigid-lattice material. This result shows that there are no large-amplitude motions such as ring flips, and few low-frequency, small-amplitude wiggles. It was therefore suggested that the mobility of the polycarbonate main-chain results in lattice distorsions which allow ring flips not permitted by the stiffer poly(2,6-dimethyl phenylene oxide) main chain [74].

As mentioned above, the strength of the ^{13}C–^{1}H dipolar coupling, $\langle b^2 \rangle$, can also be deduced from the increases in ^{13}C magnetization in cross-polarization experiments at very short contact times [64]. When ^{1}H–^{1}H homonuclear dipolar interactions are much stronger than ^{13}C–^{1}H heteronuclear dipolar interactions, the increase in magnetization as a function of the contact time is mainly exponential and its rate, $T_{CH}(SL)^{-1}$ can be calculated exactly [75, 76]. In contrast, when carbons are strongly coupled to protons, the increases in polarization at short time can no longer be described by an exponential law. At the very beginning of the contact, there occurs a coherent energy transfer between the carbon of interest and the strongly coupled protons. This oscillatory transfer is damped by the coupling with the more remote protons. For very short contact times, the carbon and its n strongly coupled protons can be considered as an isolated CH_n system. At the end of a very short contact, t, the magnetization $M(t)$ can be described by an oscillatory time-dependent function whose expression is deduced from results obtained for a C–H group [77] and extended to CH_2 groups by using the calculation for the magnetization in liquid AX_n systems [78]:

$$M(t)/M(\infty) = \sin^2 \sqrt{n\langle b^2 \rangle}\,\frac{t}{4} \tag{6.19}$$

Therefore $\langle b^2 \rangle$ can be estimated in a very simple way by measuring the contact time $t_{1/2}$ necessary to obtain half of the maximum polarization $M(\infty)$:

$$\sqrt{\langle b^2 \rangle} = \frac{\pi}{\sqrt{n}\,t_{1/2}} \tag{6.20}$$

For $r_{CH} = 1.09$ Å, $t_{1/2} = 28\,\mu s$ for a rigid CH group and $t_{1/2} = 20\,\mu s$ for a rigid CH_2 group. Experimental $t_{1/2}$ values longer than these rigid-lattice values are evidence for a reduction of the ^{13}C–^{1}H dipolar coupling by motional processes whose frequencies are higher than 10^5 Hz [64].

As an example, these $t_{1/2}$ measurements have been employed to investigate the local dynamics of the aliphatic units of the following mesomorphic polyester-ether [79]:

—(OC—⟨benzene⟩—⟨benzene⟩—⟨benzene⟩—CO—O—$(CH_2$–CH_2—$O)_4)_n$

At 25°C, the $t_{1/2}$ value for the methylene carbon a, adjacent to the carboxyl group, is 21 μs. This $t_{1/2}$ value is close to the expected value for the rigid lattice and shows that the carbon a is frozen on the timescale of the experiment, 10^5 Hz. By contrast, the $t_{1/2}$ of the other methylene carbons are longer, indicating substantial but incomplete motional narrowing of the dipolar interaction between the carbon of interest and its directly bonded protons. In the case of the methylene carbon b, next to carbon a, the 27 μs $t_{1/2}$ value can be interpreted in terms of oscillations on the valence cone of approximately 20° about an equilibrium conformation. For the other methylene carbons, located in the middle of the aliphatic sequence, the 37 μs $t_{1/2}$ value indicates either oscillations of larger amplitude or jumps between two equilibrium conformations. These motional modes of the central methylene carbons are likely related to the γ-transition exhibited by this polyester-ether.

6.3.5 Conclusion

Due to its selectivity and to the large number of available parameters, high-resolution solid-state ^{13}C NMR is a very powerful tool for investigating local motions below T_g. However, to obtain a detailed description of local dynamics in a polymer system, one should not limit oneself to one type of experiment; the results of various ^{1}H, ^{13}C and eventually 2-D solid-state NMR measurements have to be collected and compared with data obtained from other techniques such as mechanical or dielectric relaxation. The value of a combination of solid-state NMR techniques has been illustrated by Jelinski *et al.* [80] for poly(butylene terephthalate) and poly(butylene terephthalate)-containing segmented copolymers. In the example of poly(ethylene terephthalate) studied in [81], results obtained from a number of ^{1}H and ^{13}C experiments indicate that four separate motional processes can be distinguished: small angular fluctuations whose amplitude grows as a function of temperature, a slower specific process of the benzene rings that correlates with the β relaxation, and an even slower motion of the methylene groups whose activation energy correlates with that of the α relaxation, are observed in the crystalline and less mobile amorphous regions of the polymer. There also exist some very mobile amorphous segments in which the rates of the above processes are higher. These very mobile segments also undergo an almost isotropic reorientation.

6.4 ^{13}C NMR study of the molecular organization of some solid heterogeneous polymer systems

In heterogeneous polymer systems, nuclei with different mobilities and environments may have different relaxation times. This property has been widely exploited to study the partitioning of semi-crystalline polymers in their different phases. For example, measurements of the ^{13}C spin–lattice relaxation times of the crystalline portion of a set of polyethylenes have shown that the main factors determining the $T_1(^{13}C)$ values are the crystallite thickness and the nature of the interfacial structure [82]. The greater the degree of branching and the more homogeneous the branching distribution, the shorter the $T_1(^{13}C)$ values [83]. With regards to the amorphous phase, at temperatures well above the glass-transition temperature, its behaviour is liquid-like with ^{13}C spin–lattice relaxation times much shorter than those of the crystalline regions (see chapter 5) [82–85].

As a consequence of these properties, partially relaxed ^{13}C NMR spectra can be used as a test for the assignment of the different resonances of polyethylene [86]. Results thus obtained have been analysed in terms of a three-phase model comprising lamellar crystallites, crystalline–amorphous interface and isotropic amorphous phase, and motions occurring in these different regions [86].

As discussed in section 6.3, relaxation times of solid polymers are not only determined by dynamic phenomena. There may exist a contribution from the static mechanism of spin diffusion. The probability for a flip-flop between two opposite 1/2 spins having the same precession frequency is a decreasing function of the distance between the spins. Moreover, for heterogeneous solids, the spin diffusion mechanism manifests itself in a particular way. When two proton populations have different spin temperatures at a given time, they will tend to a common spin temperature by spin diffusion. (The mechanism is described in more detail in chapter 7). Such a situation occurs during a $T_1(^1H)$ or $T_{1\rho}(^1H)$ determination in a heterogeneous system. For example, if the spin–lattice relaxation times of the protons of the two constituents are quite different, for small values of the interval, t, of the $180°–t–90°$ inversion-recovery sequence, protons having the shorter $T_1(^1H)$ will have recovered most of their equilibrium magnetization: they will have a low spin temperature, whereas protons with the longer $T_1(^1H)$ will be far from equilibrium magnetization; they will have a high spin temperature. Consequently, magnetization transfer will occur from the low spin temperature spins to the high spin temperature spins by the spin-diffusion mechanism during the inversion recovery experiments and the measured spin–lattice relaxation times will no longer be the intrinsic values for the two regions.

For a diffusion duration, $t = 1/(1/R_A) - (1/R_B)$, where R_A and R_B are the relaxation rates in regions A and B, the maximum spin-diffusion length through the material during the course of the relaxation experiment, L, may be written as [87]

$$L \sim \sqrt{6D_s t} \tag{6.21}$$

where D_s is the spin-diffusion coefficient. For solid polymers, D_s, $T_1(^1H)$ and $T_{1\rho}(^1H)$ are of the order of $10^{-11}\,cm^2/s$, 1 s and 10 ms, respectively. One can thus estimate that, in polymer systems, spin diffusion will be efficient at length scales of the order of 100 Å and 10 Å during $T_1(^1H)$ and $T_{1\rho}(^1H)$ determinations, respectively.

The fact that spin-diffusion experiments can be used to determine information on the size of morphological domains having different relaxation times has often been exploited to study heterogeneous systems by wide-band ^{1}H NMR. However, the main difficulty of such experiments is the detailed analysis of relaxation curves. In this respect, determinations of proton relaxation times through high-resolution solid-state ^{13}C NMR may be of considerable interest since the selectivity of the technique allows the relaxation of the different constituents of the system to be investigated independently. The first example of a determination of ^{1}H spin–lattice relaxation times in the rotating frame by means of high-resolution solid-state ^{13}C NMR has been performed by Stejskal *et al.* [88] in the case of polymer blends. It is based on the fact that the decrease in the ^{13}C magnetization as a function of the contact time in a cross-polarization experiment, for contact times long enough, or as a function of the delay in a delayed contact pulse sequence, is a function of the $T_{1\rho}(^{1}H)$ of the protons close to the observed carbon (see chapter 5). Therefore, if, in a polymer blend, polymers are incompatible and form domains whose size is larger than 1 μm, then the selective measurements of $T_{1\rho}(^{1}H)$ for each of the components in the blend yield the $T_{1\rho}(^{1}H)$ value of the pure component. In contrast, if the polymers are miscible at the NMR spatial scale of 10 Å, the $T_{1\rho}(^{1}H)$s associated with the different species may become equal. In the example described in [88] of atactic polystyrene-poly(2,6-dimethyl phenylene oxide) blends, the decreases in ^{13}C magnetization of the PS and PMPO components (Figure 6.10) show that, although the material under study is mainly made of an intimate blend of the two homopolymers, there are also small regions where the polystyrene is not uniformly mixed with the other polymer.

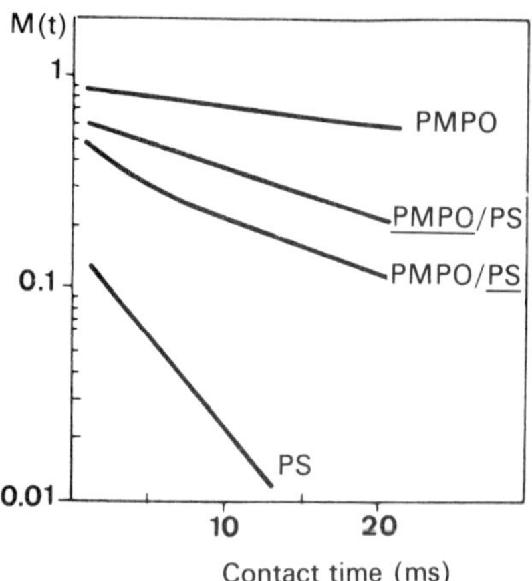

Figure 6.10 Plots of carbon magnetization, $M(t)$, generated by long-term matched spin-lock CP transfers yielding $T_{1\rho}(^{1}H)$ for protons attached to main-chain carbons of poly(2,6-dimethylphenylene oxide) (PMPO), polystyrene (PS) and 75/25 PMPO/PS blends. After [88].

Very similar experiments have been carried out to probe the miscibility of a number of polymer blends [89–94] and mixtures of polymers and additives [95, 96]. In the case of poly(methyl methacrylate) (PMMA) – poly(vinylidene fluoride) (PVF2) blends, this technique has revealed the existence of heterogeneities occurring during the preparation of the samples [89]. When the PVF2 concentration in the blend is high enough, PVF2 crystallizes. During the crystallization process, the PMMA molecules that are intermixed before crystallization occurs are rejected from the growing crystallites. If the blend has enough time to re-equilibrate, the decrease in the magnetization of the PMMA carbons as a function of contact duration is exponential. In contrast, if the re-equilibration of PMMA in the amorphous phase is not completed, the decrease is more complex and its shape indicates the existence of an excess of PMMA at the borders of the crystalline domains. Figure 6.11 shows the dependence of the ^{13}C magnetization of two blends, submitted to a fast and slow cooling, respectively, as a function of contact duration in a $T_{1\rho}(^1H)$ determination experiment. For the slowly cooled sample, the exponential character of the $T_{1\rho}(^1H)$ decrease points out the homogeneity of the amorphous phase, in contrast with data observed for the rapidly cooled sample [89].

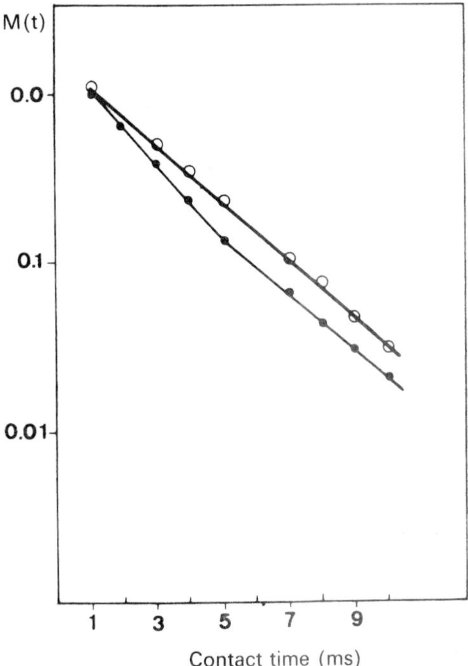

Figure 6.11 Dependence of carbon-13 atom magnetization, $M(t)$, as a function of contact duration ($T_{1\rho}(^1H)$ measurement) for a 50:50 PMMA/PVF2 blend. (○) blends cooled from 190°C to room temperature at a cooling rate of 0.2°/min; (●) blends cooled from 190°C to room temperature at a cooling rate of 10°/min. After [89].

Information on the local structure of polymers is not limited to proton–proton spin diffusion experiments. Spin diffusion among the carbons can also be used [97]. However, because of the low gyromagnetic ratio and natural abundance of the carbon-13 nuclei, it is less efficient than proton–proton spin diffusion. At natural ^{13}C abundance, the rate of spin diffusion is usually too low to compete with the rate of spin–lattice relaxation in most polymers. However, an interesting exception is that of semi-crystalline polymers such as linear polyethylene and cellulose which have very long longitudinal relaxation times $T_1(^{13}C)$ and for which natural-abundance ^{13}C–^{13}C spin diffusion has been observed [98]. Otherwise, in most cases, in order to be able to observe spin exchange among carbons, one has to use isotopically enriched samples [97]. The applicability of ^{13}C exchange measurements for investigation of polymer miscibility has been demonstrated in the case of enriched poly(ethylene terephthalate) (PET)–bisphenol A polycarbonate) (PC) blends [92, 97]. Figure 6.12 shows the pulse sequence used by Linder *et al.* to detect carbon–carbon spin diffusion in this system by two-dimensional NMR. In the absence of exchange, peaks are observed only along the diagonal in the 2-D spectrum. The presence of cross-peak intensity indicates spin exchange between sites with different chemical shifts during the mixing time τ_m. In the example shown in Figure 6.13 of a PET/PC sample heated at 260°C in four steps for 27 min, the arrow marks the cross-peak between the PET labelled methylene carbon and the PC carbonyl carbon, which is evidence for mixing of the two homopolymers at the molecular level [92].

Carbon-13 spin diffusion has also been used to study the local structure of the system PC/di-n-butyl phthalate (DBP) [99]. DBP was enriched at one of the carbonyl sites. PC was studied at natural abundance. The schemes and principles of the pulse sequences are shown in Figure 6.14. The presence of spin diffusion is indicated by an accelerated decay rate of PC ^{13}C magnetization in the DANTE (selective excitation) experiment relative to the decay rate in the simple spin–lattice relaxation experiment. Results thus

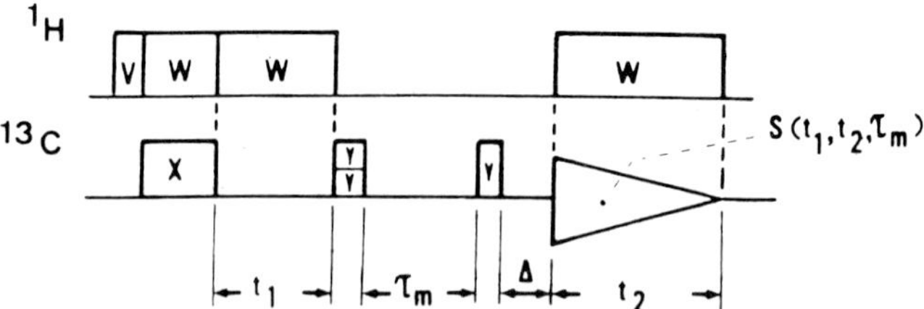

Figure 6.12 Pulse sequence used to detect carbon–carbon spin diffusion. It corresponds to a two-dimensional experiment which involves cross-polarization in the preparation period to generate ^{13}C magnetization. In the evolution period, t_1, each ^{13}C isochromat is labelled with its own resonance frequency. Spin diffusion occurs during the mixing time τ_m. Finally, in the detection period, t_2, the observed signals reflect how the labelled magnetization is redistributed during the mixing period. After [97].

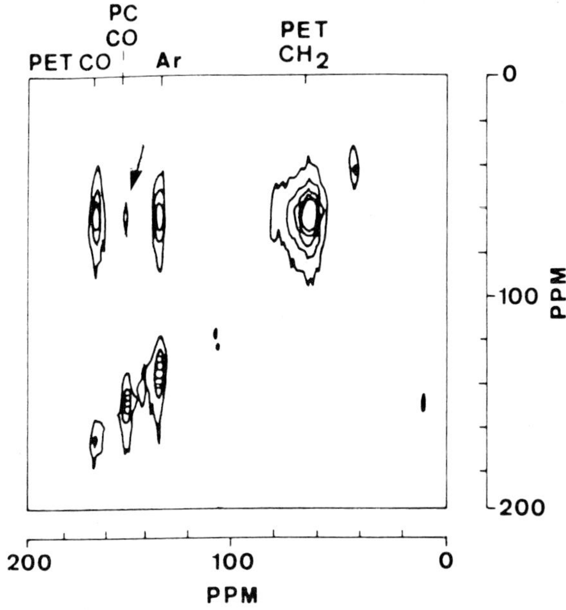

Figure 6.13 2-D ^{13}C spin-exchange spectrum for a PET/PC blend after it was heated in four steps for 27 min. The arrow marks the exchange peak. After [92].

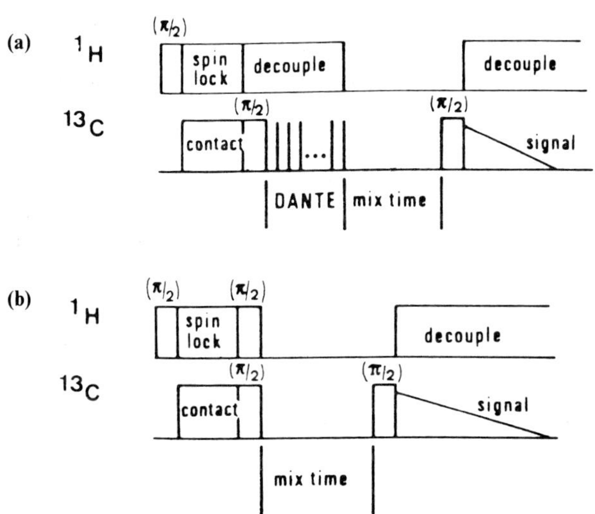

Figure 6.14 (a) Solid-state ^{13}C MAS pulse sequence for the measurement of spin diffusion. The pulse sequence begins with a normal cross-polarization procedure. The first 90° carbon pulse creates longitudinal magnetization and the DANTE sequence selectively inverts the labelled magnetization of the DBP enriched carbon. The mixing time allows for spin diffusion between the inverted peak and the remaining resonances. The last 90° carbon pulse places magnetization in the xy plane for detection. (b) Solid-state ^{13}C MAS pulse sequence for the measurement of spin–lattice relaxation. After [99].

obtained show that the relative spatial positioning of DBP and PC depends on the DBP concentration. At 10 wt%, the carbonyl ester of DBP must be near the PC phenylene groups but removed from the quaternary aliphatic site. At 25 wt%, this intermolecular structural specificity is lost and the carbonyl is near both the phenylene and quaternary aliphatic PC sites [99].

6.5 Conclusion

The large number of examples quoted in this chapter is a clear indication that high-resolution ^{13}C NMR is a very powerful tool for studying local dynamics and molecular organization in bulk polymers. Of course, other experimental techniques can provide information in these fields. However, what is specific to ^{13}C NMR is its high selectivity allied with the natural abundance of carbon-13 nuclei. A typical example, probing the particular interest of this method, is the study of secondary relaxations in polymer systems; dynamic mechanical measurements lead to a description of these phenomena in terms of temperatures and frequencies, which can be conveniently summarized on a relaxation map. However, the precise identification of the motional processes that are involved in the secondary relaxations can only be deduced from a comparison of the mechanical behaviour of different systems. High-resolution solid-state ^{13}C NMR used in the temperature and frequency ranges of interest, which can tell how the linewidths and relaxation times of identified carbons are modified by the motion, is thus able to provide an unambiguous identification of the local modes responsible for these transitions. Similar examples showing the power of the technique could also have been chosen among the numerous studies dealing with local dynamics in bulk polymers at temperatures well above the glass-transition temperature or with solid-state organization, for which the behaviour of the different parts of the molecules in the different phases to which they belong, can be investigated separately. Even more precise information can be obtained from two-dimensional NMR. At the present time, it seems that only deuterium NMR using selectively deuterated samples can compete in these areas with high-resolution ^{13}C NMR.

References

1. J.P. Cohen-Addad and J.P. Faure, *J. Chem. Phys.* **61** (1974) 2440.
2. A.D. English and C.R. Dybowski, *Macromolecules* **17** (1984) 446.
3. A.D. English, *Macromolecules* **18** (1985) 178.
4. C.K. Hall and E. Helfand, *J. Chem. Phys.* **77** (1982) 3275.
5. J.L. Viovy, L. Monnerie and J.C. Brochon, *Macromolecules* **16** (1983) 1845.
6. J.T. Bendler and R. Yaris, *Macromolecules* **11** (1978) 650.
7. A.A. Jones and W.H. Stockmayer, *J. Polym. Sci., Polym. Phys. Ed.* **15** (1975) 847.

8. J.L. Viovy, L. Monnerie and F. Merola, *Macromolecules* **18** (1985) 1130.
9. V. Veissier, Thesis, Université Pierre et Marie Curie, Paris (1987).
10. Y.Y. Lin and A.A. Jones, *J. Polym. Sci. Polym. Phys. Ed.* **22** (1984) 2195.
11. R. Dejean de la Batie, F. Lauprêtre and L. Monnerie, *Macromolecules* **21** (1988) 2045.
12. J.A. Pople and M. Gordon, *J. Am. Chem. Soc.* **89** (1967) 4253.
13. R. Dejean de la Batie, F. Lauprêtre and L. Monnerie, *Macromolecules* **21** (1988) 2052.
14. R. Dejean de la Batie, F. Lauprêtre and L. Monnerie, *Macromolecules* **22** (1989) 122.
15. R. Dejean de la Batie, F. Lauprêtre and L. Monnerie, *Macromolecules* **22** (1989) 2617.
16. U. Buchenau, M. Monkenbusch, M. Stamm, C.F. Majkrzak and N. Nucker, Workshop on Polymers in Dense Systems, Grenoble (1987).
17. O.W. Howarth, *J. Chem. Soc., Faraday Trans. 2* **76** (1980) 1219.
18. F. Lauprêtre, L. Bokobza and L. Monnerie, *Polymer* **34** (1993) 468.
19. M.L. Williams, R.F. Landel and J.D. Ferry, *J. Am. Chem. Soc.* **77** (1955) 3701.
20. J.D. Ferry, in *Viscoelastic Properties of Polymers*, 3rd edition, Wiley, New York (1980).
21. J.P. Faivre, Thesis, Université Pierre et Marie Curie, Paris (1985).
22. D.W. Van Krevelen, *Properties of Polymers. Correlations with Chemical Structure*, Elsevier, Amsterdam (1972).
23. F. Bueche, *J. Chem. Phys.* **20** (1952) 1959.
24. T. Kanaya, K. Kaji and K. Inoue, *Macromolecules* **24** (1991) 1826.
25. J. Schaefer, E.O. Stejskal and R. Buchdahl, *Macromolecules* **10** (1977) 384.
26. J. Herzfeld and A. Berger, *J. Chem. Phys.* **73** (1980) 6021.
27. E.O. Stejskal, J. Schaefer and R.A. McKay, *J. Magn. Reson.* **25** (1977) 569.
28. M. Maricq and J.S. Waugh, *Chem. Phys. Lett.* **47** (1977) 327.
29. J.S. Waugh, M. Maricq and R. Cantor, *J. Magn. Reson.* **29** (1978) 183.
30. A. Bax, N.M. Szeverenyi and G.E. Maciel, *J. Magn. Reson.* **52** (1983) 147.
31. A. Bax, N.M. Szeverenyi and G.E. Maciel, *J. Magn. Reson.* **55** (1983) 494.
32. G.E. Maciel, N.M. Szeverenyi and M. Sardashti, *J. Magn. Reson.* **64** (1985) 365.
33. E. Lippmaa, M. Alla and T. Tuherm, *Proc. 19th Congress Ampere*, Heidelberg (1976).
34. Y. Yarim-Agaev, P.M. Tutunjian and J.S. Waugh, *J. Magn. Reson.* **47** (1982) 51.
35. A. Bax, N.M. Szeverenyi and G.E. Maciel, *J. Magn. Reson.* **51** (1983) 400.
36. T. Terao, T. Fujii, T. Onodera and A. Saika, *Chem. Phys. Lett.* **107** (1984) 145.
37. T. Terao, H. Miura and A. Saika, *J. Chem. Phys.* **85** (1986) 3816.
38. W.P. Aue, D.J. Ruben and R.G. Griffin, *J. Magn. Reson.* **43** (1981) 472.
39. W.P. Aue, D.J. Ruben and R.G. Griffin, *J. Chem. Phys.* **80** (1984) 1729.
40. A.C. Kolbert, D.P. Raleigh, M.H. Levitt and R.G. Griffin, *J. Chem. Phys.* **90** (1989) 679.
41. P. Sergot, F. Lauprêtre, C. Louis and J. Virlet, *Polymer* **22** (1981) 1150.
42. J.F. O'Gara, A.A. Jones, C. Chung and P.T. Inglefield, *Macromolecules* **18** (1985) 1117.
43. A.K. Roy, A.A. Jones and P.T. Inglefield, *Macromolecules* **19** (1986) 1356.
44. H.T. Edzes and J.P.C. Bernards, *J. Am. Chem. Soc.* **106** (1984) 1515.
45. A.P.M. Kentgens, A.F. de Jong, E. de Boer and W.S. Veeman, *Macromolecules* **18** (1985) 1045.
46. A.P.M. Kentgens, E. de Boer and W.S. Veeman, *J. Chem. Phys.* **87** (1987) 6859.
47. S. Swanson, S. Ganapathy, S. Kennedy, P.M. Henrichs and R.G. Bryant, *J. Magn. Reson.* **69** (1986) 531.
48. A.N. Garroway, W.M. Ritchey and W.B. Moniz, *Macromolecules* **15** (1982) 1051.
49. A.N. Garroway, W.B. Moniz and H.A. Resing, *Faraday Symp. Chem. Soc.* **13** (1979) 63.
50. A.N. Garroway, D.L. VanderHart, W.L. Earl, *Philos. Trans. R. Soc. London, Ser. A* **299** (1981) 609.
5.1 E.O. Stejskal, J. Schaefer and T.R. Steger, *Faraday Discuss. Chem. Soc., Faraday Symp.* **13** (1979) 56.
52. J. Schaefer, E.O. Stejskal, T.R. Steger, M.D. Sefcik and R.A. McKay, *Macromolecules* **13** (1980) 1121.
53. D.L. VanderHart and A.N. Garroway, *J. Chem. Phys.* **71** (1979) 2773.
54. J. Schaefer, M.D. Sefcik, E.O. Stejskal and R.A. McKay, *Macromolecules* **17** (1984) 1118.
55. J. Schaefer, M.D. Sefcik, E.O. Stejskal and R.A. McKay, *Macromolecules* **14** (1981) 280.
56. J.R. Garbow and J. Schaefer, *Macromolecules* **20** (1987) 819.
57. J. Schaefer, E.O. Stejskal and R. Buchdahl, *J. Macromol. Sci. Phys. B* **13** (1977) 665.
58. L.C. Dickinson, P.L. Morganelli, W.J. MacKnight and J.C.W. Chien, *Makromol. Chem. Rapid Commun.* **8** (1987) 425.

59. L.C. Dickinson, P. Morganelli, C.W. Chu, Z. Petrovic, W.J. MacKnight and J.C.W. Chien, *Macromolecules* **21** (1988) 338.
60. S. Ganapathy, V.P. Chacko and R.G. Bryant, *Macromolecules* **19** (1986) 1021.
61. J. Schaefer, M.D. Sefcik, E.O. Stejskal, R.A. McKay, W.T. Dixon and R.E. Cais, *Macromolecules* **17** (1984) 1107.
62. T.R. Steger, J. Schaefer, E.O. Stejskal and R.A. McKay, *Macromolecules* **13** (1980) 1127.
63. M.D. Sefcik, J. Schaefer, E.O. Stejskal and R.A. McKay, *Macromolecules* **13** (1980) 1132.
64. F. Lauprêtre, L. Monnerie and J. Virlet, *Macromolecules* **17** (1984) 1397.
65. L.A. Belfiore, P.M. Henrichs, D.J. Massa, N. Zumbulyadis, W.P. Rothwell and S.L. Cooper, *Macromolecules* **16** (1983) 1744.
66. M.D. Sefcik and J. Schaefer, *J. Polym. Sci., Polym. Phys. Ed.* **21** (1983) 1055.
67. M.D. Sefcik, J. Schaefer, F.L. May, D. Raucher and S.M. Dub, *J. Polym. Sci., Polym. Phys. Ed.* **21** (1983) 1041.
68. D.L. VanderHart, W.L. Earl and A.N. Garroway, *J. Magn. Reson.* **44** (1981) 361.
69. D. Suwelack, W.P. Rothwell and J.S. Waugh, *J. Chem. Phys.* **73** (1980) 2559.
70. W.P. Rothwell and J.S. Waugh, *J. Chem. Phys.* **74** (1981) 2721.
71. F. Lauprêtre, J. Virlet and J.P. Bayle, *Macromolecules* **18** (1985) 1846.
72. J. Schaefer, R.A. McKay, E.O. Stejskal and W.T. Dixon, *J. Magn. Reson.* **52** (1983) 123.
73. J. Schaefer, J.R. Garbow, E.O. Stejskal and J.A. Lefelar, *Macromolecules* **20** (1987) 1271.
74. J. Schaefer, E.O. Stejskal, D. Perchak, J. Skolnick and R. Yaris, *Macromolecules* **18** (1985) 368.
75. D. Demco, J. Tegenfeldt and J.S. Waugh, *Phys. Rev. B* **11** (1975) 4133.
76. M. Mehring, in *High Resolution NMR Spectroscopy in Solids*, eds. P. Diehl, E. Fluck and R. Kosfeld, Springer-Verlag, Heidelberg (1976).
77. L. Muller, A. Kumar, T. Baumann and R.R. Ernst, *Phys. Rev. Lett.* **32** (1974) 1402.
78. G.C. Chingas, A.N. Garroway, R.D. Bertrand and W.B. Moniz, *J. Chem. Phys.* **74** (1981) 127.
79. F. Lauprêtre, C. Noël, W.N. Jenkins and G. Williams, *Faraday Discuss. Chem. Soc.* **79** (1985) 191.
80. L.W. Jelinski, J.J. Dumais, P.I. Watnick, A.K. Engel and M.D. Sefcik, *Macromolecules* **16** (1983) 409.
81. A.D. English, *Macromolecules* **17** (1984) 2182.
82. D.E. Axelson, L. Mandelkern, R. Popli and P. Mathieu, *J. Polym. Sci., Polym. Phys. Ed.* **21** (1983) 2319.
83. D.E. Axelson, *J. Polym. Sci., Polym. Phys. Ed.* **20** (1982) 1427.
84. G. Hempel and H. Schneider, *Polym. Bull.* **6** (1981) 7.
85. D.L. VanderHart, *Macromolecules* **12** (1979) 1232.
86. R. Kitamaru, F. Horii and K. Murayama, *Macromolecules* **19** (1986) 636.
87. A. Abragam, *The Principles of Nuclear Magnetism*, Oxford University Press, New York (1961).
88. E.O. Stejskal, J. Schaefer, M.D. Sefcik and R.A. McKay, *Macromolecules* **14** (1981) 275.
89. P. Tekely, F. Lauprêtre and L. Monnerie, *Polymer* **26** (1985) 1081.
90. C.W. Chu, L.C. Dickinson and J.C.W. Chien, *Polym. Bull.* **19** (1988) 165.
91. L.C. Dickinson, H. Yang, C.W. Chu, R.S. Stein and J.C.W. Chien, *Macromolecules* **20** (1987) 1757.
92. P.M. Henrichs, J. Tribone, D.J. Massa and J.M. Hewitt, *Macromolecules* **21** (1988) 1282.
93. N. Parizel, Thèse de l'Université Paris 6 (1989).
94. J.F. Parmer, L.C. Dickinson, J.C.W. Chien and R.S. Porter, *Macromolecules* **22** (1989) 1078.
95. C.M. Sultany, *Polym. Bull.* **20** (1988) 463.
96. M.K. Gupta, J.A. Ripmeester, D.J. Carlsson and D.M. Wiles, *J. Polym. Sci., Polym. Lett. Ed.* **21** (1983) 211.
97. M. Linder, P.M. Henrichs, J.M. Hewitt and D.J. Massa, *J. Chem. Phys.* **82** (1985) 1585.
98. D.L. VanderHart, *J. Magn. Reson.* **72** (1987) 13.
99. A.K. Roy, P.T. Inglefield, J.H. Shibata and A.A. Jones, *Macromolecules* **20** (1987) 1434.

7 Solid-state proton NMR studies of polymers

A.M. KENWRIGHT and B.J. SAY

7.1 Introduction

Solid-state proton NMR provides a way of investigating the motional characteristics of polymers. In particular, it can give information about the motional heterogeneity present in most solid polymer samples, which is of crucial importance in determining physical properties at the macroscopic level. It has the great advantage of requiring no special sample preparation, and can be applied to solid polymers in virtually any physical form (slugs, films, powders, fibres, etc.) as well as to melts.

The criterion we use for deciding whether a sample is sufficiently solid to fall within the remit of this chapter is based on the NMR characteristics of the sample, and is that the form of the free induction decay (FID) following a 90° pulse on a static sample be dominated by the homonuclear dipolar interaction, rather than by chemical shift effects. While this may sound somewhat esoteric, it simply means that if you can see effects in the spectrum due to having more than one type of chemically distinct proton in the sample, or if you have a Lorentzian line (exponential FID) of the sort of width that might be associated with a viscous, but isotropic, liquid (say 100 Hz or less), then you are outside the region covered by this chapter. On the other hand, if only a single line is observable, with a linewidth greater than about 100 Hz, and particularly if the FID deviates from being exponential, then you need to take account of at least some of the considerations given in this chapter. This definition does not have a sharp boundary. It actually takes us at least part way into the melt for polymers having little chemical shift dispersion, such as polyethylene and poly(dimethylsiloxane) (PDMS), but since such systems clearly lie outside the realm of traditional high-resolution NMR, it seems best to include them here.

The observation of ^{1}H and ^{19}F resonance in the solid state was among the first applications of NMR to polymers [1], and provided information on the motional heterogeneity present in semi-crystalline polymers. Subsequent development of the technique and development of NMR relaxation methods have allowed considerable insight into the structure and dynamics of polymer systems in the solid state. It is probably true to say, however, that for many years, the difficulty of obtaining results was surpassed by the difficulty of interpreting them. This was due in large part to the fact that in the classic broad-line ^{1}H NMR technique, the signal recorded is the sum of the signals from all the component regions of the sample, and therefore in order to extract information about the intrinsic properties of a particular region, it is first necessary to establish the correlation between the components of the NMR behaviour and the regions of the sample, which often turns out to be a non-trivial problem. Another problem was that molecular motion in polymers was poorly understood, which rendered theoretical interpretation of the observed results very difficult, and the analysis was further complicated by the fact that the intrinsic relaxation behaviour of the components of the system was often modified by the process of spin diffusion, which has not always been fully appreciated.

A considerable amount of valuable information was nevertheless provided by solid state ^{1}H NMR over many years, and the progress of the technique has been charted by periodic reviews of the area [2–5]. In the last decade however, there have been developments that have significantly increased the information that can be extracted. Among these have been the development of a much greater understanding of the way spin diffusion affects the observed relaxation behaviour in heterogeneous systems, and the use of experimental techniques that allow the manipulation and detection of the magnetisation on the basis of chemical shift (either ^{1}H chemical shift or ^{13}C chemical shift via cross-polarisation). These have effectively added a new dimension to the classic relaxation techniques.

Our intention in this chapter is to provide an overview of the state of the art in solid-state ^{1}H NMR of polymers with particular emphasis on those areas of recent development which we feel are most significant. The work cited reflects this, and does not represent an exhaustive catalogue of the available literature. Our intention throughout has been to attempt to explain the underlying physics in a way that should be accessible to the reader who is not a specialist in solid-state NMR, and for this reason we have occasionally sacrificed theoretical rigour in favour of simplicity of explanation. This should not present a problem since rigorous explanations are available to the interested reader from the original literature and in standard NMR texts.

7.2 Heterogeneity in solid polymers

The vast majority of solid polymers are heterogeneous, in that they contain regions of different molecular mobility, or regions of different chemical composition, or both (as introduced in chapters 5 and 6). Solid polymer systems that are not heterogeneous in some sense are quite rare, and usually not practically useful because of their poor mechanical properties.

The intimate nature of the relationship between this heterogeneity and the mechanical properties of the system is related to the motional characteristics of the different regions present in the sample. The practical implications of this are well known and widely used in, for example, block copolymer systems where the mechanical strength of a polymer with a high glass-transition temperature (T_g) is amalgamated with the impact resistance of a low T_g polymer. However, the observed mechanical properties depend not only on the motional properties of the component regions, but also on their size, shape and spatial distribution. These are all parameters of the system that polymer scientists wish to study.

The question of the molecular mobility of the different regions is usually addressed by measurements such as mechanical relaxation, dielectric loss, or NMR relaxation. While, in general, it is not possible to determine the exact nature of the molecular motion from NMR relaxation data, it nevertheless provides a sensitive probe of rates of molecular motion, and by judicious choice of experiment can be made to cover a wide range of motional frequencies. It is worth emphasising that the classic NMR relaxation experiments are sensitive to the frequencies of molecular motion. Where multiple relaxation components are observed in a sample that is known to be structurally heterogeneous, there is often the temptation to make a simple correlation between the two. This results in people referring to relaxation components as 'the crystalline component' or 'the amorphous component', but it must be borne in mind that they are really referring to 'the less mobile component' or 'the more mobile component'. There may not necessarily be a 1:1 correspondence between regions of molecular mobility and, for example, regions of crystalline order or chemical composition.

The optimum technique for investigating the size and morphology of different regions depends on the size of the regions involved. This can range from greater than a micrometre in the case of long chain block copolymers and blends of immiscible polymers, to a few nanometres or less in some semi-crystalline homopolymer systems and in blends of miscible polymers. Obviously, for the larger scale heterogeneities, optical microscopy and NMR imaging microscopy (see chapter 9) are possibilities, while at smaller scales, electron microscopy can provide useful information if the contrast between the regions can be made suitable. However, for regions with dimensions of the order of 10 nm or less, even electron microscopy begins to run into problems. On the other hand, the direct study of the NMR phenomenon of spin diffusion can yield valuable information about heterogeneity on this scale, and even down to the molecular level. Solid-state ^{1}H NMR is therefore a valuable tool for studying motional heterogeneities in solid polymer samples.

7.3 Underlying theory: the dipolar interaction

The Zeeman interaction of nuclear dipoles with an applied magnetic field (B_0) gives rise to the splitting that is the basis of NMR. The energy difference is proportional to the field and the populations of the energy levels conform to a Boltzmann distribution. The next strongest interaction is the one between the dipoles themselves. Since the applied field is usually several orders of magnitude larger than the dipolar field of a nuclear spin, it is appropriate to treat the dipole–dipole interaction as a perturbation of the Zeeman interaction.

Classically, the energy of interaction between two point dipoles μ_1 and μ_2 separated by a distance r is given by

$$E_D = \frac{\mu_0}{4\pi}\left[\frac{\mu_1 \cdot \mu_2}{r^3} - \frac{3(\mu_1 \cdot \mathrm{r})(\mu_2 \cdot \mathrm{r})}{\mathrm{r}^5}\right] \tag{7.1}$$

where μ_0 is the permeability of free space.

To obtain the equivalent quantum mechanical expression for the interaction between the dipole moments of two nuclei (i.e. the dipolar Hamiltonian), it is merely necessary to substitute the dipole moment operator, $\hat{\mu}$, for the classical quantity μ_i and then further substitute according to the equation

$$\hat{\mu} = \gamma \hat{I} \hbar \tag{7.2}$$

where γ is the magnetogyric ratio of the nucleus and $\hat{I}$ is the spin angular momentum operator of the nucleus.

After substitution and rearrangement, this gives [6]

$$\hat{\mathscr{H}}_D = \frac{\mu_0}{4\pi}\gamma^2\hbar^2\left[\frac{\hat{I}_1 \cdot \hat{I}_2}{r^3} - \frac{3(\hat{I}_1 \cdot r)(\hat{I}_2 \cdot r)}{r^5}\right] \tag{7.3}$$

which can be expressed as

$$\hat{\mathscr{H}}_D = \frac{\mu_0}{4\pi}\frac{\gamma^2\hbar^2}{r^3}[A + B + C + D + E + F] \tag{7.4}$$

where

$$A = \hat{I}_{1z} \cdot \hat{I}_{2z}(1 - 3\cos^2\theta)$$
$$B = -\tfrac{1}{4}(\hat{I}_{1+} \cdot \hat{I}_{2-} + \hat{I}_{1-} \cdot \hat{I}_{2+})(1 - 3\cos^2\theta)$$
$$C = -\tfrac{3}{2}(\hat{I}_{1z} \cdot \hat{I}_{2+} + \hat{I}_{1+} \cdot \hat{I}_{2z})\sin\theta\cos\theta\exp(-i\phi)$$
$$D = -\tfrac{3}{2}(\hat{I}_{1z} \cdot \hat{I}_{2-} + \hat{I}_{1-} \cdot \hat{I}_{2z})\sin\theta\cos\theta\exp(i\phi)$$
$$E = -\tfrac{3}{4}\hat{I}_{1+} \cdot \hat{I}_{2+}\sin^2\theta\exp(-2i\phi)$$
$$F = -\tfrac{3}{4}\hat{I}_{1-} \cdot \hat{I}_{2-}\sin^2\theta\exp(2i\phi)$$

In the above expressions $\hat{I}_+ = \hat{I}_x + i\hat{I}_y$, $\hat{I}_- = \hat{I}_x - i\hat{I}_y$, and z is taken to be the reference direction, coincident with the direction of the applied magnetic field, B_0. θ is the angle between the inter-nuclear vector and B_0, and ϕ is a phase angle perpendicular to B_0. The terms A to F are often referred to as the 'dipolar alphabet'.

The A term can be visualised as a field shift at a nucleus. The B term couples spins of opposite polarisation, enabling them to exchange polarisation states ('flip-flop' transitions). For homonuclear spin pairs, neither of these change the overall polarisation of the system and hence are energy conserving, forming the 'static' or 'secular' part of the dipolar Hamiltonian. The remaining terms, the 'non-secular' part, involve a change in the polarisation and therefore are not energy-conserving. They are only 'allowed' in the presence of suitable fluctuations in the magnetic field to compensate for the energy difference.

7.3.1 *The static part*

The lineshape of a dipolar-coupled system is governed by the A and B terms of the dipolar Hamiltonian. In order to compute the behaviour of the spin system, it is necessary to consider all the spins in it (typically Avogadro's number), since the total dipolar Hamiltonian is the sum of all the pairwise interactions. While the response of isolated systems of two or three spins is analytically soluble, the general case is not because for more than three spins the A and B terms do not commute. This can be thought of in terms of the FID following a pulse, where the precession frequency of a nucleus will depend on the spin states of its neighbours. Within the timescale of the FID, these neighbours will participate in flip-flop transitions with other spins thereby 'scrambling' the local field.

It was to surmount this problem of the inability to calculate the effects of the dipolar Hamiltonian that Van Vleck [7] proposed the method of moments. We can define the nth moment (M_n) of a resonance absorption as

$$M_n = \int I_\omega(\omega - \omega_0)^n \, \mathrm{d}\omega \tag{7.5}$$

where the reference frequency ω_0 is chosen so that $M_1 = 0$.

Effectively, we divide the lineshape into vertical strips and take the integral, weighting each strip by the nth power of its deviation from the centre of the line. For a symmetric line, the odd moments are therefore all 0. The signal, $S_{(t)}$, following a pulse can then be defined in terms of the moments [8]

$$S_{(t)} = S_{(0)}\left[1 - M_2\left(\frac{t^2}{2!}\right) + M_4\left(\frac{t^4}{4!}\right) - M_6\left(\frac{t^6}{6!}\right) + \cdots\right] \tag{7.6}$$

The signal at $t = 0$, $S_{(0)}$, is proportional to the integral of the whole spectrum, M_0.

This series converges slowly, which implies that the moments of the system must be known to high order if the FID and lineshape is to be completely determined. Van Vleck gives the method for calculating the moments from the structure of a rigid solid, using traces of commutators that reduce to a series of sums over internuclear vectors. The second moment involves summing the square of an interaction term for every pair of spins. The fourth moment involves the square of that sum, the sum of the fourth power of the interaction terms, and a triple sum over every possible combination of three spins. Successive moments become even more complex. Adding to this the necessity of taking into account molecular reorientations (such as methyl group rotation) and the 'powder averaging' necessary for non-oriented systems, it is clear that the dipolar lineshape remains, in effect, incalculable.

However, since the dipolar interaction is a through space effect and thus independent of bonding (unlike J coupling), given a crystal structure for a material it is, in principle, possible to calculate the lower moments, particularly the second, taking into account the averaging effects of molecular motions. When these are fast compared to the lifetime of the magnetic coherence (say 10 μs) then the spins involved in that motion can be regarded as being static at the mean position. This calculated second moment can then be compared with experiment. While the second moment is not usually sufficiently sensitive to allow discrimination between possible structures, it is often capable of revealing the kinds of molecular motions present.

If we were to take a more pragmatic approach to the problem of the NMR lineshape, we could consider that the probability of a neighbouring spin lying parallel or antiparallel to the applied field is effectively equal. Then, considering a sphere of nearest neighbours, there would be a roughly binomial distribution of fields at a given nucleus. This is quite close to a Gaussian distribution which is the usual first approximation to a solid lineshape. This simple picture ignores the B term, which would lead to an underestimation of the second moment by a factor of 4/9 [9], emphasising that the B term contributes to the linewidth. Abragam [9] discusses the problem of the mathematical representation of the FID shape in a rigid solid for the case of calcium fluoride, and concludes that the use of a Gaussian-broadened sync function whose parameters are adjusted to give the second and fourth moments calculated from the known structure, is a valid representation of the FID.

In liquids, where molecular motion is rapid and isotropic, the effects of the static part of the dipolar Hamiltonian are averaged to zero, giving rise to an exponential FID and hence a Lorentzian lineshape. In this case, as introduced in chapter 1, we can define a transverse relaxation time, T_2, by

$$M_{(t)} = M_{(0)} \exp\left(\frac{-t}{T_2}\right) \tag{7.7}$$

However, in a rigid solid, the definition of a transverse relaxation time is more difficult since the shape of the decay is more complex. T_2 is therefore often taken as a 'characteristic' time of the FID, usually the time taken to decay to 1/e of its initial value, and as such can be used as an effective lifetime for the transverse magnetisation. If we consider the effects of increasing the temperature of a solid sample, we find that as various motions come into play, the second moment decreases and the FID lengthens. Eventually there comes a point where there is no residual static interaction on the timescale of the

FID. The FID is then governed by the zero frequency component of the fluctuations in the local field. This is the case with rubbers and polymer melts.

7.3.2 Suppressing the dipolar interaction

In recent years much effort has been applied to removing the static part of the dipolar interaction in solids, involving manipulation of either the angular momentum operators or the trigonometric terms of the dipolar alphabet, or both. The trigonometric terms can be changed by macroscopic sample rotation, which causes the internuclear vectors to appear to be aligned, on average, along the axis of rotation. If we chose the axis so that the angle, θ, which it makes with the applied field fulfils the condition $(1 - 3\cos^2\theta) = 0$, then the A and B terms of the dipolar alphabet become zero [10]. This angle of 54.7° ('the magic angle') is the angle that the diagonal of a cube makes with each of its edges, so rotation about this axis effectively interchanges the three Cartesian axes, averaging the Hamiltonian in a way similar to isotropic molecular tumbling (see chapter 5). The drawback with this approach is that it is not practically possible to achieve spinning speeds sufficiently high to eliminate the dipolar interaction in rigid proton-rich solids. Manipulation of the angular momentum operators is carried out by the application of specific sequences of RF pulses which average the dipolar interaction to zero. Difficulties are encountered both with the finite width of pulses and with imperfections in the pulses. The effect of these sequences can be explained to first-order as rapidly cycling the alignment of the magnetisation between the x, y and z axes for equal lengths of time, thus achieving a time average related to that described previously in the case of magic-angle spinning [11]. Unfortunately, while multiple pulse techniques can eliminate the dipolar interactions, they cannot eliminate all the other interactions that would be averaged by magic-angle spinning, such as the chemical shift anisotropy. For this reason the two techniques need to be used in combination if 'high-resolution' spectra of abundant spins in solids are to be obtained, as discussed in section 7.10.

7.3.3 The non-static part

The establishment or restoration of thermal equilibrium in a spin system requires transitions between the energy levels of the system. The probability of these transitions arising spontaneously is very low, and stimulated transitions only arise as a result of fluctuations in the magnetic field. There are a variety of interactions that can give rise to such fluctuations but, for ^1H in an insulating organic solid that does not contain paramagnetic centres, the dominant mechanism is almost certain to be fluctuations in the homonuclear dipolar interactions due to molecular motion. The remaining terms of the dipolar Hamiltonian (C to F) describe the effects of such fluctuating fields.

At thermal equilibrium, the excess of spins in the lower energy ('parallel') state is given by

$$n = N\left(1 - \exp\left(\frac{-\gamma\hbar B_0}{kT}\right)\right) \tag{7.8}$$

where n is the population difference and N is the population of the low energy state.

It should be noted that any population difference can be described using this equation by changing the temperature term. Since the spin system is only weakly coupled to the other degrees of freedom of the system ('the lattice'), it is often valid to describe population differences between the energy levels by a 'spin temperature' which may be different from the lattice temperature. The utility of this description will become apparent in later sections.

It can be shown [9] that, following a disturbance of these populations, the approach to equilibrium is essentially exponential, that is it follows first-order kinetics and can be expressed as

$$(M_{(e)} - M_{(t)}) = (M_{(e)} - M_{(0)})\exp\left(\frac{-t}{T_1}\right) \tag{7.9}$$

where $M_{(e)}$ is the equilibrium magnetisation, $M_{(0)}$ is the magnetisation at some initial time arbitrarily designated as 0, $M_{(t)}$ is the magnetisation at some subsequent time t and T_1 is a characteristic time constant.

The relaxation time constant, T_1, for a given kind of spin is related to the motional characteristics of the system by

$$\frac{1}{T_1} = \frac{9}{4}\gamma^4\hbar^2\sum_k\{J_k^1(\omega_0) + J_k^2(2\omega_0)\} \tag{7.10}$$

where the $J_k^n(n\omega_0)$ are the spectral density functions of the relative motions of another spin, k, at the frequencies ω_0 (the resonance frequency) and $2\omega_0$. These depend on both the geometry of the molecular motions and their correlation functions, as described in chapters 4 and 5.

Thus, the investigation of T_1 relaxation provides a means of probing the spectral density of motions in a sample, and hence the correlation function. Whereas in a liquid the relaxation times of chemically distinct protons in the system can differ, in the solid state this is generally not the case, because the energy conserving flip-flops make it impossible for one chemical group in a rigid molecule to relax independently of its immediate neighbours. This implies that the intrinsic relaxation properties of a motionally homogeneous phase will be uniform.

The motional frequencies probed by T_1 measurements are in the region of the resonance frequency which will generally be of the order of 5–500 MHz. There are three ways of extending the measurements to lower frequencies. One makes use of 'cross-relaxation' in the presence of ^{19}F nuclei and is discussed briefly in section 7.3.5. The second, the field-cycling method of Voigt and Kimmich [12], can generate a relaxation dispersion curve that plots the relaxation time as a function of the relaxation field and is capable of investigating motions in the range 2 kHz to 20 MHz. This is the widest range of any NMR relaxation method and, as the dispersion curve is directly related to the spectral density functions of the motions occurring, it should in principle offer a means of testing models of molecular motion and autocorrelation functions. However, the technique involves the use of different magnetic fields for the relaxation period and detection of the signal, requiring equipment that is not generally available. The third and most widely applied method of extending measurements to lower frequencies is the use of 'spin-locking'. This involves aligning the magnetisation along the magnetic vector of a RF field that is larger than the local fields experienced by the spins. The spins will evolve in relation to this field in a fashion that is analogous to their evolution in the static field in the absence of radiation, but since the RF field will be orders of magnitude smaller than the static field, the frequencies studied will be correspondingly reduced, typical to 20–200 kHz. The spin-locking is achieved by following a 90° pulse with radiation that is phase shifted by 90°, and of sufficient duration for the spins to establish a spin temperature in this environment (Figure 7.1).

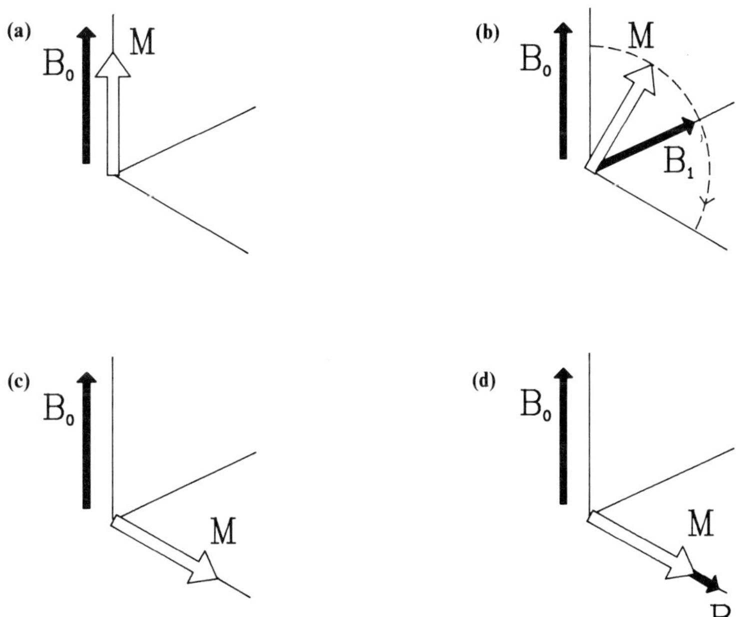

Figure 7.1 Spin-locking showing the applied field B_0, the magnetisation M and the RF field B_1. (a) At thermal equilibrium, (b) during the initial pulse, (c) following the initial pulse and (d) spin-locked.

At this point, we will have a spin-locked magnetisation that exceeds the equilibrium magnetisation in the spin-lock field by a factor of about 1000. If the radiation is maintained, this magnetisation will decay to its equilibrium value (virtually zero). In a homogeneous system, this decay will also be exponential, with a time constant $T_{1\rho}$[13]. When the spin-locking field is removed, any remaining magnetisation will generate an FID.

7.3.4 *Relaxation in heterogeneous systems*

In heterogeneous systems, the spin–lattice relaxation behaviour (T_1 and/or $T_{1\rho}$) is often found not to conform to a simple single exponential, but rather to exhibit behaviour that can be represented as a superposition of exponential processes. If we consider the simple case of a mixture of several discrete homogeneous components, each conforming to exponential relaxation as previously described, then it follows that the overall relaxation behaviour of the mixture can be described by

$$(M_{(\mathrm{e})} - M_{(t)}) = (M_{(\mathrm{e})} - M_{(0)}) \sum_i P_i \exp\left(\frac{-t}{T_i}\right) \qquad (7.11)$$

where P_i is the proportion of component i and T_i is the time constant of component i.

Analysis of behaviour of this kind can, in principle, yield information about the constituent components. However, this type of relaxation behaviour is often observed in solid polymer systems where we do not have mixtures of discrete components, but rather have chemical or structural heterogeneity on a microscopic level. In these cases, naive analysis of relaxation data in terms of constituent components as just described can be grossly misleading because of the phenomenon of 'spin diffusion'. It has been shown earlier in this section that the homonuclear dipolar Hamiltonian contains a term that allows energy conserving simultaneous radiationless transitions for a pair of adjacent nuclei (term B in the dipolar alphabet, also known as the 'flip-flop' term). This interaction is the mechanism by which spin diffusion occurs since, if different states of polarisation occur in different parts of a system, the whole of which is strongly coupled by homonuclear dipolar interactions, the 'flip-flop' interaction allows transportation of polarisation from one part of the system to another, tending to even out the polarisation (or 'spin temperature') throughout the system. It can readily be shown that under the typical conditions of, say, the proton lattice of an organic solid, this transportation process takes the form of a diffusion equation. For diffusion in one dimension [9], the equation may be written

$$\frac{\mathrm{d}M}{\mathrm{d}t} = Wa^2 \frac{\mathrm{d}^2 M}{\mathrm{d}x^2} \qquad (7.12)$$

where M is the polarisation at position x, W is the 'flip-flop' transition probability, a is the inter-nuclear distance between neighbours on the lattice and x is the positional vector. The quantity Wa^2 is often given the symbol D, and referred to as the spin-diffusion coefficient. Wolf [14] shows that for a crystal lattice, the time taken to cover nearest-neighbour distances is of the order of T_2. It is important to note that what is diffusing is magnetic polarisation, not the nuclei.

For a tightly coupled lattice in a heterogeneous system where different relaxation rates in different parts of the system would produce polarisation gradients during relaxation, spin diffusion may greatly modify the observed behaviour. This will depend not only on the intrinsic relaxation properties of the various regions, but also on their sizes, their spatial distribution and morphology, and on the rates of spin diffusion within the various regions, which is obviously a much more complicated situation than the simple superposition of intrinsic relaxation properties considered above. The general equation that governs the behaviour of the magnetic polarisation within a given region of such a system has the form

$$\frac{\mathrm{d}M_{(x,t)}}{\mathrm{d}t} = \frac{\mathrm{d}}{\mathrm{d}x}\left[D_i \frac{\mathrm{d}M_{(x,t)}}{\mathrm{d}x}\right] + \frac{(M_{(\mathrm{e})} - M_{(x,t)})}{T_i} \qquad (7.13)$$

where D_i is the effective spin diffusion coefficient in region i, $M_{(\mathrm{e})}$ is the appropriate equilibrium value of the magnetisation and T_i is the appropriate relaxation time in region i. It can be shown that the solution of this type of equation is a superposition of exponential processes [15]. Behaviour of this type is discussed in more detail in subsequent sections, and particularly in section 7.8.

7.3.5 Heteronuclear effects

Before leaving the dipolar Hamiltonian, we should give brief consideration to systems where there is a significant population of non-resonant spins. In the case of spins such as ^{13}C at natural abundance, the low natural abundance and relatively small magnetogyric ratio make the heat capacity of the spin system negligible, so that the effects are undetectable. On the other hand, in the case of partially fluorinated polymers, the non-resonant ^{19}F spins are not negligible, and the possibility of cross-relaxation adds another dimension to the problem. In the case of T_1 relaxation, multi-exponential behaviour is expected even in a motionally homogeneous system. The relaxation rates depend on the intrinsic resonance frequencies of the nuclei, and the sum and difference frequencies. It is this last that is of particular interest since it offers a means of investigating motions at a few megaHertz, a frequency otherwise only achievable at very low field, where sensitivity would be a problem [16, 17], or by the use of field-cycling methods [12].

7.4 Experimental methods

NMR can be detected either by continuous wave or pulse methods and, if carried out under appropriate conditions, these two are Fourier complements of one another. Although, in principle, any experiment can be carried out in either mode, pulse methods have several advantages and virtually all measurements are now done this way. One major advantage is the ability to create a known non-equilibrium state at a given time and then observe the evolution of the system. However, it must be stressed that where a series of measurements are made, as in a relaxation experiment, it is important to ensure that exactly the same state is achieved before each measurement. The simplest state to generate is that of thermal equilibrium which is achieved by leaving the sample in the magnetic field for a time substantially longer than the longest T_1 of the sample (typically three, five or even ten times T_1).

7.4.1 FID/lineshape measurement

The simplest measurement available is the response of the system to a pulse at or near the resonance frequency. Following the pulse, there is a signal that decays with a characteristic shape and time constant, the FID. Fourier transformation of this signal yields the spectrum. The information content of the two presentations is identical; the difference lies in our ability to extract information from the data.

In solution-state NMR or high-resolution solid-state NMR, great attention is paid to the homogeneity of the magnetic field lest the decay of the FID be dominated by this rather than the intrinsic properties of the sample. In most of the experiments discussed here, this is rarely a problem. The required homogeneity is not great except for line-narrowing experiments and the most mobile samples. A greater problem is the 'dead-time' which is the time taken for the spectrometer to recover from the overload caused by the RF pulse. This leads to a lack of information about the early part of the FID and is more important in solids than in liquids because the FID is generally shorter. When the FID is transformed, the effects of this problem are spread throughout the resulting spectrum, so there are distinct advantages in analysing the FID. The initial amplitude of the FID is proportional to the total z-magnetisation before the pulse and so is equal to the integral of the spectrum. The amplitude of the signal measured as soon as possible after the pulse is often used as a single point measure of the total magnetisation of the system. In order to properly characterise the entire proton FID of a rigid solid, digitisation rates as high as 1 MHz are necessary.

One technique that is often used to alleviate the effects of receiver recovery is the solid echo sequence (Figure 7.2(a)) [18], where a second 90° pulse (phase shifted by 90° with respect to the first) is applied to the system after an interval that is slightly longer than the dead time. This generates an echo

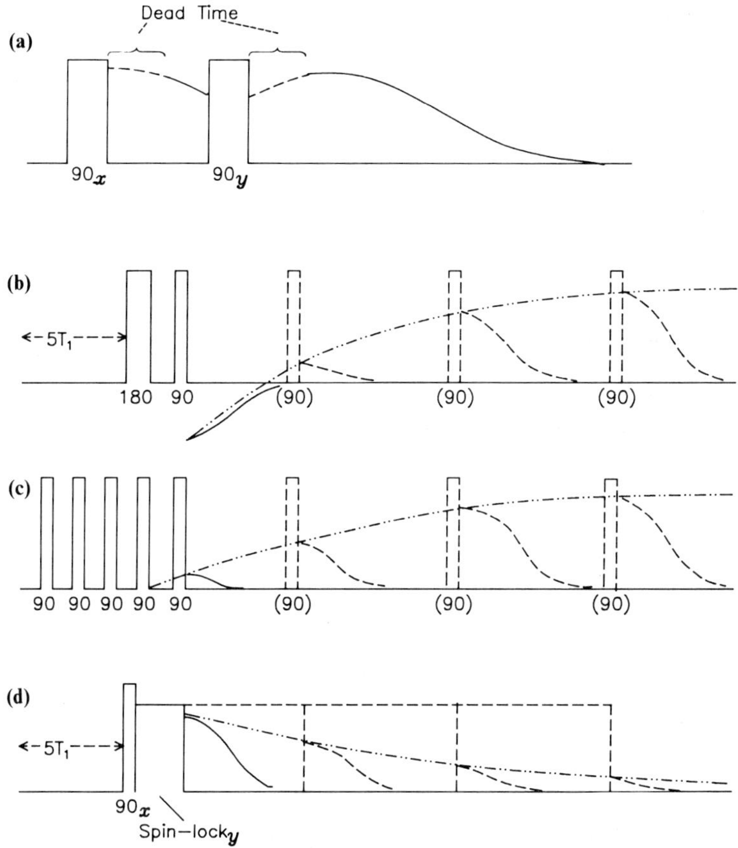

Figure 7.2 Pulse sequences: (a) the solid echo, (b) T_1 by inversion recovery, (c) T_1 by saturation recovery, (d) $T_{1\rho}$. In (b) and (c), the sampling pulse is positioned at increasing time intervals to record each data point in the relaxation curve. In (d), the duration of the spin-lock pulse is increased for each data point.

which, for sufficiently short intervals, would retain the shape of the FID, and for slightly longer intervals only distorts the shape by a small amount. The amplitude of the echo is generally a reasonable approximation to the initial amplitude of the FID, but will always discriminate against the most rigid parts of the system.

Other methods of data acquisition involve a driven element. In the case of very mobile samples where the T_2 is sufficiently long that the effects of field inhomogeneity can be observed, it may be necessary to use spin echo techniques to refocus dephased magnetisation. This can either take the form of a single pulse applied at a variable time after the initial pulse or a string of 180° pulses. There is also the possibility of using multiple pulse techniques to extend the length of the decay by removing parts of the dipolar Hamiltonian.

Since all these methods involve the use of RF radiation to modify the shape of the signal, the possible distortion of that signal by imperfections in the experiment must be borne in mind.

7.4.2 *Longitudinal relaxation,* T_1

Knowledge of T_1 is essential if production of a state of thermal equilibrium is required. Various schemes have been proposed to measure T_1 rapidly, but these are always approximate and usually unable to detect any complex, multi-exponential behaviour. The two accurate methods in common use are inversion recovery and saturation recovery. Both of these techniques involve the establishment of a known non-equilibrium population state, evolution, and sampling of the magnetisation as a function of the evolution time.

Inversion recovery (Figure 7.2(b)) needs to start with the sample at thermal equilibrium. A 180° pulse inverts the populations and, after some time, a 90° sampling pulse 'measures' the z-magnetisation. The sample must then be allowed to return to equilibrium before another measurement can be made. This is undoubtedly the most reliable method of T_1 determination but suffers from the drawback that an approximate value for T_1 must be known beforehand, in order to establish the required time between measurements. 180° and 90° pulses are normally used in order to maximise the dynamic range of the observed relaxation behaviour, but the measured relaxation times would be the same for virtually any pair of pulses.

Saturation recovery (Figure 7.2(c)) can be a more rapid method of measuring T_1 providing a condition of saturation can be reliably achieved. In the case of a rigid solid where $T_1 \gg T_2$, a simple method exists. The repeated application of 90° pulses at intervals greater than T_2 but much less than T_1 allows the preparation of a system with no polarisation. The system is allowed to evolve and the z-magnetisation sampled. Since the measurement does not start from thermal equilibrium, the saturation sequence for the next measurement can be applied immediately.

7.4.3 *Relaxation in the rotating frame,* $T_{1\rho}$

Measurement of $T_{1\rho}$ requires the preparation of a spin-locked spin system as described earlier. The magnetisation is measured as a function of the length of the spin-locking pulse (Figure 7.2(d)). The amplitude of this field can be varied, which to some extent allows selection of the frequency of motions affecting the relaxation. The spin diffusion coefficients are halved for a spin-locking field at right angles to the applied field [19]. If the RF field is not large in comparison with the local field, only a fraction of the magnetisation will be spin-locked, the remainder being stored as dipolar order. If the radiation is off-resonance the effective spin-locking field will not be perpendicular to the static field. By varying the off-resonance condition, it is possible to spin-lock at almost any angle. The $T_{1\rho}$ and the spin diffusion coefficient will vary with the angle, and in the case where the angle between the applied field and the effective spin-locking field is the magic angle, spin diffusion would be suppressed to first-order [19, 20].

All the above techniques can be combined with any of the acquisition methods discussed earlier (FID or single point detection). This allows the possibility of a full two-dimensional analysis of the data, which has been suggested by other authors but not carried out in practice [5].

7.4.4 *Experimental details*

Since we are not usually interested in chemical distinctions between protons, a low magnetic field can be useful as this avoids intrusive chemical shift effects. A further advantage of low field is that the correspondingly low frequency of operation reduces sample heating due to dielectric loss. This is a particularly insidious effect which can go almost unnoticed. If the problem were due to conduction or radiation of heat from the coil, it could be remedied by an adequate gas flow to cool the surface of the sample, thus preventing any ingress of heat. However, since the problem arises from absorption from the electric component of the RF field, it will occur uniformly throughout the sample. The maintenance of thermal equilibrium is then determined by the flow of heat from the interior of the sample to the exterior, where it can be removed by the cooling gas. This problem is particularly significant in the case of $T_{1\rho}$ where the sample is being irradiated during the relaxation period. There is no absolute cure, but low field will help and small samples intimately connected with the cooling stream will reduce the problem.

The quality factor (Q) of the probe is also of great importance in such measurements. This is a factor that governs many aspects of a tuned circuit. A high Q probe circuit will have a low bandwidth, good sensitivity, efficient use of RF power and long dead-time. In order to obtain faithful reproduction of the early part of the FID, low Q systems are necessary, but this degrades the signal-to-noise ratio and requires higher RF powers, thus increasing the sample heating problems. A compromise must be reached.

The pulse power required is not necessarily enormous. A 90° pulse of 2 μs ($v = 125$ kHz) will usually be adequate although pulses down to 1 μs are often used. Short dead-times are essential, but the definition of dead-time is rather flexible, involving a subjective measure of the amount of time that must elapse following a pulse before the signal is unaffected by that pulse. The effects of the pulse are various, possibly involving gain changes in the preamplifier, transient phase changes, and transients added to the signal. All of these will die away following the pulse, each with its own time constant. The decays are generally exponential, so the length of time for which these are significant actually depends on the amplitude of the signal being observed and the signal-to-noise ratio.

7.4.5 *Treatment of results*

At all but the simplest level, treatment of the results from a time-domain experiment involves some mathematical procedure such as non-linear least squares analysis. Least squares analysis is generally carried out by some modification of the Newton–Raphson method, that proposed by Marquardt currently being popular [21, 22]. There is a fundamental difficulty in that the 'normal equations' that must be solved as part of the procedure are often ill-conditioned. This means that rather than having a single well-defined solution, there is a group of solutions all of which are equally valid. This is particularly troublesome where there are exponential components whose time constants differ by less than a factor of about three. It is easy to demonstrate that the behaviour is multi-exponential, but much more difficult to extract reliable parameters. The fitting procedure is also dependent on the model used and it is often quite difficult to determine the number of exponentials needed to adequately represent the data. Various procedures have been suggested to overcome these difficulties, but none has yet received wide acceptance in solid-state NMR [23–26].

7.5 Lineshape/FID analysis of polymers

The analysis of lineshapes from solid polymer systems is possibly the oldest application of NMR to polymers, revealing information about the motions present in the sample without the complication of spin diffusion. It does have the drawback, however, that the mathematical form of the functions necessary to represent the data is, in general, not known from theory, and therefore the 'curve fitting' part of data analysis is likely to be more difficult than in the case of relaxation measurements such as T_1 and $T_{1\rho}$ where the decay will be a superposition of exponentials. Despite these difficulties, the technique is still widely used in its own right as a means of characterising motionally distinct components. It is also used as a 'fingerprint' method for the characterisation of samples in applications such as industrial plant control, and implicitly in other techniques such as the Goldman–Shen experiment or the 'delayed contact' ^{13}C CP/MAS experiment (cross-polarisation/magic-angle spinning).

In the following section, we consider at some length the application of the technique to semi-crystalline homopolymers, including some discussion of experimental difficulties and the problems of finding suitable functions to represent the data. In subsequent section, we discuss in less detail the application of the technique to other classes of polymer.

7.5.1 *Semi-crystalline homopolymers*

The observation of the broad line continuous wave (CW) ^{1}H spectrum of polyethylene (PE), and ^{19}F spectrum of polytetrafluoroethylene (PTFE) by

Wilson and Pake in 1953 [1] was one of the first attempts to study polymers by NMR. They observed that the line shapes of both PE and PTFE were inhomogeneous, and tentatively assigned this is to a broad component due to crystalline material, and a narrower component due to amorphous material. This stimulated a number of subsequent studies of semi-crystalline homopolymers using CW broad line techniques. It was soon recognised in the case of homopolymers that decomposition of the line into two components does not show consistent correlations with physical properties. In particular, if the broad component is assumed to arise from crystalline material, the values obtained for percentage crystallinity are strongly temperature-dependent, and do not agree with the values measured by other techniques. It was suggested [27] that more than two components might be needed to adequately describe the phase behaviour of the system. Decomposition into three components was subsequently widely used as a method for the analysis of such lines, and was particularly developed by Bergmann [28, 29]. An important part of the development of this technique was the introduction of suitable corrections for the distortions introduced by modulation of the RF (an essential part of the experiment). These are discussed by Kitamaru *et al.* [30]. Although analysis in terms of three components gave better agreement with the crystallinity measured by other techniques, there was still considerable scope for debate, with different authors favouring different functions to represent the lineshapes, while others claimed that four components were necessary to characterise the lineshape in certain samples [31, 32]. This state of uncertainty is understandable bearing in mind the following points.

Firstly, it is not possible to find theoretical expressions for the ^{1}H NMR lineshape for large, tightly coupled spin systems. The situation is further complicated by the introduction of motion that is sufficiently fast to begin averaging the homonuclear dipolar interactions, but not fast enough to average them completely. This is especially true if the motion is anisotropic as is commonly the case for the main chain motion of polymers. Despite some early theoretical work on the effects of incomplete averaging of the dipolar interaction by motion [33], and some more recent work on expressions for the ^{1}H lineshape in rigid solids [34] and polymer melts [35], it has generally been necessary to resort to purely phenomenological descriptions of the lineshape in such systems.

Secondly, there is no firm basis for choosing the number of components to represent the lineshape. The observed lineshape is determined by the distribution of spins in the sample and their mobility, which may well show more complex behaviour than would be suggested from consideration of factors such as crystallinity, morphology, and microphase separation. In this regard, the investigator can only be guided by 'Occam's razor' to take the least number of components that gives an adequate representation of the data, and by whether the behaviour of the components thus obtained correlates well with other properties of the sample.

As NMR hardware and computer technology have progressed, CW NMR has been almost totally superseded by pulse methods. The main reasons for this change are the advantages of signal averaging associated with the pulse technique, the time-saving in acquisition of a single transient, and the ability to manipulate spins using sequences of pulses. While these advantages may not be as great in the case of broad line ^{1}H measurements as for some other techniques, it is nevertheless true that very few laboratories still have the instrumental capacity to perform wide line CW experiments. Nevertheless, where such instruments do still exist, they are capable of providing valuable information [36].

Since the lineshape is the Fourier transform of the FID, the information content of the two domains must be identical, and it should therefore be possible to analyse the on-resonance FID in a way analogous to the decomposition of the lineshape to yield equivalent information on molecular mobility. There are, however, some problems in making a direct comparison between the two sets of results.

Firstly, the main source of instrumental distortion of the signal is quite different in the two domains. In the CW method, the signal is broadened due to the modulation of the applied RF, and corrections must be made for the effects of this broadening on the different components [30]. In the pulse method, the main distortion of the signal is due to the dead-time. Efforts may be made to minimise the dead-time by optimisation of instrumental characteristics, or in the case of rigid solids, some of the problem may be overcome by the application of suitable 'echo' sequences [18, 37], but it cannot be totally eliminated.

Secondly, it has not always been possible to use corresponding functions in the analyses in the two domains. For some of the functions used to represent lineshape components in the frequency domain, such as the Gaussian and Lorentzian, the corresponding FID function is known analytically, and can therefore be used in fitting FIDs. This is not the case for some functions that have been used to represent lineshape components believed to be intermediate in character between a Gaussian and a Lorentzian. This can make comparison of the results problematic. However, since none of these functions has any theoretical basis, there is no reason to prefer functions that are tractable in the frequency domain over those that are tractable in the time domain.

In the field of semi-crystalline homopolymers, an early attempt at quantitative FID analysis on polypropylene by Kluver and Ruland [38] met with only limited success largely because they attempted to represent all except the crystalline component as a sum of exponentials, even though it was already known from previous CW work that this representation was inappropriate. Later work by Bergmann [39, 40] also had problems in finding suitable functions to represent the FID, especially the component that is generally thought to be in some way intermediate between a Gaussian and an exponential. Bergmann suggested decomposition of the FID into just two components, the first a Gaussian multiplied by a sync function (as suggested by Abragam [9] for the FID of CaF_2 and hence referred to as the Abragam function) to represent the signal from rigid regions, and the second a Gaussian distribution of Kubo–Tomita [33] functions to represent the signals from more mobile regions. This not only goes against much of his earlier CW work, but also introduces the additional problem of trying to explain and justify the choice of distribution function.

The question of which functions should be used to represent the components of the FID is one that remains largely unresolved. For approximate evaluations, the representation of the FID as a sum of Gaussian and exponential decays may be adequate, but is unlikely to generate a fit of statistical significance. Our own experience, based mainly on work on polypropylene, is that the long time tail of the FID can usually be represented by an exponential or sum of exponentials, the fastest decaying component (presumably the crystalline phase) corresponds to an Abragam function, and the residual component, which is intermediate in character between a Gaussian and an exponential, is well represented by a Weibull function (see the following section). Using these functions, it has been possible to generate fits that approach statistical significance for data obtained using the solid echo technique at 40°C. However, the results obtained are quite sensitive to the measurement temperature and also to the values of experimental parameters such as the solid echo delay. Despite the difficulties of finding functions to adequately represent the complex decay of the FID, the technique has been quite widely used, particularly in studies of the effects of drawing [41–44] and annealing [45].

7.5.2 Filled elastomers and block copolymers

In the case of filled elastomers, and of some block copolymers, the difference in molecular mobilities between the two phases is much more pronounced than for homopolymers, and analysis is correspondingly easier; however, the problem of representing components that appear intermediate in character between a Gaussian and an exponential still arises. Several authors have used functions of the general form

$$M_t = M_0 \exp - \left(\frac{t}{T_2}\right)^n, \quad \text{where } 1 < n < 2$$

to try to represent such intermediate components. This function was first suggested by Weibull in a different context [46], but has been successfully used to represent part or all of the FID in several polymer systems [47–50]. Since the exponential and Gaussian decays are simply limiting cases of this function ($n = 1$ and $n = 2$, respectively), this function is a good candidate to represent components of intermediate character. It should be noted, however, that as far as we know, the Fourier transform of the Weibullian [51] does not correspond to any of the functions previously used to represent components in the analysis of CW data.

In some cases, the difference in mobility between the regions is such as to allow the decay to be adequately represented by a combination of Gaussian and exponential decays [52, 53], even to the extent of apparently revealing a component due to interfacial material. One particular problem that occurs in such systems is the difficulty of adequately recording a very short-lived component from the rigid phase and a very long-lived component from the mobile phase. This may require the use of different digitisation rates for different parts of the FID, and the use of echo techniques to overcome both 'dead-time' problems for the short component and magnet homogeneity/off-resonance effects for the long component [49, 53].

7.5.3 Glassy systems

Glassy systems are in many ways the least promising candidates for the techniques of lineshape/FID analysis since they are by definition below or around the glass-transition temperature, and amorphous, which leads one to expect undifferentiated, broad, featureless lines. Nevertheless, lineshapes for epoxy resin systems have been shown to be sensitive to molecular motions even below the glass-transition temperature [54], and also to reflect the inhomogeneous distribution of crosslink density [55, 56]. A particularly interesting example of motion in a glassy system is provided by the work of Li *et al.* [57] on a polycarbonate in which the only protons are those on the phenylene rings. Since the protons are effectively isolated in pairs, the expected lineshape in the absence of rapid molecular motion is a Gaussian broadened doublet of the kind first described by Pake in the case of gypsum [58]. At temperatures around the glass transition, however, the lineshape is well represented by the superposition of a fairly narrow Lorentzian line on a Gaussian broadened Pake doublet. This indicates considerable motional heterogeneity in the sample suggesting that some phenylene rings are reorienting quite rapidly while others are not. Similar conclusions have been reached on the basis of ^{2}H and ^{13}C measurements on a number of systems containing *para*-linked phenylene groups, and ways in which such a lineshape could arise have been analysed [59]. In this particular case, the lineshapes used do have some theoretical justification and the results obtained show how useful the technique can be in favourable circumstances.

7.5.4 Highly mobile systems: melts

Polymer melts mark the effective boundary of this chapter. It may be necessary to record the decay of the transverse magnetisation under some sort of echo sequence designed to compensate for magnet inhomogeneities if T_2 is long, but in the following discussion no distinction is made and perfect magnet homogeneity is assumed.

It has been known for some time that the FIDs of such systems often decay in a way that is well represented for the most part by a Weibullian [60, 61]. Arguments based on consideration of correlation functions suggested that the FID of high molecular weight polydimethyl siloxane in the melt should decay for the most part with a Weibullian power of between 1.25 and 1.5 [62], and the existence of residual 'static' dipolar interactions in these systems was confirmed by the existence of the 'pseudo-solid' echo [63]. This reference forms part of a much larger body of work on such systems by Cohen-Addad and co-workers which it is beyond the scope of this chapter to cover in any detail, but interested readers are directed to literature such as [64] and [65].

Further study of molecular motion in this state has been carried out by Kimmich and co-workers using the field-cycling technique [66] as well as other methods to characterise the kinds of molecular motions present [67]. These measurements, together with the work of Callaghan and co-workers [68, 69] (see chapter 9), have provided a valuable tool for evaluating the theories of polymer motion proposed by DeGennes [70] and Doi and Edwards [71], and have revealed an interesting dependence of NMR parameters on chain length. Again, this work is too wide ranging to be covered in detail in this chapter, and the interested reader is directed to the original literature.

More recently, it has begun to be possible to apply the kinds of theories of polymer chain motion developed by DeGennes [70] and Doi and Edwards [71] (see chapters 4 and 9) to predicting the theoretical form of the transverse relaxation function (FID) in polymer melts [35, 72–74], and to use experimental measurements as a test of such theories [75, 76]. These are important developments in that they provide new insights into the relationship between observed lineshapes and the motional characteristics of polymers. Brereton [35] has shown that, in the melt, motionally homogeneous homopolymers can give rise to surprisingly complex FIDs which would have been represented by more than one phenomenological function. It is interesting that the form of the FID predicted by him for motionally homogeneous viscous melts can be represented to a very good approximation as the sum of a Weibullian and one or two exponential components. This emphasises the difficulty of definitively associating motional or morphological regions with the results of phenomenological analyses.

7.6 Relaxation time analysis of polymers

The measurement of relaxation times of polymers has been an area of interest for many years, because of its direct relation to the motions in solid polymers.

These may be correlated with changes in the composition and structure of the materials as well as with external factors such as stress. The values of relaxation times depend upon the details of the motions that are occurring, and while the complexity of the problem is too great to allow a full understanding, the variation of relaxation times with temperature offers a good opportunity to obtain estimates of correlation times for various motions, as discussed in chapters 4 and 6. As the temperature of a material is increased, the frequencies of the motions occurring in the sample will increase, and when these become equal to the characteristic frequency of the relaxation time, a maximum relaxation rate (minimum relaxation time) will be observed. If the correlation function and the temperature dependence of the correlation time were known, the shape of the relaxation curve could be completely determined. Harrell *et al.* [77] have published a study of the polymerisation of trimethylolpropane triacrylate and trimethylolpropane trimethacrylate in which they were able to represent the temperature dependence of the rotating frame relaxation of the monomers by a sum of three spectral density functions (Figure 7.3). The curves were based on the assumption of an Arrhenius activation process, and good agreement was obtained. Any deviations between

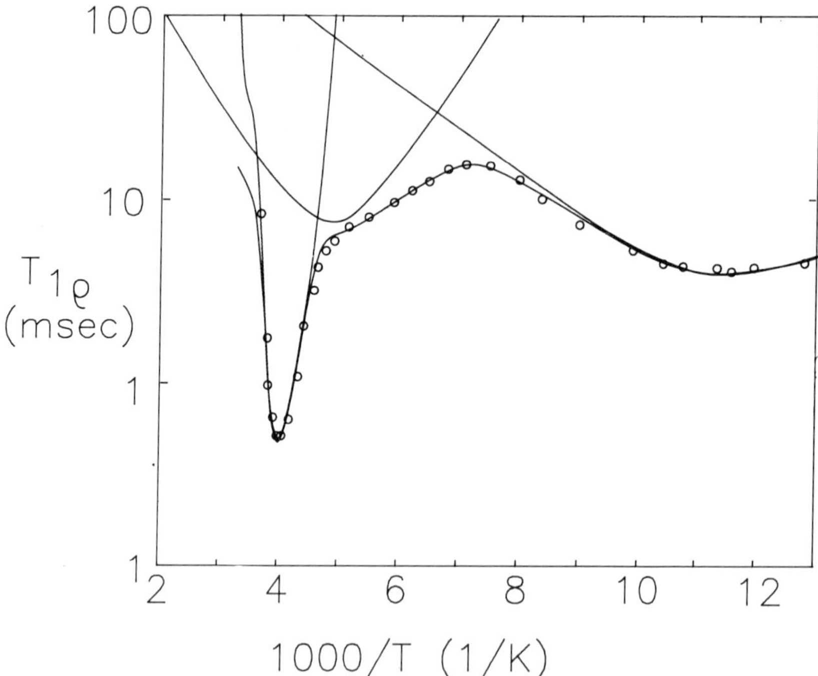

Figure 7.3 $T_{1\rho}$ versus inverse temperature for trimethylolpropane triacrylate at $B_1 = 1.8\,\mathrm{mT}$. Reprinted with permission from [77]. © (1991) John Wiley and Sons Inc.

the measurements and theory can be explained in terms of deviation from Arrhenius behaviour. However, the authors were unable to treat the results for the polymers in the same fashion, and instead analysed the effect of polymerisation by looking at the effect on the minima in the relaxation times.

A representative study of relaxation times as a function of temperature for poly(vinyl acetate) is presented in Figure 7.4 [3] where the effects of methyl group rotation and main-chain motion (glass-transition) can easily be distinguished. Correlation times are derived from the minima of T_1 or

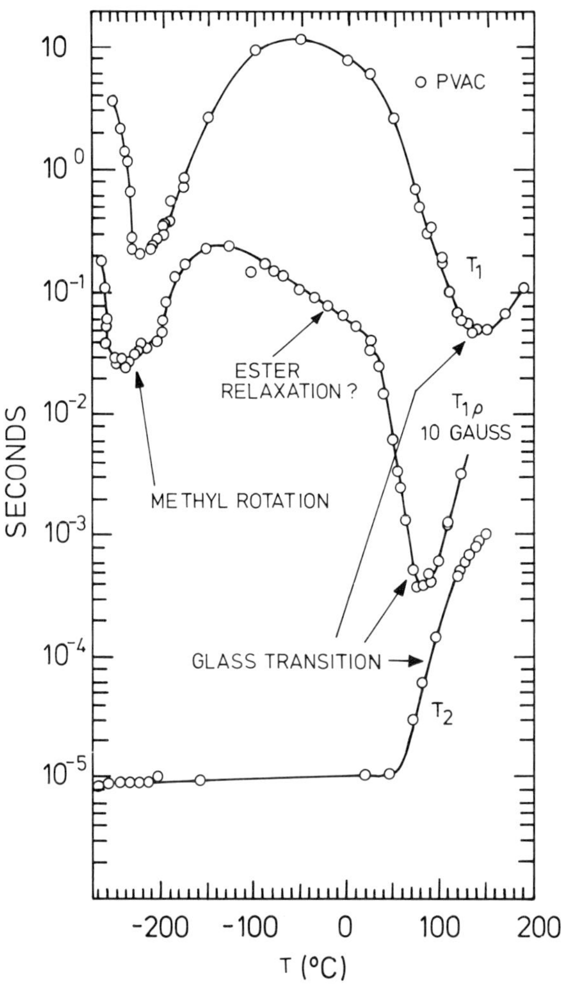

Figure 7.4 Variable temperature relaxation study of poly(vinyl acetate). Reprinted with permission from [3]. © (1971) American Chemical Society.

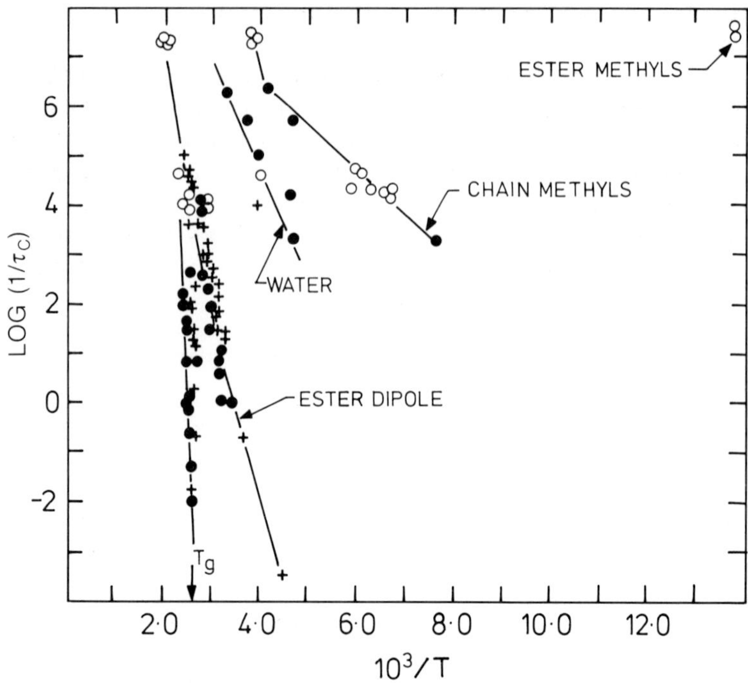

Figure 7.5 Arrhenius plot of inverse correlation times ($1/\tau_C$) measured by various techniques for poly(methyl methacrylate). (●) Mechanical; (+) dielectric: (○) NMR. Reprinted with permission from [3]. © (1971) American Chemical Society.

$T_{1\rho}$ or from the mid-point of a transition in T_2. The justification for such an approach can largely be traced to a compilation of Arrhenius plots by McCall [78], such as Figure 7.5, which show good agreement between the NMR data and mechanical relaxation and dielectric relaxation.

The review by Slichter [2] covers much of the early work in this field showing the effects of composition, molecular weight and applied stress on the T_1 minima of various polymers. It should be noted that in this early work, no mention is made of multi-exponential behaviour. In general, T_1 is less prone to exhibiting complex relaxation behaviour than $T_{1\rho}$. This is partly due to the fact that T_1 is longer than $T_{1\rho}$ allowing spin diffusional averaging to be effective over greater distances, and partly due to spin diffusion being more effective in the absence of a spin-locking field. Interpretation of relaxation data obtained from a heterogeneous system is relatively straightforward either in the case where the average behaviour of the whole system is analysed, or where the intrinsic behaviour of the constituent parts can be analysed. This is often very difficult to achieve in practice because of the effects of spin-diffusion.

7.7 Spin diffusion in relaxation methods for polymers

In this section, we look in some detail at the effects of spin diffusion in relaxation measurements on heterogeneous systems. Although the fact that spin diffusion can affect relaxation measurements has been known for many years, it was regarded for a long time simply as a slight inconvenience when doing measurements. Commonly only two cases were considered: the case of 'slow' spin diffusion, leading to the observation of intrinsic relaxation times, and the case of 'rapid' spin diffusion, leading to the observation of the population weighted rate average for the relaxation, that is

$$R_{(\mathrm{obs})} = \frac{\sum(P_i R_i)}{\sum P_i} \tag{7.14}$$

where P_i is the population of region i and R_i is the relaxation rate ($= 1/T_i$) for region i.

'Rapid' spin diffusion is commonly the case for T_1 relaxation in polymers, but not $T_{1\rho}$. In the last 10 years there have been significant developments in the understanding of the ways in which the effects of spin diffusion manifest themselves in ^{1}H relaxation measurements, particularly $T_{1\rho}$. In 1984 Packer *et al.* published a study of the effects of morphology on the observed relaxation behaviour in semi-crystalline homopolymers [79], in which they used a computer program to model the observed relaxation behaviour for a two or three region lamellar system in which each region was assumed to be homogeneous in its intrinsic values of relaxation rates and spin diffusion coefficients (Figure 7.6). The mathematical basis of the model is a set of coupled partial differential equations describing the relaxation and spin diffusion behaviour in each region, such that the behaviour of the magnetisation M, at a point x at time t is described by equation (7.13). The boundary conditions that are applicable are that there should be no net diffusion across the centres of regions A and C (by symmetry), and that the magnetisation profile must be continuous across the boundaries. Two region models may also be constructed simply by making the relaxation and spin diffusion properties of the interfacial region identical with those of one of the other regions.

They showed that such a model could reproduce the kind of behaviour observed experimentally in samples of polypropylene. The question of the kind of behaviour to be expected from such a system had previously been addressed mathematically by Cheung [80], but the results he had obtained were not accessible except in the two limits mentioned above. Although the form of the mathematical equation involved was already well known from the study of conduction of heat in solids [15], the work of Packer *et al.* [79] was the first approach that allowed the relationship between the observed relaxation behaviour and the intrinsic relaxation and spin diffusion properties of the different regions to be explored between the two limits of 'slow' and

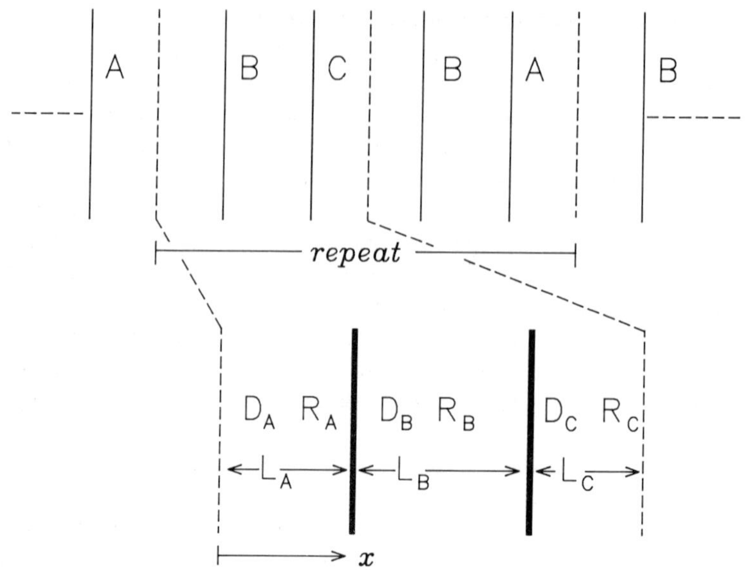

Figure 7.6 One-dimensional three-region model used in the computer simulation of relaxation in a heterogeneous system. The overall repeat distance is indicated. The calculations are carried out over the unit indicated in the lower half of the figure, comprising half of the regions A and C and the whole of B for which the dimensions, spin diffusion coefficients and relaxation rates are indicated. Redrawn and adapted with permission from [79].

'fast' diffusion. This intermediate region is of interest when the characteristic time for spin diffusion from one region to another lies between the intrinsic relaxation times of the two regions, as is commonly the case for ^{1}H $T_{1\rho}$ measurements on semi-crystalline polymer systems having minimum domain sizes up to several tens of nanometres.

Subsequent investigations using the computer model, in conjunction with more detailed analysis of the resultant behaviour, showed that two region systems in the intermediate regime were capable of generating up to four exponential components in their $T_{1\rho}$ decays [18]. This result was initially surprising since many authors had previously assumed a direct correlation between the number of components observed in $T_{1\rho}$ decays and the number of regions present in the system (see, for example, [79] or [82]). The result was confirmed by an exact analytical solution of the relaxation/spin diffusion equation [83] which showed agreement with the results of simulations to within the accuracy of the multi-component analysis. The exact solution indicated that the expected behaviour of such a system in the intermediate regime was, in principle, an infinite number of exponential components, but that only four or five at most would have significant amplitudes. It also showed that it is not possible to make any simple direct correlation between the observed parameters (populations and time constants) and the intrinsic properties of the regions. It was subsequently shown, however, that by re-casting the equation in terms of three dimensionless parameters [84], it was possible to predict to a large extent the type of observed relaxation behaviour and, more importantly, in the case of $T_{1\rho}$ relaxation close to the 'slow' diffusion limit, it was possible to extract information about the sizes of the lamellar regions from the values of the two longest observed relaxation components. This kind of size information had previously only been accessible through Goldman–Shen type experiments (see section 7.8) or from approximate calculations of the maximum possible diffusion distances on the time-scale of the longest $T_{1\rho}$ component (see, for example, [85]). Use of the model to extract morphological information (particularly distances) for semi-crystalline homopolymers has subsequently been developed [86, 87] and quite detailed information obtained which correlates well with data obtained by other techniques.

Unfortunately, detailed analysis of this kind is so far only possible for systems having regular lamellar morphologies since numerical simulations [88] have shown that for cylindrical and spherical morphologies the problem is complicated by the fact that the observed relaxation behaviour depends not only on the distance over which spin diffusion occurs, but also on the 'direction' of the diffusion. Here, the term 'direction' refers to whether the diffusion is 'out' from a slower relaxing intrusion to a faster relaxing matrix, or 'in' from a slower relaxing matrix to a faster relaxing intrusion. This is not to suggest that the spin diffusion coefficient is modified, but that the rate of magnetisation transport is morphology- and direction-dependent. This can be understood qualitatively since the rate of spin temperature transfer (magnetisation transfer) is dependent on the relative 'spin heat capacities' of spatially adjacent elements along the direction of diffusion. Consider diffusion through a point r in a model of lamellar morphology. The volume of material (and therefore the 'spin heat capacity') contained in the element between r and $r - \delta$ is the same as that in the element between r and $r + \delta$. However, for cylindrical morphology, these volumes differ by an amount proportional to $2\delta^2$. For spherical morphology, this difference rises to $6r\delta^2$. (A simple analogy may be drawn with the layers of an onion. The further you are from the centre, the more onion you get in each layer.) Thus, the rate of magnetisation transfer will be enhanced for diffusion 'out' and retarded for diffusion 'in'.

This new understanding of the role of spin diffusion in determining the observed ^{1}H relaxation behaviour, particularly $T_{1\rho}$, in structurally heterogeneous polymer systems can perhaps be briefly summarised as follows. The idea that a simple 1:1 correlation can be made between the number of observed relaxation components and the number of phases or regions present in the material has been shown to be false, but it has also been shown that in favourable cases, $T_{1\rho}$ measurements can yield valuable information about the size of regions present in the sample.

7.8 Attempts to observe spin diffusion directly

One of the fundamental questions when dealing with heterogeneous polymer systems concerns the sizes of the different regions present in the sample. If it were possible to observe the effects of spin diffusion directly and quantitatively then it should be feasible, given the applicable spin diffusion coefficient, to deduce the minimum domain dimensions of the regions from which spin diffusion is occurring. In practice, the spin diffusion coefficient, which depends on the strength of the homonuclear dipolar coupling, can be calculated to a good approximation from the second moment of the line, and the mathematics of the diffusion equation are soluble for simple geometries (lamellae, cylinders, spheres), so the problem comes down to one of making direct quantitative observations of spin diffusion.

The conceptual basis of all the experiments which have been tried in this context is a three pulse sequence (Figure 7.7) first described by Goldman and Shen [89]. In its original form, the experiment uses a 90° pulse to rotate the magnetisation into the xy plane, and then waits a time that is long compared to the decay time of the fast decaying component of the FID, but short compared to the decay time of the slow decaying component. A second 90° pulse, phase shifted 180° from the first, then returns the magnetisation from the slow decaying component to the $+z$ direction for a time τ before a third 90° pulse returns the magnetisation to the xy plane for detection. If spin diffusion couples the regions corresponding to the fast and slow decaying components of the FID, magnetisation transport during the time τ will lead to the reappearance of the fast decaying component in the detected FID. By monitoring the reappearance of the fast decaying component as a function of τ, it is possible to determine the rate of transfer of the magnetisation between

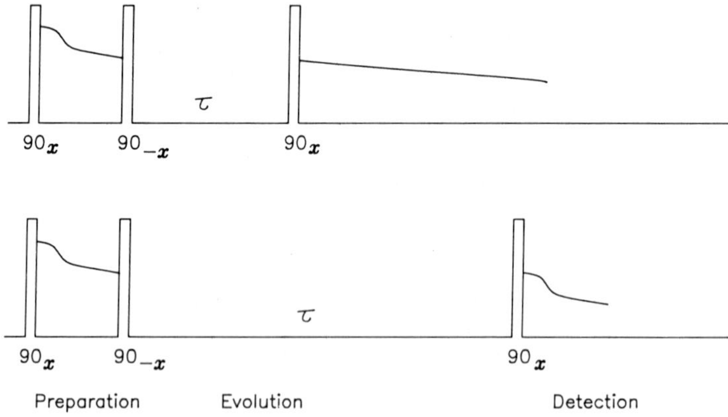

Figure 7.7 Pulse sequence for the Goldman–Shen experment. At short τ, only the magnetisation from the slower decaying component is observed, whereas at longer τ, spin diffusion has caused the magnetisation from the faster decaying component to reappear.

the regions, and therefore, given the spin diffusion coefficient, to deduce information about the distance over which spin diffusion must be taking place.

The experiment may be thought of in more general terms as consisting of three distinct phases. The first is a preparation phase in which a spatial magnetisation profile is somehow created. The second is an evolution phase in which the magnetisation profile is allowed to evolve towards an equilibrium via spin diffusion, and the third is a detection phase in which the state of the magnetisation profile is somehow detected and monitored. A remarkable range of variants consistent with this basic structure has been employed to look at polymers, particularly in the preparation and detection phases. A summary (not exhaustive) of experiments using the Goldman–Shen philosophy to study heterogeneity in various polymer systems is presented in Table 7.1. While some of the experiments use ^{13}C detection methods, they are included here because they conform to the basic Goldman–Shen philos-

Table 7.1 Goldman–Shen type experiments on polymer systems

Sample	Preparation	Detection	Reference
PVC	^{1}H FID	^{1}H FID	82
Polyurethane	^{1}H FID	^{1}H FID	90
PE	^{1}H FID	^{1}H FID	91
PE, nylon	^{1}H FID	^{1}H FID	92
Polycarbonate	^{1}H FID	^{1}H FID	93
PS/PB block copolymer	^{1}H FID	^{1}H FID (solid echo)	94
PP	^{1}H $T_{1\rho}$	^{1}H FID, ^{1}H $T_{1\rho}$	79
PET	^{1}H $T_{1\rho}$	^{1}H T_{1y} (DNCP)	95
PET	^{1}H T_{1xz} (MREV-8)	^{1}H T_{1xz} (MREV-8)	96
PS/PVME blends	^{1}H chemical shift (CRAMPS)	^{1}H chemical shift (CRAMPS) (2-D)	97
PEO/resorcinol blends	^{1}H chemical shift (CRAMPS)	^{1}H chemical shift CRAMPS) (2-D)	98
PS/PVME blends	^{1}H chemical shift (selective inversion)	^{1}H chemical shift (CRAMPS)	99
POM	^{1}H FID	^{13}C CP	100
PPS/PES blends	^{1}H $T_{1\rho}$	^{13}C CP/MAS	101
PEI/PEK blend	^{1}H chemical shift (multiple pulse)	^{13}C CP/MAS	102
Epoxy resins	'Dipolar filter'	^{13}C CP/MAS	103
PE	^{1}H $T_{1\rho}$	^{13}C CP	104
Nylon/PBZT blend	Various	Various	105
	^{1}H FID	^{1}H FID	
	^{1}H chemical shift (multiple pulse)	^{1}H chemical shift (multiple pulse)	
	Dipolar relaxation (Jeener–Broekaert	^{1}H spectrum	
		^{13}C CP/MAS	
PEI/PBI blends	^{1}H chemical shift (CRAMPS)	^{1}H chemical shift (CRAMPS)	106
PS/PBMA blend	^{1}H chemical shift (CRAMPS)	^{1}H chemical shift (CRAMPS)	107

ophy and are essentially experiments on the ^{1}H spin system. Similarly, it will be obvious to those familiar with two-dimensional NMR experiments that the description of the Goldman–Shen experiment in terms of preparation, evolution and detection phases corresponds to the generalised description of a two-dimensional experiment. Where ^{1}H chemical shift information (CRAMPS, see section 7.10) has been used in both the preparation and detection phases, these experiments have actually been performed as two-dimensional experiments, corresponding to the solution-state spin exchange experiment, but again they are included here to emphasise the fact that they are Goldman–Shen type experiments and have many of the same limitations and restrictions as Goldman and Shen's original experiment.

In practice, there are various potential problems to be considered when using the experiment. The first of these is the difficulty of associating a particular morphological phase in the polymer system with a particular component in the NMR signal. Such problems have already been discussed. Next, there is the problem of producing a magnetisation profile that leaves the magnetisation where you want it. A good example of this kind of problem is the case of semi-crystalline polymers such as polyethylene [91] and polypropylene [79], where the slowest decaying component of the FID corresponds to the amorphous material. This is an undesirable situation because the amorphous material will probably be the minor component, the spin diffusion coefficients associated with the amorphous region will be smaller than for the crystalline region, and its intrinsic spin–lattice relaxation time (T_1) will be shorter (see below). For all these reasons, it would be preferable to create a profile in which it was the crystalline (short FID) component that was left behind. This has prompted people to try selection using other criteria, such as spin–lattice relaxation times. An associated problem is ensuring that the profile produced is homogeneous within each of the different regions, and has sharp edges at the region boundaries. This problem is particularly acute for schemes in which spin diffusion is active during the selection phase, such as with $T_{1\rho}$ selection, since the spin diffusion will tend to blur the edges of the regions. In this case, the severity of the problem will depend in part on the length of the selection phase. Experiments in which spin diffusion is inactive during both the preparation and detection phases are to be preferred. The question of sharpness of selection is also particularly relevant to schemes that select on the basis of ^{1}H chemical shift, since the resolution available is generally limited. This has recently been addressed in considerable detail by Campbell and VanderHart [106].

A much more general consideration is the question of whether there are any processes other than spin diffusion that can affect the detected signal. An almost universal problem in this regard is T_1 relaxation during the evolution period, which can change not only the total amount of magnetisation detected, but also the relative populations of the regions in a manner similar to spin diffusion effects.

One common approach to this problem of relaxation has been to restrict the evolution period to times over which the effects of T_1 are hopefully negligible, say τ less than $0.1 \times T_1$, and then assume that no T_1 effects are observed. This may present a difficulty since a maximum diffusion time of $0.1 \times T_1$ may not be sufficiently long to allow equilibration by spin diffusion, and therefore it may not be possible to accurately determine the long time behaviour of the system.

A second approach to the problem of T_1 effects in the evolution period has been to try to design an experiment in which T_1 effects are eliminated by suitable manipulation of the magnetisation. Such a scheme, in which the magnetisation is stored along the $+z$ and $-z$ axes on alternate transients during the evolution period has been proposed [79, 108], and used in experimental measurements [79, 106]. Unfortunately, it has subsequently been shown [109] that such schemes are fundamentally flawed in that while they tend to eliminate or cancel bulk changes in the magnetisation due to T_1 processes, they do not eliminate the tendency of T_1 processes to flatten out the magnetisation profile in a manner that resembles the action of spin diffusion. Such schemes therefore seem undesirable in that they tend to replace a known distortion of the data with an unknown one. The idea of storing the magnetisation in the $-z$ direction during the evolution period is a useful one, however, since, if the Goldman–Shen experiment is performed twice, once storing the magnetisation in the $+z$ direction and the second time storing it in the $-z$ direction, and the results of the two experiments are the same to within experimental error, this is absolute proof that T_1 effects are negligible on the timescale of the experiment. Despite the simplicity of this check, it has rarely been used.

A third approach is to accept the intimately coupled nature of T_1 and spin diffusion in such systems, and attempt a complete analysis of the combined relaxation/spin diffusion behaviour. This is undoutedly a more rigorous approach, but requires the determination somehow of the intrinsic T_1s of the various regions! Recently, an attempt has been made to implement such an approach [104], but because of the number of assumptions and approximations used in the work, the results obtained must be regarded as rather optimistic. Nevertheless, future developments using this approach may well increase the range of heterogeneities accessible by Goldman–Shen type experiments.

A rather more minor problem with regard to distortion of the detected signal is change in the shape of the FID as a function of τ. This was first reported by Cheung [92] for values of τ less than 500 μs. He mistakenly ascribed the effect to modulation of the FID by spin diffusion, but it was subsequently shown to be due to the formation of multiple quantum coherences [110]. It can be eliminated by suitable phase cycling.

Despite these potential problems, the techniques have been successfully applied to a number of systems, often yielding information that would be difficult or impossible to obtain in other ways. It is perhaps at its most useful in investigations of miscibility/heterogeneity in polymer blends, where the size of regions is often only a few nanometres, or less, and the associated time for equilibration by spin diffusion is correspondingly very short. In this regard, selection on the basis of chemical shift is usually the technique of choice, and such experiments have revealed information about heterogeneity on distance scales smaller than could be readily detected either by bulk techniques such as DSC, or by microscopy.

7.9 Attempts to suppress spin diffusion in relaxation measurements

In the preceding sections, we have seen how spin diffusion can affect the observed values in relaxation measurements and how it can provide information on the size of regions present in a sample. However, we have also seen that spin diffusion is often an unwanted complication, obscuring the intrinsic relaxation information from the different regions. In this section, we will look at ways of trying to measure intrinsic relaxation information for the different regions by suppressing spin diffusion. Some of the techniques discussed below have already been mentioned in connection with the preparation and/or detection phases of the Goldman–Shen type experiments, but are dealt with in this section as experiments in their own right. Such experiments are most commonly used as a way of obtaining information about the number of components present in a sample, and their relative amounts. In this sense they are providing similar information to that provided by FID analysis (section 7.5). They have the advantage that the form of the decay for each component is generally known to be exponential, but have the disadvantage that they are 'driven' techniques and therefore the results obtained can depend critically on how well the experiment is set up.

Spin diffusion can be suppressed by manipulating the spins in such a way that the flip-flop term of the dipolar Hamiltonian is averaged to zero. Because of the geometrical factors in the dipolar Hamiltonian, this always involves manipulation in which the principal axis of the modulation is at the magic angle to the applied field. The modulation may be either by RF or by physical rotation of the sample (magic-angle spinning), although spinning speeds sufficiently high to completely eliminate the ^{1}H–^{1}H flip-flop term are rarely practical for organic solids.

In terms of RF modulation, there are essentially two possibilities. The first is applying the RF pulse off-resonance by an amount such that the effective field in the rotating reference frame of the applied RF is at the magic angle to the external field, thereby causing precession about the magic angle. This technique was first suggested by Lee and Goldburg [111, 112] as a way of suppressing dipolar interactions with a view to obtained high-resolution solid-state spectra, but it was subsequently modified by Tse and Hartmann [19] to spin-lock magnetisation at the magic angle, thereby providing an experiment similar to the on-resonance $T_{1\rho}$ experiment, but without the complications of spin diffusion. Of course, it is also possible to arrange to spin-lock magnetisation at angles other than the magic angle, and the effects of such experiments on the effective spin diffusion coefficient and intrinsic relaxation times have been analysed in detail [20].

The second possibility is the use of multiple pulse techniques to effectively achieve the same sort of precession. There are several multiple pulse sequences available to achieve this, of varying complexity, but all consist of sequences of 90° pulses with cyclic phase permutations. At its simplest, the sequence can be thought of as being designed such that the time average precession of the magnetisation is about the [1 1 1] direction, and therefore at the magic angle to the external field, thus giving modulation properties similar to the Lee–Goldburg experiment. This may be achieved conceptually by cyclically applying 90° pulses with phases x, z, and y, at a rate that is fast compared to the size of the interaction to be averaged (in this case the dipolar coupling). In practice, it is not possible to apply a 90° pulse of phase z, so longer pulse sequences must be used to achieve the same objective. The simplest of these would be to use two 90° pulses of phases $-x$ and $-y$ to move the magnetisation from the y axis to the x axis, in place of the 'conceptual' 90° z pulse. This gives a cycle of four phases, x, $-x$, $-y$, y which is the basis of the WAHUHA pulse sequence [113], but higher order permutations of 8 (MREV-8) [114, 115] and 24 (BR-24) [116] pulses have subsequently been developed which have better compensation for imperfections, and are more efficient.

If the magnetisation is first made colinear with the [1 1 1] modulation axis then these sequences can be used to 'spin-lock' the magnetisation, and a time constant for the decay of the spin-locked magnetisation can be measured. The value of the measured time constant depends on the experiment being performed, and therefore each experiment has an associated time constant. The time constant for relaxation under the MREV-8 sequence is usually denoted by T_{1xz}, while relaxation under BR-24 is denoted by $T_{1\mathrm{BR}}$. Although some understanding of the relaxation phenomena under these sequences has been derived from average Hamiltonian theory [117, 118], comparison between time constants obtained using different sequences is not straightforward. A further development of these techniques has been the introduction of a measurement in which the relaxation of magnetisation is monitored under multiple pulse conditions, with an additional 180° pulse being used in the middle of the sequence to eliminate chemical shift information. This has been termed the dipolar narrowed Carr–Purcell sequence (DNCP) [119], and its associated time constant is designated T_{1y}.

We have seen in section 7.8 that these techniques have been widely used in the preparation and detection phases of spin-diffusion experiments, but they have also found application elsewhere. Spin-locking at the magic angle ($T_{1\rho}^{\mathrm{MA}}$) has been used to derive estimates for the lower limits of intrinsic relaxation times in polypropylene [79], and in morphological investigations of PET fibres [96]. Spin-locking at other angles ($T^{\theta}_{1\rho}$) has been used to reduce the effective rate of spin diffusion, thereby extending downwards the range of lamellar thicknesses that could be determined in samples of polyethylene by detailed analysis of multi-exponential relaxation behaviour [86]. Determination of $T^{\theta}_{1\rho}$ as a function of θ has also been used as part of a detailed study of the phase behaviour of a copolymer of vinylidene fluoride and trifluoroethylene [17]. Although off-resonance spin-locking is a conceptually appealing experiment, it is experimentally very demanding and there is always the problem of assessing to what extent spin diffusion has actually been suppressed.

Proton relaxation under multiple pulse conditions has also been used to characterise phase composition in, for example, PET [95, 96], and polyethylene [120]. The technique is particularly useful in the case of PET because the phases present generally do not show large differences in the decay times of their FID components, so FID analysis would be particularly problematic. Although the problem of assessing the extent to which spin diffusion is suppressed also applies to relaxation under multiple pulse conditions, the available experimental evidence suggests that they may be more effective in practice than the corresponding off-resonance spin-locking experiment [96]. This is almost certainly due to practical considerations rather than theoretical ones.

7.10 High-resolution proton methods for polymers, MAS and CRAMPS

In the preceding section, we discussed methods of suppressing dipolar interactions as a means of eliminating spin diffusion in relaxation measurements, but dipolar interactions are also the principal source of line-broadening in solid-state ^{1}H spectra, and therefore suppression of dipolar interactions also offers the possibility of obtaining 'high resolution' ^{1}H spectra of solids in which chemical shift information is discernible. It is this, rather than the suppression of spin diffusion that has been the motivation in the development of these techniques, particularly the multiple pulse sequences.

We have already mentioned that, in principle, magic-angle spinning (MAS) alone will remove dipolar broadening, provided sufficiently high spinning speeds can be achieved. The spinning speeds that can currently be achieved in practice still fall considerably short of this goal for organic polymers. Dec *et al.* [121] have shown that, for a rigid organic solid, even at spinning speeds around 20 kHz, there is still substantial line broadening due to dipolar interactions, and it should be borne in mind that at the time of writing a realistic estimate for routinely achievable spinning speed would be more like 15 kHz. The problem of residual line-broadening is particularly acute in ^{1}H spectra, since the chemical shift range is relatively small. In ^{19}F spectra, on the other hand, the chemical shift dispersion tends to be much greater and so residual line-broadening effects are proportionally less important, allowing useful information to be obtained using high speed MAS alone [122].

Even for ^{1}H, magic-angle spinning alone can still be a useful technique if the size of the homonuclear dipolar interactions in the static sample has already been substantially reduced, either by the effects of molecular motion, or by chemical or isotopic dilution. Schneider *et al.* [123] demonstrated in 1972 that the signal due to the amorphous phase of polyethylene could be substantially narrowed at quite modest spinning speeds, although their aim was to facilitate easier lineshape analysis rather than to obtain chemical shift information. VanderHart *et al.* [124] used magic-angle spinning to obtain high-resolution ^{1}H spectra of the (isotopically dilute) residual protons in 98% deuterated atactic polystyrene, and used the observation of re-introduced line-broadening when the material was blended with 30% per-proteo material as evidence of intimate mixing at the molecular level. There are also numerous examples of the ability of magic-angle spinning to improve the resolution in ^{1}H spectra of highly mobile systems such as swollen gels and amorphous systems above the glass-transition temperature (see, for example, [125, 126]), although such materials, while definitely not liquids, are often on or even over the border line of what might normally be considered solid.

We have seen in the previous section that multiple pulse techniques offer the most efficient way of suppressing the A and B terms of the dipolar alphabet, which are the origin of the line-broadening. If we apply such a pulse sequence to a solid sample, allowing the magnetisation to precess freely from axis to axis and sample it 'stroboscopically' every time it is coincident with, say, the y axis, then a 'pseudo FID' will be constructed, the Fourier transform of which will be a scaled version of the spectrum without dipolar broadening, but retaining the chemical shift information. (The distinction between this and the previously described relaxation methods is that in the relaxation experiments the magnetisation was first made colinear with the modulation axis and 'spin-locked' there.) The spectra obtained using such multiple pulse techniques alone will still have unwanted broadening because the multiple pulse sequences average only the dipolar interactions (unlike MAS), and so chemical shift anisotropy (the variation in chemical shift as a function of molecular orientation, which is averaged to a single isotropic value by rapid molecular tumbling in solution) will still be present in the spectra. This anisotropy is much smaller than the size of the dipolar interactions, but may still be large compared to the chemical shift range (e.g. up to 20 ppm for ^{1}H). However, it does fall well within the range of interactions that may be effectively eliminated by magic-angle spinning. The technique of combined rotation and multiple pulse spectroscopy (CRAMPS) was first proposed by Gerstein *et al.* [127] and demonstrated on the ^{19}F spectrum of poly(1,1,2-trifluoro-2-chloroethene), in which it was shown to be possible to resolve the two chemically distinct types of fluorine. Since then there has been considerable development of the technique and fairly widespread application of it to ^1H systems. It has been shown [121, 128] that in favourable cases with crystalline organic solids, ^1H linewidths of less than 1 ppm are achievable, and in the case of crystalline polymers, comparable linewidths may be achieved. However, in amorphous polymers, the linewidths obtained are usually considerably greater, being typically of the order of 2–4 ppm. This is not due to any deficiency in the technique as applied to polymers, but rather is due to the distribution of physical environments experienced by a chemically distinct type of proton in a polymer sample.

Nevertheless, in certain favourable cases, it has been possible to obtain composition information about polymer mixtures [128], and configuration information relating to relatively simple crystalline polymer systems [129]. The latter case is illustrated (Figure 7.8) by the spectrum of poly(ethylene adipate) where the observed spectrum is more complex than would be expected in the absence of conformational effects.

^1H CRAMPS is an extremely demanding technique requiring a spatially homogeneous irradiation field and a very high degree of spectrometer stability, both in timing and phase. For a recent survey of the specific experimental requirements and some of the measures to overcome them; see the article by Maciel *et al.* [130]. In common with other solid-state techniques, development of pulse sequences that are more tolerant of spectrometer imperfections [131]

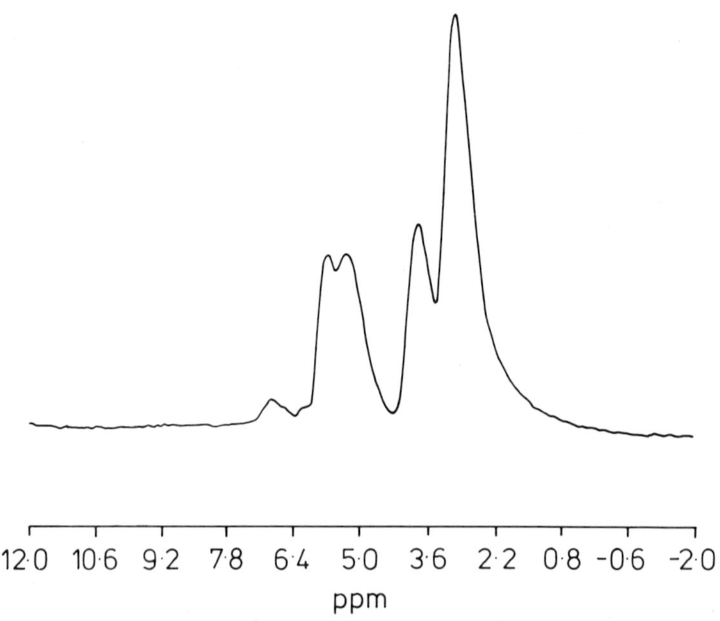

Figure 7.8 300 MHz ^1H CRAMPS spectrum of poly(ethylene adipate). 64 transients, relaxation delay 10 s. Reprinted with permission from [129]. © (1992) American Chemical Society.

and improved spectrometer hardware may greatly increase the applicability of the technique. In the case of ^{19}F, the experimental difficulties are increased by the typically greater spectral width needed which requires a shorter cycle time for the multiple pulse sequence [132, 133].

^{1}H CRAMPS has not been widely applied to polymers as a technique in its own right, but, as we have already seen in section 7.8, the available resolution has been sufficiently high to enable it to be successfully used in both the preparation and detection phases of spin diffusion experiments. There has been considerable development of ways of selecting magnetisation on the basis of its chemical shift under CRAMPS, including selective inversion [134] using a version of the CRAMPS sequence effectively incorporating the frequency selective DANTE [135] sequence, as well as other sequences based on selecting nulling of spectral components [102]. There has also been work done on ways of optimising the polarisation gradients produced by such selection procedures [106]. It seems likely that the use of the CRAMPS sequence as part of Goldman–Shen type experiments will prove to be its major use in polymer science.

7.11 Carbon detection of proton magnetisation

The subject of high-resolution ^{13}C methods as applied to polymers has been dealt with in chapters 5 and 6, but we feel it would be wrong to leave this chapter without pointing out explicitly that there exist a whole range of experiments which have been applied to polymers where the magnetisation is detected as a ^{13}C signal although a major part of the experiment is essentially being performed on the ^{1}H spin system, and as such is subject to all the interactions discussed in this chapter. These range from simple variations on the standard CP/MAS sequence such as delayed spin-locking (selection on the basis of the ^{1}H FID), and delayed contact (selection on the basis of the ^{1}H $T_{1\rho}$) to much more sophisticated spin diffusion experiments employing ^{13}C detection, and even two-dimensional heteronuclear correlation experiments which use multiple pulse sequences on both channels [136].

A recent example from the literature which is particularly illustrative of this approach is the so-called WISE experiment [137], in which the ^{1}H lineshape is detected as a function of ^{13}C chemical shift in a two-dimensional experiment. This sequence has also been modified to permit observation of the effects of ^{1}H spin diffusion in what is essentially a heteronuclear two-dimensional variant of the Goldman–Shen type experiment. This has allowed clear observation of the effects of 'micro heterogeneity' in 'compatible' blends of PS/PVME.

The use of ^{13}C detection obviously adds to these experiments by allowing separate analysis for the different chemically resolved peaks in the ^{13}C spectrum, but it is important to note that this does not imply that what is detected is the intrinsic behaviour of the protons directly bonded to the carbons being detected. In particular, there is no justification for assuming that the decay of the signal associated with a particular type of carbon as a function of some proton parameter will necessarily be a single exponential process. The magnetisation detected in such experiments originates in the ^1H spin system, and is subject to the same laws within that spin system whether ^{13}C detection is used or not. While this point should be fairly obvious, it has nevertheless been a source of confusion with some authors.

A useful analysis of the utility of such experiments in heterogeneous polymer systems has been made by Tekely *et al.* [100], and several other authors have made extensive use of such measurements (see, for example [138]). While the gain in chemical selectivity is somewhat balanced by the accompanying loss in sensitivity, there are occasions when detection of the ^1H magnetisation via the ^{13}C chemical shift allows analyses which would otherwise be difficult or impossible [102].

7.12 Oriented polymers

So far we have assumed, at least implicitly, that the samples under consideration have an isotropic distribution of molecular orientations with respect to the applied field. This situation arises where there is an absence of crystalline order, or the crystallites within the sample are themselves randomly oriented. When linear polymers are stretched, drawn, or extruded, some degree of molecular alignment occurs leading to a preferred direction of molecular orientation, which changes both the macroscopic properties of the polymer, and the NMR parameters. Hyndman and Origlio [27] first reported such measurements on polyethylene and polypropylene fibres. The spectra they presented clearly change with the orientation of the fibres to the magnetic field, demonstrating that the probability distribution of molecular orientations has been affected by the drawing process and is no longer isotropic. Study of the ^1H FID has subsequently been used to determine the effects of sample drawing [41–44].

In such systems, the changes in the NMR properties are twofold. Firstly, the introduction of orientation can simplify the spectra and secondly, the variation of the NMR parameters with sample orientation is itself a source of information. Further, it is also possible to study the effects of applied stress on the sample by the change in the NMR parameters. The study of such oriented samples can yield information both about the structure of the materials and the orientation process.

All NMR parameters are, in principle, dependent on the angle between the orientation axis and the applied field, although it is generally lineshape information such as the second and fourth moments, and T_2 that are most amenable to quantitative analysis. McBrierty and Ward [139] investigated the second moment of the ^1H spectrum of polyethylene as a function of draw ratio and orientation and found that the second moment alone was insufficient to characterise the orientation distribution, but contained enough information to show that a single phase model of polyethylene was inadmissible. A later publication [140], extending the work to a study of the fourth moments, allowed calculation of the distribution of molecular orientations to two further orders, and predicted the tensile modulus of the samples far more accurately. Figure 7.9, taken from the work of Kretz *et al.* [141], shows the kind of results obtainable by this method. This study of poly(tetramethylene oxide) correlates the results obtained from ^{1}H FIDs and ^{2}H echoes. The quadrupolar splitting observed in the deuterium spectrum is another orientation-dependent

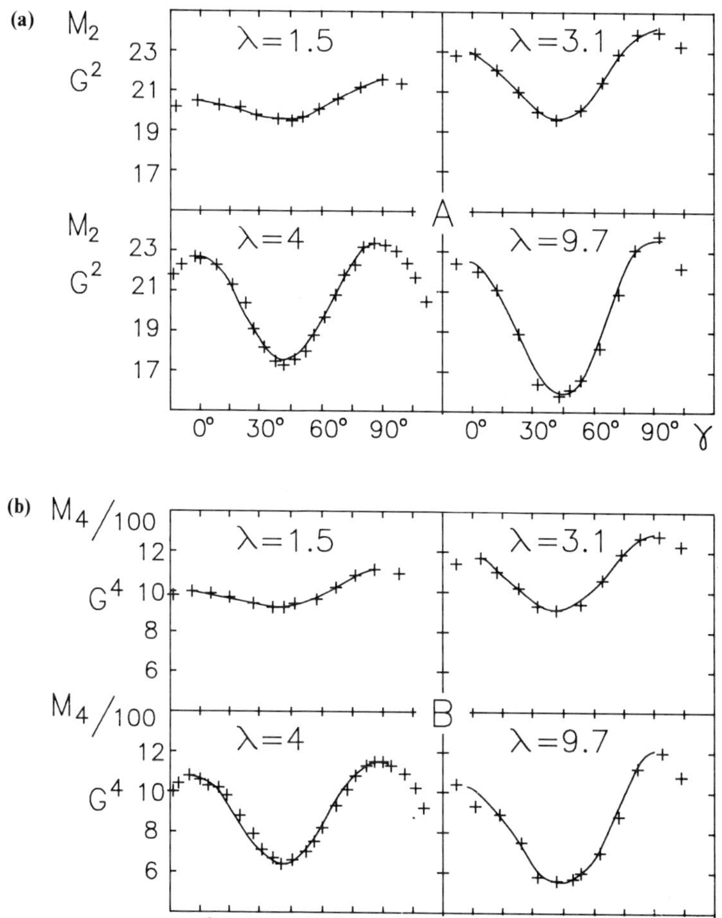

Figure 7.9 Example of ^{1}H moment anisotropy in drawn samples of poly(tetramethylene oxide) at 150 K. (a) Second moment, (b) fourth moment. λ is the draw ratio, γ is the sample orientation with respect to B_0. Reprinted with permission from [141]. © (1988) John Wiley and Sons Inc.

parameter, and offers great possibilities in the study of oriented samples as discussed in chapter 8. An alternative approach by Brandolini and Dybowski [142] used the chemical shift anisotropy of the ^{19}F resonance of PTFE observed under multiple pulse NMR to measure the orientation distribution of crystallites. Although this was suggested as a general method of obtaining the orientation distribution, there does not appear to be much likelihood of its being used in proton NMR where both the isotropic dispersion and the anisotropy of the chemical shift are small. The use of ^{13}C chemical shift anisotropy information is described in chapter 5.

In an isotropic sample, the distribution of orientations gives a 'powder doublet' of the kind described by Pake [58] for pairs of spins, which is then broadened by the inter-pair interactions. This renders the spectra difficult to analyse since many possible models will reflect the gross features, leaving the discrimination to rather minor 'features'. When this powder averaging is removed, the nearest-neighbour dipolar couplings for preferred orientations become dominant in the spectrum. Smith *et al.* [143] utilised this in a study of polyethylene [143], decomposing the spectra in terms of a three-region model and observing the variation in the composition with the draw ratio. Further, analysis of the spectra obtained at differing orientations allowed evaluation of the orientation distribution in the intermediate region.

A recent publication by Clements *et al.* [36] describes the effect of applied tensile stress on the proton spectra of polyvinylidene fluoride. The spectra consist of a narrow singlet superimposed on a broad doublet, which are assigned to amorphous and crystalline material, respectively. When stress is applied, the proportions of these components change, the fractional crystallinity increasing when the stress is parallel to the draw direction and decreasing when perpendicular. These measurements were all carried out with the draw direction parallel to the magnetic field direction because the crystalline doublet was most apparent in this case. The authors offer this crystallinity change as a partial cause of the piezoelectric response of the material.

Since many commercial applications of polymers involve their use in an oriented form, investigation of the nature and distribution of molecular orientation in such systems is of considerable interest to polymer scientists. In this regard, solid-state ^{1}H NMR is a valuable tool and complements results obtained from diffraction methods, and from other spectroscopies which are sensitive to molecular orientation [144].

7.13 Conclusions

We have seen that the homonuclear dipolar interactions between protons in a solid polymer sample are sensitive to the molecular motions present. This makes ^{1}H NMR particularly useful for studying the motional heterogeneity which plays such an important role in determining the macroscopic properties of solid polymers. Recent developments in the technique have seen particular emphasis on its use to determine the minimum dimensions of small scale heterogeneities, where the information it provides is often not available from other sources. We have also seen that under appropriate conditions solid state ^{1}H NMR can provide useful information on polymer configuration and orientation.

References

1. C.W. Wilson and G.E. Pake, *J. Polym. Sci.* **10** (1953) 503–505.
2. W.P. Slichter, *NMR Basic Principles and Progress* **4** (1961) 209–231.
3. D.W. McCall, *Acc. Chem. Res.* **4** (1971) 223–232.
4. V.J. McBrierty and D.C. Douglass, *J. Polym. Sci. Macromol. Rev.* **16** (1981) 295–366.
5. V.D. Fedotov and H. Schneider, in *Structure and Dynamics of Bulk Polymers by NMR Methods*, Springer-Verlag, Berlin (1989).
6. N. Bloembergen, *Physica* **15** (1949) 386–426.
7. J.H. Van Vleck, *Phys. Rev.* **74** (1948) 1168–1183.
8. P. Mansfield, *Prog. NMR Spectrosc.* **8** (1971) 41–101.
9. A. Abragam, in *Principles of Nuclear Magnetism*, Oxford University Press, Oxford (1961).
10. E.R. Andrew, *Prog. NMR Spectrosc.* **8** (1971) 1–40.
11. B.C. Gerstein and C.R. Dybowski, in *Transient Techniques in NMR of Solids*, Academic Press, Orlando (1985).
12. G. Voigt and R. Kimmich, *Polymer* **21** (1980) 1001–1008.
13. T.M. Connor, *NMR Basic Principles and Progress* **4** (1971) 247–270.
14. D. Wolf, in *Spin-temperature and Nuclear-spin Relaxation in Matter*, Oxford University Press, Oxford (1979).
15. H.S. Carslaw and J.C. Jaeger, in *Conduction of Heat in Solids*, Oxford University Press, Oxford (1959).
16. D.C. Douglass and V.J. McBrierty, *Macromolecules* **11** (1978) 766–773.
17. J. Hirschinger, B. Meurer and G. Weill, *J. Phys. (Paris)* **50** (1989) 563–582.
18. J.G. Powles and J.H. Strange, *Proc. Phys. Soc. (London)* **82** (1963) 6–7.
19. D. Tse and S.R. Hartmann, *Phys. Rev. Lett.* **21** (1968) 511–514.
20. G.P. Jones, *Phys. Rev.* **148** (1966) 332–335.
21. D.W. Marquardt, *J. Soc. Ind. Appl. Math.* **11** (1963) 431–441.
22. W.H. Press, B.P. Flannery, S.A. Teukolsky and T. Vetterling, in *Numerical Recipes–the Art of Scientific Computing*, Cambridge University Press, Cambridge (1986).
23. S.W. Provencher, *Comput. Phys. Commun.* **27** (1982) 213–227.
24. S.W. Provencher, *Comput. Phys. Commun.* **27** (1982) 229–242.
25. E. Yeramian and P. Claverie, *Nature* **326** (1987) 169–174.
26. C. Tellier, M. Guillou-Charpin, D. Le Botlan and F. Pelissolo, *Magn. Reson. Chem.* **29** (1991) 164–167.
27. D. Hyndman and G.F. Origlio, *J. Polym. Sci.* **39** (1959) 556–558.
28. K. Bergmann, *J. Polym. Sci., Polym. Phys.* **16** (1978) 1611–1634.
29. K. Unterforsthuber and K. Bergmann, *J. Magn. Reson.* **33** (1979) 483–495.
30. R. Kitamaru, F. Horii and S.H. Hyon, *J. Polym. Sci., Polym. Phys.* **15** (1977) 821–836.
31. J. Loboda-Cackovic, H. Cackovic and R. Hosemann, *J. Macromol. Sci.* **B16** (1979) 127–144.
32. F. Horii and R. Kitamaru, *J. Polym. Sci., Polym. Phys.* **19** (1981) 109–120.
33. R. Kubo and K. Tomita, *J. Phys. Soc. Jpn.* **9** (1954) 888–919.
34. G.E. Karnaukh, A.A. Lundin, B.N. Provotorov and K.T. Summanen, *Sov. Phys. JETP* **64** (1986) 1324–1331.
35. M.G. Brereton, *J. Chem. Phys.* **94** (1991) 2136–2142.
36. J. Clements, G.R. Davies and I.M. Ward, *Polymer* **32** (1991) 2736–2740.
37. S. Zhang and X. Wu, *Chem. Phys. Lett.* **156** (1989) 82–86.
38. W. Kluver and W. Ruland, *Prog. Colloid Polym. Sci.* **64** (1978) 255–256.

39. K. Bergmann, *Polym. Bull.* **5** (1981) 355–360.
40. K. Bergmann, H. Schmiedberger and K. Unterforsthuber, *Colloid Polym. Sci.* **262** (1984) 283–293.
41. M. Ito, H. Serizawa, K. Tanaka, W.P. Leung and C.L. Choy, *J. Polym. Sci., Polym. Phys.* **21** (1983) 2299–2309.
42. H. Tanaka, *J. Appl. Polym. Sci.* **28** (1983) 1707–1715.
43. M. Ito, T. Kanamoto, K. Tanaka and R.S. Porter, *J. Polym. Sci., Polym. Phys.* **23** (1985) 59–71.
44. L.B. Liu, N. Murakami, M. Sumita and K. Miyasaka, *J. Polym. Sci., Polym. Phys.* **27** (1989) 2427–2440.
45. H. Tanaka, F. Kohrogi and K. Suzuki, *Eur. Polym. J.* **2** (1989) 449–453.
46. W. Weibull, *J. Appl. Mech.* **18** (1951) 293–297.
47. S. Kaufman, W.P. Slichter and D.D. Davis, *J. Polym. Sci.* **9** (1971) 829–839.
48. S. Kaplan and J.J O'Malley, *Polymer* **22** (1981) 221–225.
49. K. Fukumori, T. Kurauchi and O. Kamigaito, *J. Appl. Polym. Sci.* **38** (1989) 1313–1334.
50. E. von Meerwall and T. Stone, *J. Polym. Sci., Polym. Phys.* **27** (1989) 503–522.
51. F. Oberhettinger, in *Fourier Transforms of Distributions and their Inverses*, Academic Press, New York (1973).
52. R.A. Assink and G.L. Wilkes, *Polym. Eng. Sci.* **17** (1977) 606–612.
53. H. Tanaka and T. Nishi, *J. Chem. Phys.* **82** (1985) 4326–4331.
54. L. Banks and B. Ellis, *J. Polym. Sci., Polym. Phys.* **20** (1982) 1055–1067.
55. A.C. Lind, *ACS Polym. Preprints* **21** (1980) 241–242.
56. C.G. Fry and A.C. Lind, *Macromolecules* **21** (1988) 1292–1297.
57. K.L. Li, P.T. Inglefield, A.A. Jones, J.T. Bendler and A.D. English, *Macromolecules* **21** (1988) 2940–2944.
58. G.E. Pake, *J. Chem. Phys.* **16** (1948) 327–336.
59. J.I. Kaplan and A.N. Garroway, *J. Magn. Reson.* **49** (1982) 464–475.
60. R. Folland, J.H. Steven and A. Charlesby, *J. Polym. Sci., Polym. Phys.* **16** (1978) 1041–1057.
61. R. Folland and A. Charlesby, *Polymer* **20** (1979) 207–210.
62. Y. Martin-Borret, J.P. Cohen-Addad and J.P. Messa, *J. Chem. Phys.* **58** (1973) 1700–1709.
63. J.P. Cohen-Addad and R. Vogin, *Phys. Rev. Lett.* **33** (1974) 940–943.
64. J.P. Cohen-Addad and C. Schmit, *Polymer* **29** (1988) 883–893.
65. J.P. Cohen-Addad, *Macromolecules* **22** (1989) 147–151.
66. H. Koch, R. Bachus and R. Kimmich, *Polymer* **21** (1980) 1009–1016.
67. G. Schnur and R. Kimmich, *Chem. Phys. Lett.* **144** (1988) 333–338.
68. P.T. Callaghan, *Polymer*, **29** (1988) 1951–1959.
69. T.M. Huirua, R. Wang and P.T. Callaghan, *Macromolecules* **23** (1990) 1658–1664.
70. P.G. DeGennes, *J. Chem. Phys.* **55** (1971) 572–579.
71. M. Doi and S.F. Edwards, in *The Theory of Polymer Dynamics*, Oxford University Press, Oxford (1986).
72. M.G. Brereton, *Macromolecules* **22** (1989) 3667–3674.
73. H. Schneider and W. Hiller, *J. Polym. Sci., Polym. Phys.* **28** (1990) 1001–1014.
74. M.G. Brereton, *Macromolecules* **23** (1990) 1119–1131.
75. M.G. Brereton, I.M. Ward, N. Boden and P. Wright, *Macromolecules* **24** (1991) 2068–2074.
76. R. Kimmich, M. Kopf and P. Callaghan, *J. Polym. Sci., Polym. Phys.* **29** (1991) 1025–1030.
77. J.W. Harrell, M. Choudhury, S. Ahuja and W. Walker, *J. Polym. Sci., Polym. Phys.* **29** (1991) 1039–1046.
78. D.W. McCall, *Nat. Bur. Stand. (U.S.) Spec. Publ.* **310** (1969) 475–492.
79. K.J. Packer, J.M. Pope, R.R. Yeung and M.E.A. Cudby, *J. Polym. Sci., Polym. Phys.* **22** (1984) 589–616.
80. T.T.P. Cheung, *Phys. Rev. B* **23** (1981) 1404–1418.
81. A.M. Kenwright, K.J. Packer and B.J. Say, *J. Magn. Reson.* **69** (1986) 426–439.
82. V.J. McBrierty, *Farad. Discuss. Chem. Soc.* **68** (1979) 78–86.
83. A.D. Booth and K.J. Packer, *Mol. Phys.* **62** (1987) 811–828.
84. A.D. Booth and K.J. Packer, *ACS Polym. Preprints* **29** (1988) 19–20.
85. M.E.A. Cudby, K.J. Packer and P.J. Hendra, *Polym. Commun.* **25** (1984) 303–305.
86. K.J. Packer, I.J.F. Poplett and M.J. Taylor, *J. Chem. Soc., Faraday Trans. 1* **184** (1988) 3857–3863.

87. K.J. Packer, I.J.F. Poplett, M.J. Taylor, M.E. Vickers, A.K. Whittaker and K.P.J. Taylor, *Makromol. Chem., Macromol. Symp.* **34** (1990) 161–170.
88. A.M. Kenwright, *unpublished work.*
89. M. Goldman and L. Shen, *Phys. Rev.* **144** (1966) 321–331.
90. R.A. Assink, *Macromolecules* **11** (1978) 1233–1237.
91. T.T.P. Cheung and B.C. Gerstein, *J. Appl. Phys.* **52** (1981) 5517–5528.
92. T.T.P. Cheung, *J. Chem. Phys.* **76** (1982) 1248–1254.
93. K.L. Li, A.A. Jones, P.T. Inglefield and A.D. English, *Macromolecules* **22** (1989) 4198–4204.
94. H. Tanaka and T. Nishi, *Phys. Rev. B* **33** (1986) 32–42.
95. T.T.P. Cheung, B.C. Gerstein, L.M. Ryan, R.E. Taylor and D.R. Dybowski, *J. Chem. Phys.* **73** (1980) 6059–6067.
96. J.R. Havens and D.L. VanderHart, *Macromolecules* **18** (1985) 1663–1676.
97. P. Caravatti, P. Neuenschwander and R.R. Ernst, *Macromolecules* **18** (1985) 119–122.
98. L.A. Belfiore, T.J. Lutz, C.M. Cheng and C.E. Bronnimann, *J. Polym. Sci., Polym. Phys.* **28** (1990) 1261–1274.
99. P. Caravatti, P. Neuenschwander and R.R. Ernst, *Macromolecules* **19** (1986) 1889–1895.
100. P. Tekely, D. Canet and J.J. Delpuech, *Mol. Phys.* **67** (1989) 81–96.
101. X. Zhang and Y. Wang, *Polymer* **30** (1989) 1867–1871.
102. K. Schmidt-Rohr, J. Clauss, B. Blumich and H.W. Spiess, *Magn. Reson. Chem.* **28** (1990) s3–s9.
103. N. Egger, K. Schmidt-Rohr, B. Blumich, W.D. Domke and B. Stapp, *J. Appl. Polym. Sci.* **44** (1992) 289–295.
104. T. Kimura, K. Neki, N. Tamura, F. Horii, M. Nakagawa and H. Odani, *Polymer* **33** (1992) 493–497.
105. D.L. VanderHart, *Makromol. Chem., Macromol. Symp.* **34** (1990) 125–159.
106. G.C. Campbell and D.L. VanderHart, *J. Magn. Reson.* **96** (1992) 69–93.
107. G.C. Campbell, D.L. VanderHart, Y. Feng and C.C. Han, *Macromolecules* **25** (1992) 2107–2111.
108. S. Zhang and M. Mehring, *Chem. Phys. Lett.* **160** (1989) 644–646.
109. A.M. Kenwright and K.J. Packer, *Chem. Phys. Lett.* **173** (1990) 471–475.
110. K.J. Packer and J.M. Pope, *J. Magn. Reson.* **55** (1983) 378–385.
111. W.I. Goldburg and M. Lee, *Phys. Rev. Lett.* **11** (1963) 255–258.
112. M. Lee and W.I. Goldburg, *Phys. Rev. A* **140** (1965) 1261–1271.
113. J.S. Waugh, L.M. Huber and U. Haeberlen, *Phys. Rev. Lett.* **20** (1968) 180–182.
114. P. Mansfield, *J. Phys. C* **4** (1971) 1444–1452.
115. W.K. Rhim, D.D. Elleman and R.W. Vaughan, *J. Chem. Phys.* **58** (1973) 1772–1773.
116. D.P. Burum and W.K. Rhim, *J. Chem. Phys.* **71** (1979) 944–956.
117. A.J. Vega and R.W. Vaughan, *J. Chem. Phys.* **68** (1978) 1958–1966.
118. A.J. Vega, A.D. English and W. Mahler, *J. Magn. Reson.* **37** (1980) 107–128.
119. C. Dybowski and R.G. Pembleton, *J. Chem. Phys.* **70** (1979) 1962–1966.
120. J.R. Havens and D.L. VanderHart, *J. Magn. Reson.* **61** (1985) 389–391.
121. S.F. Dec, C.E. Bronnimann, R.A. Wind and G.E. Maciel, *J. Magn. Reson.* **82** (1989) 454–466.
122. S.F. Dec, R.A. Wind and G.E. Maciel, *Macromolecules* **20** (1987) 2754–2761.
123. B. Schneider, H. Pivcova and D. Doskocilova, *Macromolecules* **5** (1972) 120–124.
124. D.L. VanderHart, W.F. Manders, R.S. Stein and W. Herman, *Macromolecules* **20** (1987) 1724–1726.
125. A.D. English and C.R. Dybowski, *Macromolecules* **17** (1984) 446–449.
126. H.D.H. Stover and J.M.J. Frechet, *Macromolecules* **22** (1989) 1574–1576.
127. B.C. Gerstein, R.G. Pembleton, R.C. Wilson and L.M. Ryan, *J. Chem. Phys.* **66** (1977) 361–362.
128. C.E. Bronnimann, B.L. Hawkins, M. Zhang and G.E. Maciel, *Anal. Chem.* **60** (1988) 1743–1750.
129. N. Zumbulyadis, J.M. O'Reilly and D.M. Teegarden, *Macromolecules* **25** (1992) 3317–3319.
130. G.E. Maciel, C.E. Bronnimann and B.L. Hawkins, *Adv. Magn. Reson.* **14** (1990) 125–150.
131. D.G. Cory, *J. Magn. Reson.* **94** (1991) 526–534.
132. R.K. Harris and P. Jackson, *Chem. Rev.* **91** (1991) 1427–1440.
133. P. Jackson, *J. Magn. Reson.* **90** (1990) 391–396.
134. P. Caravatti, M.H. Levitt and R.R. Ernst, *J. Magn. Reson.* **68** (1986) 323–334.

135. G.A. Morris and R. Freeman, *J. Magn. Reson.* **29** (1978) 433–462.
136. A. Bielecki, D.P. Burum, D.M. Rice and F.E. Karasz, *Macromolecules* **24** (1991) 4820–4822.
137. K. Schmidt-Rohr, J. Clauss and H.W Spiess, *Macromolecules* **25** (1992) 3273–3277.
138. C.W. Chu, L.C. Dickinson and J.C.W. Chien, *J. Appl. Polym. Sci.* **41** (1990) 2311–2325.
139. V.J. McBrierty and I.M. Ward, *J. Phys. D* **1** (1968) 1529–1542.
140. V.J. McBrierty, I.R. McDonald and I.M. Ward, *J. Phys. D* **4** (1971) 88–101.
141. M. Kretz, B. Meurer, P. Spegt and G. Weill, *J. Polym. Sci., Polym. Phys.* **26** (1988) 1553–1568.
142. A.J. Brandolini and C. Dybowski, *J. Polym. Sci., Polym. Lett.* **21** (1983) 429–431.
143. J.B. Smith, A.J. Manuel and I.M. Ward, *Polymer* **16** (1975) 57–65.
144. I.M. Ward, in *Structure and Properties of Oriented Polymers*, Elsevier Applied Science, London (1975).

8 Deuterium NMR of synthetic polymers

D.M. RICE

8.1 Introduction

The development of deuteron NMR for the characterization of the segmental dynamics and orientation of polymers has taken place during just over the last decade. The particular technological advance that allowed this development was the availability of high-speed (greater than 1 MHz) signal averagers and economical high-power pulse amplifiers, which have made it possible to obtain the broad deuterium, Fourier transform lineshapes associated with crystalline materials and amorphous materials below their glass transition temperature, T_g. Much of the early development of deuteron NMR methods for materials research took place in the laboratory of Spiess and co-workers in the early 1980s [1–4]. There is also a large parallel literature involving the use of deuteron NMR for the study of biopolymers [5, 6], and biomembranes [7], areas which are also certainly important, but beyond the purposes of this review. A comprehensive review of early applications in materials research was presented in 1986 by Jelinski [8]. At that time, the field was described as at 'an early stage of development'. In the intervening years, the technical ability to obtain a deuteron quadrupole echo lineshape has become more commonplace and nearly all commercial instrument vendors offer some type of deuterium 'wide-line' option. There now exists a body of solid-state deuteron lineshape and spin-relaxation experiments [9] which can be employed to study the mechanistic details of a particular chain motion, or to obtain structural details from oriented materials.

The unique capabilities of solid-state deuteron NMR (as well as experiments with ^{13}C, ^{15}N and ^{31}P) result from the ability to assess ordering and mobility of individual bonds in a solid material. One can specifically examine the molecular mechanisms that determine bulk properties. This ability has motivated deuteron studies of such fundamental issues as the nature of the glass-transition [10–18] interactions in blends and mixtures [19–26], the molecular ordering and dynamics of crystalline materials [10, 11, 27–31] dynamics of elastomers [32–45], dielectric properties [46, 47] and mechanical spectroscopy [48–51]. New experiments continue to be tied to methods development, particularly in the areas of multi-dimensional NMR [10, 11, 46, 52–61]. Recently, deuteron NMR has been employed in the characterization of new materials, for example for new liquid crystalline polymers [62–84], conducting polymers [85–91] polyamides [92–99] and other materials [100–105]. Deuteron NMR still remains one of the smaller areas of NMR materials research, principally due to the expense associated with a necessary program of isotopic enrichment. The need to label also prevents the routine characterization of samples from the production line or when processing methods cannot be reproduced at the laboratory scale. However, with modern instrumentation, NMR technology no longer limits this approach and to a certain extent the deuteron is just another nucleus for study, particularly for the multi-dimensional exchange methods, described later in this chapter.

The objectives in this chapter are to describe, with examples, the group of experiments that is now available for study of a new material with deuteron NMR. The focus is upon methods for the characterization of order and mobility in the crystalline state or below T_g, which is our own interest. To be comprehensive, the literature of experiments associated with networks and liquid crystalline materials have been referenced. Examples are taken from the study of the conducting polymer poly(*p*-phenylene vinylene) carried out in the laboratory of F.E. Karasz at the University of Massachusetts [85–89]. Special attention is also given to several recent experiments from the laboratories of G. Zachmann [31], A. English [92–97] and H.W. Spiess [10, 11, 46, 53]. The focus is upon the means and criteria that motivate one to choose deuterium NMR for characterization.

8.2 Experimental theory

The advantages obtained from the NMR study of a deuterium-labelled polymer result from the fact that its lineshape and relaxation are determined principally by the quadrupole interaction, the coupling between the deuterium nuclear quadrupole moment and the electric field gradients at the nucleus that are due to the electronic distribution of a particular bonding arrangement [9]. The deuterium quadrupole spectrum of a C–D or N–D bond spans a breadth of 250–300 kHz. Other interactions are of about 100 times lesser magnitude, including the deuteron chemical shift, along with the deuteron–proton and deuteron–deuteron dipolar interactions [8]. With spin $\boldsymbol{I} = 1$, the high field spectrum of deuterium consists of two transitions (-1 to 0 and 0 to $+1$) which are separated by a quadrupole frequency splitting, $\Delta\nu$ (Figure 8.1(a)) [106]:

$$\Delta\nu = Q_{xx}\,\boldsymbol{x} + Q_{yy}\,\boldsymbol{y} + Q_{zz}\,\boldsymbol{z} \tag{8.1}$$

The quadrupole splitting depends upon the relative orientation of a principal axis system (PAS, unit vectors $\boldsymbol{x}$, $\boldsymbol{y}$ and $\boldsymbol{z}$) fixed in the C–D bond and the magnetic field vector, $\boldsymbol{B}_0$. The values Q_{xx}, Q_{yy}, Q_{zz} are quadrupole tensor frequency components ($Q_{xx} + Q_{yy} + Q_{zz} = 0$) associated with each principal axis and they are the quadrupole splittings, $\Delta\nu$, which result when $\boldsymbol{B}_0$ is aligned

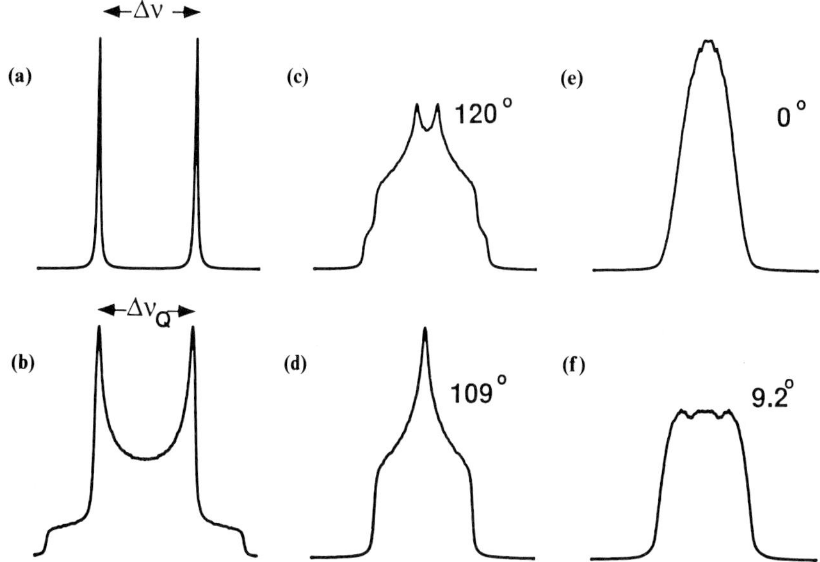

Figure 8.1 (a) The quadrupole splitting of an individual C–D bond orientation, $\Delta\nu$, is defined by equation (8.1) or (8.2); (b) a typical 'powder' spectrum for isotropic distribution static C–D bonds with $\eta \sim = 0.0$. Two-site jumps lead to apparent axial asymmetry ($\eta' > 0$); (c) a powder spectrum for fast phenylene ring-flips (jumps angle 120°, $\eta' = 0.6$); (d) a powder spectrum for fast two-site jumps on a tetrahedral lattice (jump angle 109°, $\eta' = 1.0$). Spectra of a phenylene ring on a partially oriented chain (10° Gaussian uncertainty) (e) whose 1,4 axis is aligned with the chain axis ($\psi = 0.0$) and (f) whose 1,4 axis is tilted at an angle ($\psi = 9.2°$) to the chain axis.

with each of x, y and z. For a C–D bond, the electronic distribution has approximate axial symmetry about the bond direction (the z axis). Q_{zz} is therefore unique and $Q_{xx} = Q_{yy} = -Q_{zz}/2$. An isotropic distribution of static C–D bonds yields a distribution of quadrupole splittings and a lineshape that is approximately a 'Pake doublet' (Figure 8.1(b)) [8, 9]. The frequencies Q_{xx}, Q_{yy} are obtained from the frequency splitting between peaks of Figure 8.1(b) and Q_{zz} is obtained from that of the outer edges.

When (x, y, z) are expressed in spherical coordinates, Θ and Φ, a more familiar equation (8.2) results [8]:

$$\Delta\nu = \Delta\nu_Q(3\cos^2\Theta - 1 - \eta\sin^2\Theta\cos 2\Phi) \qquad (8.2)$$

$\Delta\nu_Q = Q_{zz}/2$. The value of $\eta = (Q_{yy} - Q_{xx})/Q_{zz}$ (with $Q_{zz} < Q_{yy} < Q_{xx}$) quantifies the deviation from axial symmetry. For C–D bonds, $\Delta\nu_Q$ is typically 120–130 kHz and the asymmetry parameter is $\eta < 0.05$. Equation (8.2) also establishes the conventional nomenclature for labelling tensor axes according to the magnitude of the components. However, this convention is not necessary for the calculation of a lineshape.

Deuteron NMR can be used to study motion. Fast motions that reorient the C–D bond characteristically alter the lineshape. Figure 8.1(c) shows the spectrum for an isotropic distribution of phenylene rings which undergo fast 180° rotational jumps ('ring flips') about the 1,4 axis [107, 108]. The jumps average the quadrupole splittings of the C–D bonds at the 2 and 6 positions (or 3 and 5). The motionally averaged lineshape depends upon the orientation of $\boldsymbol{B}_0$ relative to an average principal axis system (x', y' and z') through equation (8.1). A two-site jump spectrum has a motionally averaged or apparent asymmetry parameter η' (defined by equation (8.2)) that may not be equal to zero. The parameter η' also depends upon the angle between the jump sites ($\eta' \approx 0.60$ for an angle of 120°). Figure 8.1(d) shows a second lineshape that is expected for two-site reorientation with equal populations on a tetrahedral lattice ($\eta = 1.0$ for an angle of 109°) [10]. This particular lineshape has been obtained for reorienting polymethylene chains [109]. The jump spectrum also depends upon the a priori populations of the jump sites, which might be unequal for a group without symmetry, such as the polymethylene chain.

Deuteron NMR can also be used to characterize orientation. An oriented film or fiber with chain alignment has a spectral shape characteristic of a modified distribution of C–D bond orientations [4]. Figure 8.1(e) shows the lineshape for a phenylene ring whose 1,4 axis is approximately aligned with $\boldsymbol{B}_0$ (the figure is calculated with a 10° Gaussian uncertainty of alignment) [80]. Because the C–D bonds of the phenylene ring are at 60° and 120° to the 1,4 axis, only C–D bond orientations at about these angles to $\boldsymbol{B}_0$ contribute to this spectrum. The probable quadrupole splittings are therefore $\Delta\nu \approx \Delta\nu_Q\,(3\cos^2(60° \text{ or } 120°) - 1)$, values near zero. Orientation about an axis other than the 1,4 axis, resulting either from a tilt of the ring relative to the orientation axis or a change in the uncertainty of alignment, will alter the pattern and lead to a new distinct lineshape. For example, Figure 8.1(f) is a spectrum of a phenylene ring which is tilted by 9.2° relative to the axis of alignment [87]. Note the substantial broadening from the additional orientations that result from the tilt.

Data about chain alignment and reorientation must usually be obtained by modelling or simulation of the lineshape, and usually, data from several different experiments are required to yield an unambiguous conclusion. An exception involves the more recent multi-dimensional experiments, described in section 8.5 [53, 59], that are designed to yield the appropriate data graphically or with simple calculation. Even with 2-D spectroscopy, if multiple motions are present or if diffusion is also present, there remains a requirement to model the data.

8.2.1 Basic pulse sequences

Several basic deuteron pulse sequences are shown in Figure 8.2. These pulse sequences form the basis of a group of deuteron experiments that allow one to obtain the spectra of oriented materials and provide for characterization

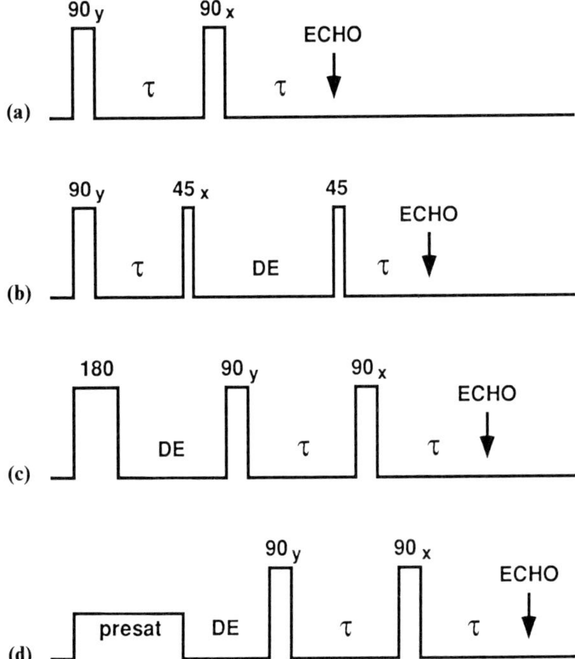

Figure 8.2 Basic 1-D pulse sequences: (a) the quadrupole echo; (b) the spin-alignment echo; (c) inversion-recovery quadrupole echo; and (d) quadrupole echo with pre-saturation for amorphous/crystalline selection. τ is the quadrupole echo pulse spacing in each sequence, DE is a delay for spin–lattice relaxation, of either (b) quadrupolar order or (c) and (d) Zeeman order. Flip angles are labelled above the pulses.

of motion over a wide range of correlations times ($10^0 < \tau_c < 10^{-10}$ s). The two-pulse quadrupole echo [110] (90_y–τ–90_x–τ–echo, Figure 8.2(a)) is most often used to obtain the solid-state deuteron spectrum. It is impractical to obtain a deuterium spectrum with a single pulse, due to probe ringing in the first 10 μs following the pulse. For the quadrupole echo, magnetization is refocused at a time τ following the second pulse, allowing complete observation after the ringing has decayed. When motion is present, the magnitude and shape of the quadrupole echo spectrum can depend upon the pulse spacing, τ, resulting from intermediate rate exchange ($10^{-7} < \tau_c < 10^{-3}$ s) due to molecular reorientation [107]. This effect has been labelled 'echo-distortion' [9], although it is a fundamental property of a motionally averaged line shape. In the absence of intermediate-exchange motion, the decay of the quadrupole echo is governed by a relaxation time T_{2E}, analogous to T_2 of the Hahn spin echo for spin 1/2 nuclei. For the deuteron, T_{2E} is usually a few milliseconds and is also affected by the static chemical shift anisotropy and dipolar coupling. In contrast, decay due to echo distortion can be very rapid (<20 μs).

The three-pulse spin-alignment echo [111, 112] (90_y–τ–45_x–DE–45–τ–echo, Figure 8.2(b)) provides a second method with which to obtain a spectrum. For DE = 0, the experiment is similar to a quadrupole echo and the DE dependence is determined by slow reorientation ($10^{-3} < \tau_c < 10^0$ s). In the absence of slow motion, this dependence is determined by a spin–lattice relaxation time T_{1Q} (see section 8.2.2), the decay of quadrupole order. The value of T_{1Q} is determined by faster motions, $\tau_c = 1/\omega_0$, where ω_0 is the Larmor frequency. Typically, both T_1 (the usual Zeeman spin–lattice relaxation time) and T_{1Q} are 'anisotropic' [113] (i.e dependent upon orientation) and the anisotropy is specific to a particular mechanism of motion. Kothe and co-workers have presented a 2-D method which allows separation of the effects of relaxation in a second dimension [114].

Figure 8.2(c) is an inversion-recovery quadrupole echo pulse sequence, which is used to measure the Zeeman spin–lattice relaxation time, T_1, with quadrupole echo detection [8, 9, 115]. Pre-saturation (Figure 8.2(d)) or progressive saturation (variation of the delay between transients) are also used to measure T_1. Notably, pre-saturation with spectral subtraction can separate the spectra of domains with different T_1 and is used to obtain the individual spectra of the amorphous and crystalline regions of semicrystalline polymers [8]. Also, Vold and co-workers have recently presented methods involving selective inversion for the measurement of slow molecular reorientation, which provide an alternative to spin alignment or multidimensional methods [116].

8.2.2 A vector model for spin 1 dynamics

The spin dynamics of the deuteron (spin $\boldsymbol{I} = 1$) are more complex than those of the spin 1/2 nuclei, and the simple vector model used in other chapters, derived from the Bloch equations, provides no particular insight into deuteron spin dynamics. However, some of the geometric simplicity of the Bloch equations is present in a product–operator formalism, used to describe spin 1 NMR [117]. This formalism can provide a visual understanding of the deuteron pulse sequences in terms of simple precession and pulse rotations, albeit among a greater number of coordinate axes. The formalism can be used to understand the production of quadrupole order and the T_{1Q} relaxation time (Figure 8.2(b)) and the two-dimensional deuteron exchange experiment (section 8.5).

For spin 1 nuclei the three components of the spin I_x, I_y, and I_z are replaced with nine product operators I_{xm}, I_{ym} and I_{zm} ($m = 1, \ldots, 3$). Three of these are directly proportional to the observable components of the rotating frame spin angular momentum ($I_{x1} = (1/2)I_x$, $I_{y1} = (1/2)I_y$ and $I_{z1} = (1/2)I_z$). Each of these three components can be grouped with two others ($m = 2$ and 3, which are non-observable) to yield three, three-dimensional subsystems (for example I_{x1} with I_{x2} and I_{x3}, etc).

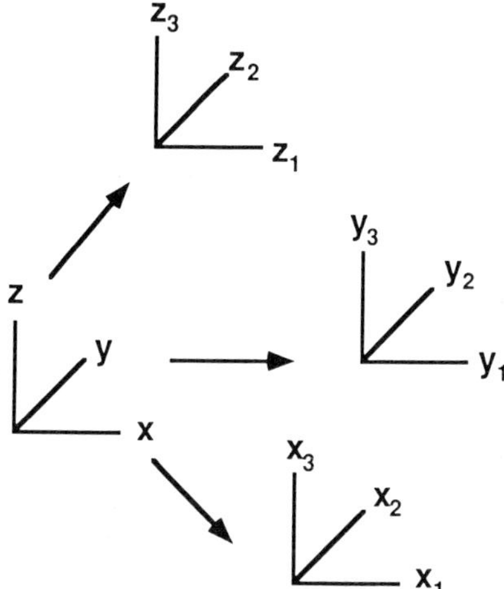

Spin 1 dynamics is described by three rules: (1) pulses with a rotation strength of ω_1 cause the usual rotations among the observable components (I_{x1}, I_{y1} and I_{z1}); (2) precession due to the quadrupole interaction is confined to a particular subsystem. For example magnetization along I_{x1} (created by a 90° pulse along I_{y1}) precesses in the I_{x1}, I_{x2} plane (or I_{y1} in the I_{y1}, I_{y2} plane, etc); (3) pulses with a strength ω_1 rotate the non-observable components within a subsystem with twice the rate, $2\omega_1$. In a subsystem, a 45° pulse becomes a 90° pulse, a 90° pulse becomes a 180° pulse, etc.

For example, after the first pulse of the quadrupole echo, rotation occurs among I_{x1}, I_{x2}. The second 90° pulse transforms I_{x2} to $-I_{x2}$ and refocusing occurs at I_{x1} after a delay, τ. These rotations are analogous to those that occur among I_x, I_y and I_z during the normal 90_x–τ–180_y Hahn spin echo, and the τ dependence of the quadrupole echo lineshape is therefore analogous to that of the Hahn spin echo. T_{2E} defines relaxation of the vector in the I_{x1}, I_{x2} plane, analogous to T_2 in the I_x, I_y plane for the Hahn echo.

For the spin alignment echo experiment, the first 45° pulse creates alignment along I_{x3}. I_{x3} is the sum of two components, the first proportional to the quadrupolar order, which is affected only by T_{1Q} spin–lattice relaxation processes (the second component, double quantum coherence, decays rapidly and can be neglected). The second 45° pulse returns the altered magnetization to the I_{x1}, I_{x2} plane where the resulting intensity is observed as a quadrupole echo. Slow exchange or T_{1Q} relaxation during DE alters the detected magnetization.

8.2.3 Experimental methods

The equipment and special methods of deuteron NMR provide the means to accurately represent the deuteron spectrum, which is among the broadest

NMR spectra that are routinely obtained [118, 119]. Deuteron spectra should be obtained with the quadrupole echo (Figure 8.2(a)) in almost all cases where the linewidth is greater than about 10–20 kHz. A minimum convenient pulse spacing is $\tau = 20\,\mu s$, a longer τ must be used for dilute samples (when the echo amplitude is small compared to the ring-down) or with an NMR instrument equipped with a solution-state pre-amplifier. There is usually no penalty for longer values, unless T_{2E} is short or echo distortion dominates the lineshape. When using a pulse spacing near $20\,\mu s$, one must beware of distortions caused by a phenomenon known as virtual signal from the first pulse and residual probe or pre-amplifier ringing [118]. Virtual signals are minimized by using high power and a short 90° pulse ($< 2.5\,\mu s$) or a large value of τ. The effects of probe ringing are minimized by phase cycling.

A 90° pulse of less than $2.5\,\mu s$ is also necessary for reasonable excitation of the spectral bandwidth, particularly for the three-pulse sequences (Figures 8.2(b), (c)) where roll-off at the edge of the bandwidth and the effects of virtual signals are magnified. A typical deuterium probe is a 5 or 7 mm horizontal solenoid and for this configuration, pulse power should be about a kilowatt. Several commercial amplifiers supply this power level. Composite pulse sequences [120] extend the spectral bandwidth but are less useful when echo distortion is present. Noticeable roll-off is always present with any pulse sequence and correction of simulations for finite-pulse should always be used [121].

Above liquid helium temperatures, the intensity pattern of a pure deuterium quadrupole echo spectrum is always symmetric about the resonance position (a large chemical shift anisotropy can, however, cause minor spectral asymmetry) [122]. Otherwise, the lack of spectral symmetry is an artifact which has two sources, 'phase-glitch', resulting from a finite phase rise-time of pulses, and an asymmetric probe bandwidth for detection. Phase-glitch is minimized with tuning procedures developed for ^1H multiple-pulse NMR. It should be recognized that the detection bandwidth of the NMR coil is usually just greater than the width of the deuterium spectrum and is also slightly asymmetric. Correction for this distortion is required in addition to the finite-pulse correction. For careful work, one might calibrate the off-resonance response of the probe with a liquid sample such as D_2O [115] (or D_2SO_4 at high temperature) [95] and apply this function to simulations in addition to the pulse-width correction. More simply, the total distortion (probe plus finite-pulse) might be determined with a known lineshape. Addition of a capacitor and a Q-damping resistor will symmetrize and broaden the probe bandwidth, although at the expense of power loss and sensitivity. Symmetrized and Q-damped NMR coils are available from instrument vendors.

Although deuteron spectra are symmetric, they should be acquired with quadrature detection (acquisition of two orthogonal receiver channels) in order to evaluate the artifactual lack of spectral symmetry. Some investigators prefer to fold a lineshape before simulation. Before folding, the spectrum should be known to have been acquired on resonance, and contain no dispersive features. To obtain correct phase, acquisition should be begun before the echo and the receiver phase should be set to obtain all signal in the real channel. The first point is shifted to the echo top. It is useful to over-digitize the echo (a 2–5 MHz digitization rate is common) to ensure that a point is present at the peak. Alternatively, one can interpolate between the points with a spline fit of the echo.

Curve-fitting deuteron spectra is still an art. Individual models are often distinguished by small features of the lineshape and the judgement of the investigator is required. For a multi-parameter model, it is best to have visual understanding of the multi-dimensional array of acceptable simulations as well as the standard deviation for each fit. It should be recognized that the $\pm$ error associated with asymmetry and phasing will usually be greater than that due to the signal-to-noise ratio.

8.3 Polymer orientation

Deuteron NMR is used to characterize the chain alignment of drawn films and fibers [4, 27, 31, 87, 89], or for materials that have been aligned in other ways, such as through poling [46] or with a magnetic field [80]. Methods for the study of orientation [3, 4] were developed for the study of high density polyethylene and NMR remains a useful characterization method for this material [27]. The goal in this type of experiment is to characterize the distribution of chain orientation angles about the draw axis. The result is often expressed as an 'order parameter' or 'Hermans orientation function' [123], $f = \langle 3/2 \cos 2\beta - 1/2 \rangle$ where f is an average over the chain orientation distribution $P(\beta)$. The results are often compared with X-ray diffraction data [123] or that from spectroscopic dichroism [124]. An advantage of NMR is that it will provide the complete orientation distribution $P(\beta)$ [125]. Spectroscopic dichroic methods yield only the orientation function. NMR methods also do not depend upon long-range crystalline order (as do diffraction methods) and it is possible to study poorly ordered or amorphous materials. However, the NMR experiment is also more difficult than these other methods and probably only justified in special cases if the orientation function is the only goal.

A second, and perhaps more unique, goal of the orientation experiment is the determination of chain conformation. The NMR experiment can reveal details of conformation which may be below the resolution of the fiber diffraction experiment. Two examples of the second approach are described, a study of the conducting polymer poly(*p*-phenylene vinylene) (PPV) [87, 89] and of poly(ethylene terephthalate) (PET) [31].

*8.3.1 Stretched films of poly(*p*-phenylene vinylene)*

Films of the conducting polymer, poly(*p* phenylene vinylene), (PPV) are obtained from a water-soluble sulfonium-salt precursor polymer. Chain-aligned

PPV films of high crystallinity are obtained by elimination through simultaneous heating and stretching of the precursor film, followed by high temperature annealing. PPV possesses neither a glass-transition nor a crystalline melting temperature. Doping by both oxidation and reduction produces a conducting material. Simpson and co-workers have used deuteron NMR to examine the structure and chain alignment of these films [87, 89].

As discussed in section 8.2.1, the lineshape of a deuteron spectrum of an oriented material is determined by the distribution of C–D bonds relative to the magnetic field, and can be calculated from equation (8.1) with knowledge of the distribution of coordinates $\boldsymbol{x}$, $\boldsymbol{y}$, $\boldsymbol{z}$ or equivalently of Θ and Φ for equation (8.2). This distribution is obtained through a series of rotational coordinate transformations:

$$\text{PAS} \Longleftrightarrow \text{CHAIN} \Longleftrightarrow \text{FILM} \Longleftrightarrow \text{LAB}$$

$$\boldsymbol{R}_1(\alpha_1\beta_1\gamma_1) \quad \boldsymbol{R}_2(\alpha_2\beta_2\gamma_2) \quad \boldsymbol{R}_3(\alpha_3\beta_3)$$

where α_n, β_n and γ_n are the Euler angles for the transformation between coordinate systems. Figure 8.3(a) shows the coordinate transformations that define the orientation of C–D bonds in drawn PPV. $\boldsymbol{R}_1$ relates the principal axis system (PAS, $\boldsymbol{x}$, $\boldsymbol{y}$, and $\boldsymbol{z}$) of the C–D bond to a CHAIN or molecular-frame axis system. For a static material, $\boldsymbol{R}_1$ depends upon the bond lengths and angles of a segment. If motion is present, the PAS of the C–D bond is replaced by a motionally averaged PAS ($\boldsymbol{x}'$ $\boldsymbol{y}'$ and $\boldsymbol{z}'$). For static PPV, with a repeating *trans*-stilbene-like structure, the 1,4 ring axis is expected to be oriented at an angle $\psi = 9.2°$, where ψ, the 'ring-tilt' angle, is the angle between the ring, $\boldsymbol{r}$, and the unique chain axis, $\boldsymbol{c}$ (Figure 8.3(a)). The two C–D bonds are therefore oriented at $60 \pm \psi°$ to the unique chain axis. If the chain axis were perfectly aligned with $\boldsymbol{B}_0$, two quadrupole splittings would result.

$\boldsymbol{R}_2$ relates the CHAIN axis sytem to a macroscopic FILM or 'molecular director axis' (MD) [31], fixed in the drawn PPV film. A film might also contain an additional anisotropic chain orientation distribution about an axis normal to the film. This axis, ND, and a transverse axis, TD are also defined. The transformation $\boldsymbol{R}_2$ contains information about the distribution of chain orientations. The full lineshape is obtained by integration over the angles, α_2, β_2 and γ_2 with the chain orientation distribution $P(\alpha_2\beta_2\gamma_2)$. For PPV, chain alignment is uniaxial and $\boldsymbol{c}$, the unique CHAIN axis, is distributed isotropically about MD. Also the normal to the ring plane ($\boldsymbol{x}$ in Figure 8.3) is distributed isotropically about the CHAIN axis. Therefore $P(\alpha_2, \gamma_2) = 1.0$ and the orientation distribution is simply $P(\beta_2)$. For a reasonably symmetric orientation distribution, one can usually reduce the number of angles of integration from three to two [80]. Spiess has also presented analytical expressions for sub-spectra that can be used when echo distortion is absent [4]. $\boldsymbol{R}_3$ describes the orientation of the FILM axis system relative to $\boldsymbol{B}_0$, the unique axis of the LAB frame. β_3 is the angle between the draw axis and the magnetic field. An additional sample rotation of α_3 about the draw axis is required to obtain information about anisotropic chain orientation.

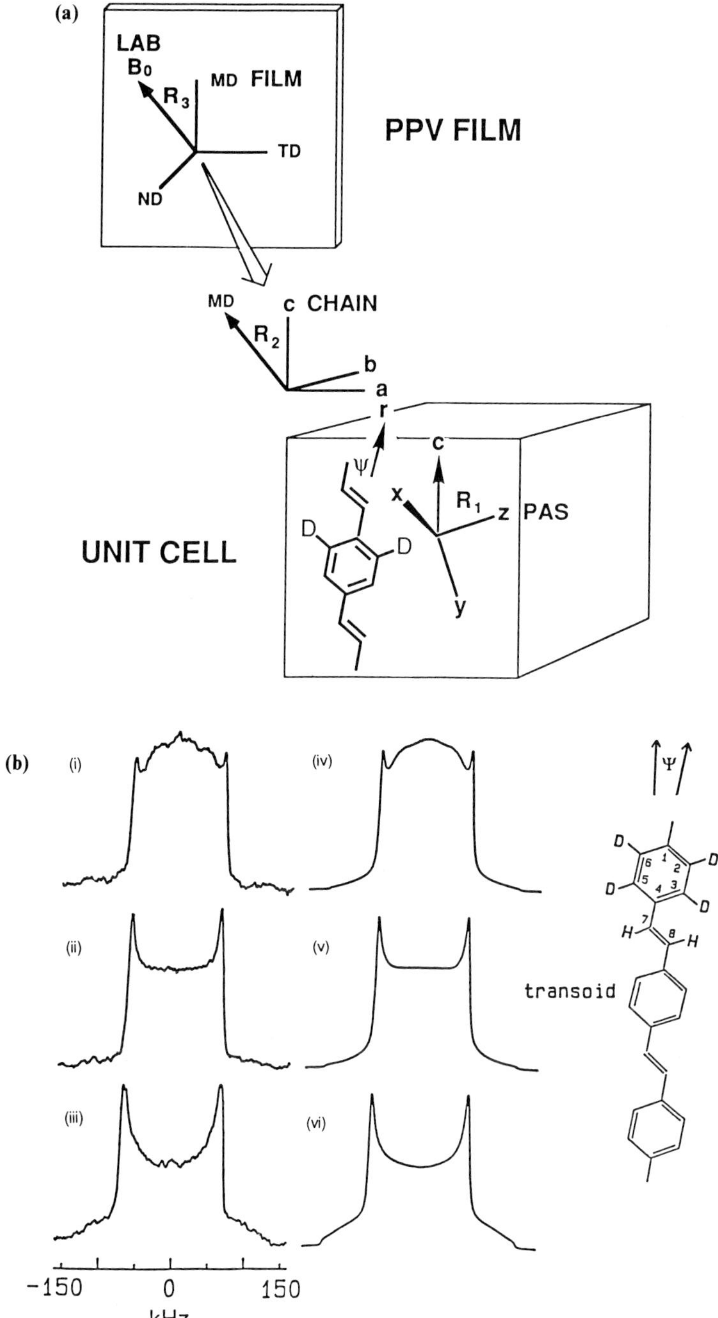

Figure 8.3 (a) The four coordinate frames (LAB, FILM, CHAIN and PAS) that are used to describe the distribution of C–D bond orientations of an oriented film or fiber, as described in the text. (b) Experimental deuteron spectra of drawn PPV with (i) $\beta_3 = 0°$, (ii) $\beta_3 = 15°$ and (iii) $\beta_3 = 90°$. The corresponding modelled lineshapes are shown in (iv), (v) and (vi), respectively. (Reprinted from J.H. Simpson, D.M. Rice and F.E. Karasz, *Macromolecules* **25** (1992) 2099–2016 by permission of the publishers, American Chemical Society.)

The description of deuterium data from an oriented sample is inherently a modelling process. The three transformations contain eight angles which can all potentially represent parameters in a simulation. One's model for the CHAIN axis system, i.e. the conformation of the segment, and $P(\alpha_2, \beta_2, \gamma_2)$ interact and influence the final result. NMR data and diffraction data should be used together, and ideally both should be modelled with similar assumptions. Convenient commercial software exists for the modelling of diffraction data, but to our knowledge as yet no similar programs are available for deuteron NMR.

Figure 8.3(b) shows a set of three spectra of drawn PPV for $\beta_3 = 0°$, 15° and 90°, together with corresponding modelled data. Simpson *et al.* [87] showed that the well-aligned central component of these spectra was best fit with a uniaxial Gaussian distribution of chain (or crystallite orientations) with a 10° width (at the 1/e points) and with a ring-tilt, $\psi = 7.7°$. For a complete fit, a second broader distribution was also required; the second component represented either a poorly oriented part of the sample or part of a non-Gaussian orientation distribution $P(\beta_2)$ for crystallites. The experimental value of $\psi = 7.7°$ supported a picture of PPV as a repeating *trans*-stilbene structure ('transoid'). An alternating *trans*-stilbene structure ('cisoid') with $\psi = 0.0°$, was inconsistent with the experimental data. Such a distinction was possible because the width of $P(\beta_2)$ was of the same order as the ring-tilt. The difference between 9.2° and 7.7° was significant and was attributed to a small amount of disorder within crystallites. Simpson *et al.* [89] also examined a vinylene labelled PPV-d_2 and obtained the same conclusion. This model for the PPV structure was consistent with that proposed from diffraction data. With NMR data Simpson *et al.* were able to evaluate the applicability of the pseudo-affine model for PPV orientation.

8.3.2 Planar orientation of polyethylene terephthalate

Zachmann and co-workers have examined the deuterium spectra of PET that has been oriented through necking at 45°C followed by annealing at 240°C [31]. They also examined a non-annealed sample which was shown to have a different structure. PET presents a somewhat more complex problem for simulation than PPV because of the possibility for planar orientation of the PET chains in addition to the axial orientation. PET also possesses a complex morphology including crystalline domains, amorphous domains and boundary regions, and the alignment of each of these domains could potentially be different. T_1 selection methods (Figure 8.2(d)) were used to separate the spectra of amorphous and crystalline domains. Some of the issues are similar to those of PPV, the breadth of the orientation distribution and the ring-tilt angle. In addition, it was shown that annealing of PET causes parallelization of the phenylene rings relative to the ND axis, i.e. chain alignment about two axes. The structure of a PET segment is drawn in Figure 8.4, as des-

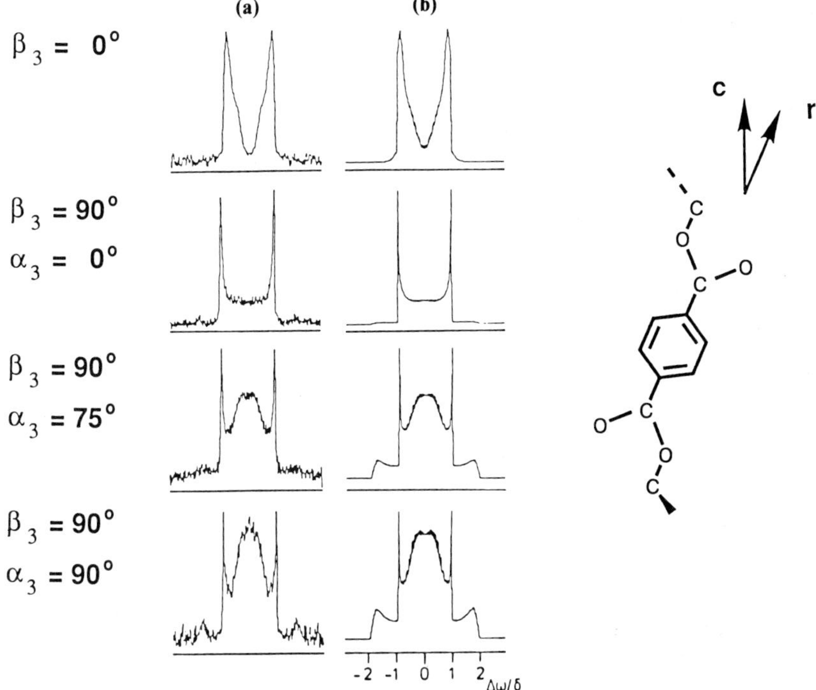

Figure 8.4 (a) Measured and (b) calculated deuteron spectra of the crystalline regions of oriented PET and a drawing of PET. The spectra are dependent upon α_3 (the angle of rotation around the draw axis) for β_3 (the angle between the draw axis and the magnetic field) = 90°. (Adapted from S. Rober and H.G. Zachmann, *Polymer* **10** (1992) 2061–2075 by permission of the publishers, Butterworth–Heinemann Ltd.)

cribed by Rober and Zachmann [31] (for PET $\psi \approx 24°$). Other angles that describe the orientation of PET are similar to those of Figure 8.3. The chain alignment is described with two uniaxial Gaussian functions. The first describes the distribution of $\boldsymbol{c}$ about MD and the second for the phenylene-ring normal axis, $\boldsymbol{x}$ relative to ND, following the use of X-ray pole-figures. Hermans functions, f are determined for each distribution.

For drawn PET which has not been annealed it is the 1,4 axis of the phenylene rings that are preferentially oriented in the draw direction with a broad orientation distribution. The crystallization which results from annealing, however, clearly tilts the ring axis so that the chain axis becomes preferentially oriented in the draw direction. It also causes the parallelization of the phenylene rings. Figure 8.4 shows the spectra of the crystalline regions of a ring-deuterated PET obtained by Rober and Zachmann [31] which support this conclusion. Information about planar orientation is clearly evident for rotation of α_3 when $\beta_3 = 90°$.

It should be noted that other NMR methods are available for the study of orientation, as discussed in chapters 5 and 7. Harbison *et al.* [126] have presented a 2-D ^{13}C magic-angle spinning (MAS) method, based upon the orientation dependence of the anisotropic ^{13}C chemical shift (see chapter 5) [126]. This method possesses a significant advantage over deuteron NMR, because it allows the study of orientation from all ^{13}C sites simultaneously with a natural abundance sample. In a 1-D slow-spinning MAS experiment, side-bands are present that retain information about the original chemical shift anisotropy. For a material that has been oriented in the MAS rotor, the two-dimensional method re-introduces sufficient information to reconstruct the original line shape. Harbison *et al.* [126] and Tzou *et al.* [127] have used this method to study PET.

It is useful to compare the ^{13}C method with the deuteron method—should one spin or label? We suggest that to do both is desirable. Notably, Harbison *et al.* [126] were able to strongly establish that the normal to the ring plane was perpendicular to MD, however, the uncertainty of the ring tilt angle was larger. In contrast, Rober and Zachmann [31] were able to strongly establish the ring-tilt angle as well. With the ^{13}C method, it was possible to observe a hint of the planar orientation while with ^2H, it was possible to measure the orientation function. Angular resolution of the orientation experiment is determined by the orientation and relative magnitude of quadrupole tensor components, the anisotropic chemical shift components σ_{xx}, σ_{yy} and σ_{zz} for ^{13}C (chapter 5), and the intrinsic linewidth. For rotation of the ring normal, the two methods are quite comparable; this rotation involves the unique tensor axes for both ^2H and ^{13}C (for ^{13}C σ_{zz} is perpendicular to the ring plane and for $^2H Q_{zz}$ is in the ring plane). Both changes are large. For characterization of ring-tilt angle, the deuteron method is preferable; this rotation involves σ_{xx} and σ_{yy} for ^{13}C whose difference is smaller. For methylene groups, the ^{13}C anisotropy is small and deuteron NMR is clearly the preferable method. Both methods should be equally suited for the study of planar orientation, although the ^{13}C method has a minor disadvantage; one must reposition the sample in the MAS rotor and get it to spin.

It appears that 2-D MAS side-band patterns are easier to simulate than the static lineshape. The information of a deuteron lineshape is usually found in regions of high slope (i.e in the vicinity of peaks and edges). Harbison *et al.* [126] note that MAS transforms this information into intensity data for side-bands for which a $\pm$ error is more readily determined. The use of the 2-D MAS method with deuterium is feasible and is a worthwhile alternative approach to simulation of the non-spin lineshape. The two approaches should yield the same answer but would be affected differently by experimental error.

8.4 Polymer motion

Deuteron NMR is used to obtain mechanistic information about local segmental motions in the solid state. The results have had an impact upon several issues, among them the study of chain motions that are associated with the glass-transition temperature, T_g [10–18] or other mechanical or dielectric relaxations of the amorphous state [3, 48–51, 80, 128, 129]. Amorphous polymers, including atactic-polypropylene, a-PP [10, 11] and polycarbonate [48–51] continue to be of interest, along with polymer blends and other mixtures [19–26]. Miura *et al.* [94] have examined the amorphous domains of Nylon-6,6. Interest in the crystalline state is twofold. One is interested in mechanisms associated with mechanical or dielectric spectroscopy, but also characterization of a motional mechanism will reveal details about conformation that are unavailable from diffraction experiments. Two experiments provide good examples, a 2-D study by Hirschinger *et al.* [46] of poly(vinylidene fluoride), PVF_2 and a 1-D study by Simpson *et al.* [88, 89] of PPV. Two other examples are described. Hirschinger *et al.* [95] have used deuteron NMR to reveal the nature of the Brill transition of Nylon-6,6 and Hagemeyer *et al.* [10] and Schaefer *et al.* [11] have examined motion in isotactic polypropylene, i-PP.

It has been shown in chapters 4 and 6 that the individual mechanisms describing polymer chain motion in the bulk are often complex. An individual segment may simultaneously execute small-angle libration or rotational diffusion ($10^{-12} < \tau_c < 10^{-9}$ s) and rotational jump motion with a broad distribution of correlation times ($10^{-7} < \tau_c < 10^{-3}$ s). An amorphous material may also undergo slow, large-amplitude chain motions ($10^{-3} < \tau_c < 10^{0}$ s), associated with the glass transition. Crystalline materials often permit slow jumps. Data from a single experiment are usually inadequate. A combination of lineshape and relaxation experiments should be used, based upon the pulse sequences of Figure 8.2 and more recently 2-D spectroscopy. The examples show the use of different combinations of methods.

*8.4.1 Ring-flip motion of poly(*p*-phenylene vinylene)*

The 180° rotational jump of the phenylene ring (the 'ring-flip', Figure 8.1(c)) has been a motion frequently studied by deuteron NMR, due to a quite distinctive lineshape and the ubiquity of this group in materials of interest. Ring-flips are present in annealed PPV films. Figure 8.5 shows a set of deuterium spectra and simulations, obtained by Simpson *et al.* [88] with the quadrupole echo (Figure 8.2(a)) for a ring deuterated PPV-d_4. At low temperature, the spectrum is a rigid lattice Pake doublet, while at high temperature, the spectrum is dominated by the ring-flip motion (Figure 8.1(c)). The result was initially surprising, due to the high crystallinity and conjugated structure of PPV. The activation energy for the flips was of interest to provide information about bond rotation energy and potential packing disorder associated with the PPV crystalline structure.

Quadrupole echo lineshapes for the phenylene ring-flip motion (or any jump model) are simulated with an *n*-site exchange calculation; one must also include the exchange-caused relaxation between the pulses (echo distor-

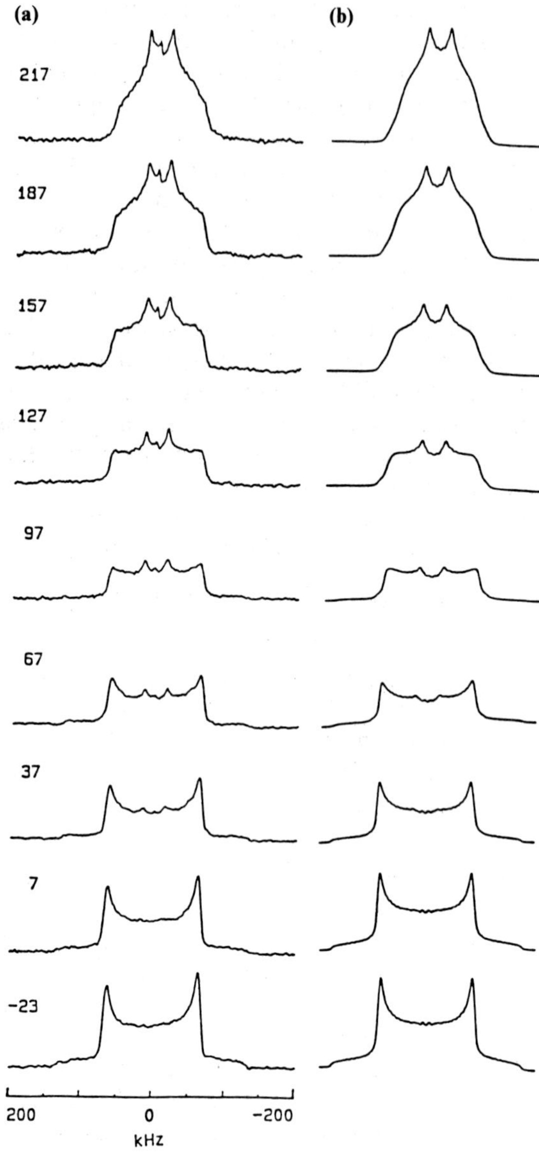

Figure 8.5 (a) Deuteron spectra of PPV-d_4 at different temperatures in °C and (b) simulations, showing the effect of a distribution of ring-flip rates. (Reprinted from J.H. Simpson, D.M. Rice and F.E. Karasz, *J. Polym. Sci. B, Polym. Phys. Ed.* **30** (1992) 11–18, John Wiley and Sons, Inc.)

tion, see section 8.2.1) [9, 113]. The correctness of this picture has been verified with studies of crystalline model compounds [108, 115]. Notably, spectra that are affected by intermediate jumps ($\tau_c = 10^{-5}$–10^{-4}) have a strong shape and intensity dependence upon either temperature or the value of the pulse spacing, τ, and this distortion can be used to evaluate the distribution of jump rates. The experimental intensity is best obtained from the time-domain echo height. The probe response must be calibrated for various temperatures, for a material with constant echo intensity, for example crystalline alanine with a CD_3 group [88]. For the intermediate-rate ring-flip, up to 80% of echo intensity is lost within 30–100 μs. Crystalline peptides show nearly this full loss, suggesting a narrow distribution of jump rates. In contrast, an amorphous polymer such as a poly(ethersulfone) [129] or poly-(butylene terephthalate) (PBT) [128] can undergo the full transition from static spectrum to jump spectrum with little loss of intensity. The latter effect is attributed to broad distribution of jump rates which might span seven orders of magnitude in an amorphous polymer [8].

Spectra can be simulated with a superposition of ring-flip spectra whose rates, k, are defined by a log-Gaussian distribution, $P(\log(k))$:

$$P(\log(k)) \propto \mathrm{e}^{-[(\log(k) - \log(k_0)/\Delta \log(k)]^2} \tag{8.3}$$

Each spectrum yields two parameters, a median jump rate, k_0 and a log-Gaussian width $\Delta \log(k_0)$. An activation energy, E_a, can be obtained from an Arrhenius plot of the jump rate. Typically, the distribution of jump rates is attributed to a distribution of activation energies and

$$\Delta \log(k) = \frac{\Delta E_a}{2.303RT} \tag{8.4}$$

Notably, a small distribution of activation energies can cause a large distribution of rates.

Spectra of PPV (Figure 8.5) show a 50% drop of intensity at temperatures of about 100°C. The activation energy was determined to be $E_a = 15$ kcal/mol and the width of the log-Gaussian function of ± 2 (four orders of magnitude between the 1/e points of the distribution) explained the intensity change. A 2–3 kcal distribution of activation energies would be sufficient to cause this distribution. The gas-phase bond rotation energy for the single bonds of stilbene has been estimated to be 5–10 kcal/mol [130] (potentially 10–20 kcal/mole for two bonds). Therefore, intramolecular effects might explain the entire activation energy of the PPV ring-flip and packing energies must be small or cooperative motion must be present. Doping of PPV, to produce the conducting state, is expected to increase the conjugation length of PPV and raise the activation energy for flips. This effect has been observed with H_2SO_4 doped PPV [86]. Doping, however, also converts the herringbone packing of rings to parallel packing, and this change may also affect the activation energy [131].

Kaplan *et al.* [91] have examined the conducting polymer polyaniline (PA). The morphology of PA is considerably more heterogeneous than PPV. Echo distortions are not present and some rings remain static at all temperatures. Nevertheless, doping was shown to reduce the population of flipping rings. English and co-workers have made measurements with ring deuterated poly(phenylene terephthalamide) PPTA [29, 30]. The structure of PPTA is quite comparable to PPV (an amide group replaces vinylene group). They find a slightly lower activation energy for flips (12 kcal) and a broader distribution 4 kcal. Packing interactions must determine this activation energy.

For amorphous materials, the ring-flip has often been associated with a β-mechanical or dielectric transition. The flip is not necessarily the cause of the transition itself, but activation energies are coincident. It is tempting to replace equation (8.3) with a theoretical function more meaningful for mechanical analysis [132], but to accomplish this experiment precise measurements of the ring-flip rate distribution are needed. Also, with deuterium NMR only one frequency, the quadrupole coupling frequency, is available and direct time-temperature correlation is not possible [133]. Nevertheless Wherle *et al.* [50] have made precise measurements of the ring-flip rate distribution for the β-transition of polycarbonate. For polycarbonates, the distribution is distinctly non-Gaussian, and antiplasticizers increase the population of slowly flipping rings [49–51].

Jumps are also characterized redundantly from spin–lattice relaxation, T_1, data using the pulse sequence of Figure 8.2(c) [88]. Note that a jump with a correlation time that affects the quadrupole echo lineshape must be predominantly in the non-extreme narrowing region (where T_1 is linearly proportional to τ_c; see chapter 4). Therefore, the distribution of T_1 values should be consistent with the distribution of jump rates. Deviation from this behaviour would suggest an additional mechanism, for example small-angle libration. For PPV, a log-plot of an inversion recovery decay is quite non-linear, indicating a distribution of rates. Figure 8.6 shows a plot of T_1 versus inverse temperature for PPV-d_4, and $E_a = 7$ kcal/mol with a two-order of magnitude distribution of rates. This level of agreement with lineshape data is sufficient to establish the jump model. The difference may result from the effects on T_1 of small angle diffusion at high temperature. Small-angle diffusion is also evident in the lineshape as a slight narrowing of the total pattern at high temperature.

For a jump mechanism, T_1 is also anisotropic. The anisotropy, measured with an inversion recovery experiment near the null-point, can provide a signature pattern for a particular mechanism. Figure 8.6 also shows an inversion recovery spectrum, obtained at 220°C and a simulation [88]. Wittebort *et al.* have described the theoretical calculation of T_1 anisotropy for a jump model [113, 134]. For C–D bond orientations close to the PAS axes, T_1 has a larger value and therefore these signals are inverted. This particular pattern is unique to the ring-flip mechanism and distinguishes it from other motions which might coincidently cause a spectrum of the same shape.

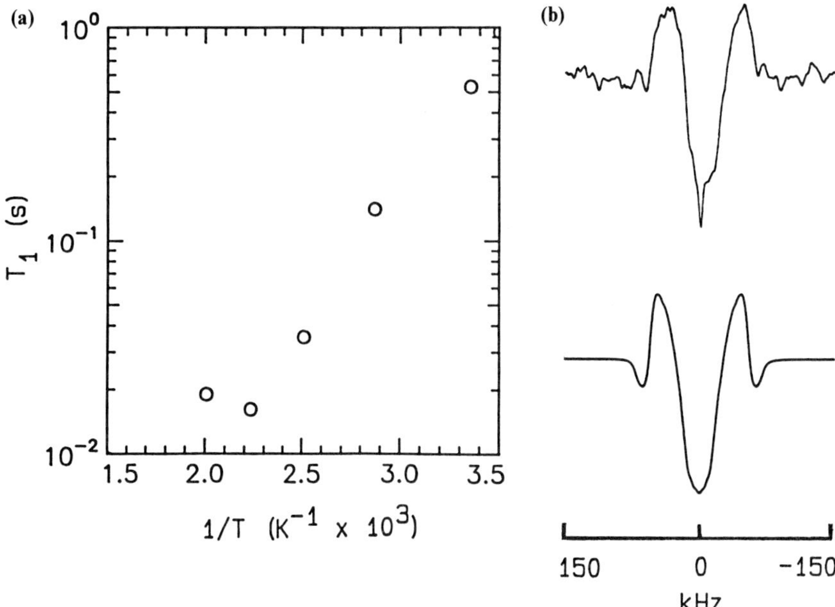

Figure 8.6 (a) A plot of T_1 versus inverse temperature for PPV-d_4. This temperature dependence is typical of slow, large-amplitude motion. (b) Experimental and (c) simulation inversion recovery spectra near the null-point showing the expected T_1 anisotropy for a ring-flip. (Reprinted from J.H. Simpson, D.M. Rice and F.E. Karasz, *J. Polym. Sci. B, Polym. Phys. Ed.* **30** (1992) 11–18, John Wiley and Sons, Inc.)

*8.4.2 Chain motion of poly(*p*-phenylene vinylene)*

Information about PPV chain motion could be obtained from a vinylene labelled material PPV-d_2 [89]; the presence of ring motion obscures data about chain motion from measurements of PPV-d_4. The interest here was to understand the ring-flip mechanism and determine whether there might exist a particular kink-propagation mechanism for the sp^2 hybridized chain, analogous to the three-bond motions of vinyl polymers. At ambient temperature PPV-d_2 possesses a static Pake doublet spectrum. Figure 8.7(a) shows the spectrum at 220°C; the linewidth is reduced and there is a large apparent asymmetry parameter ($\eta = 0.14$).

Clearly motion is present, but the mechanism is ambiguous. The shape might be explained by one of two models, 180° jumps of the PPV chain about its axis (Figure 8.7(b)) (similar to a ring-flip, except that both vinylene C–D bonds remain with $\beta_1 = 72°$ to the chain axis), or by small angle diffusion (Figure 8.7(c)) within a single potential well. Study of an oriented sample provides a way to distinguish the two mechanisms. Figures 8.7(d), (g) show spectra from stretched samples aligned at $\beta_3 = 0°$ and 90° to the magnetic

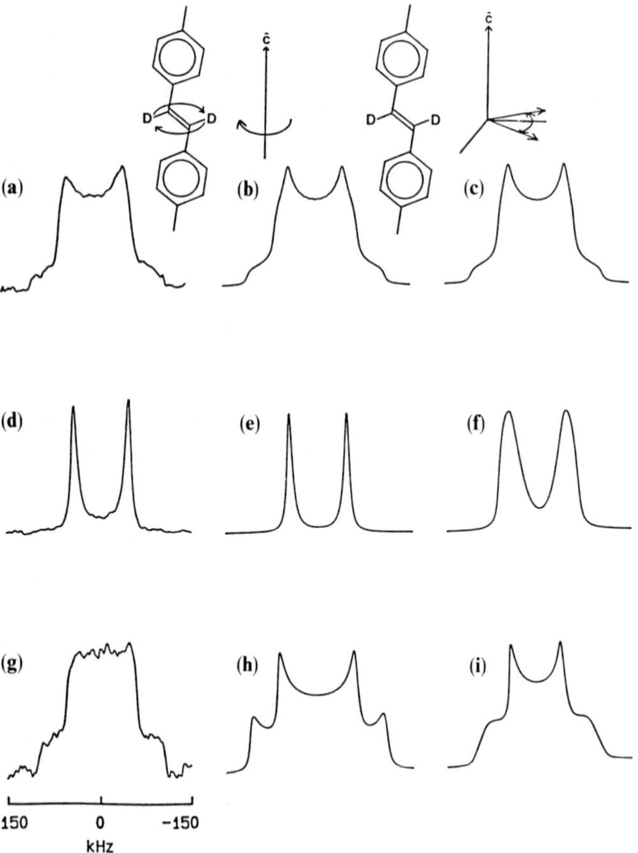

Figure 8.7 Deuteron spectra of oriented PPV-d_2 at 220°C. Simulations distinguish jumps from libration as described in the text. (Reprinted from J.H. Simpson, W. Liang, D.M. Rice and F.E. Karasz, *Macromolecules* **25** (1992) 3068–3074 by permission of the publishers, American Chemical Society.)

field. The spectra are consistent with the chain-jump model (Figures 8.7(c), (h)), although some smaller angle diffusion must also be present. Chain jumps might be attributed to cooperative rotation of multiple segments, probably due to chain defects at crystallite boundaries. A kink propagation mechanism is less likely unless it also involves simultaneous readjustment of the chain axis. The orientation experiment works, while the powder experiment fails, because by using an oriented sample, one additionally determines the orientation of the motionally averaged principal axis system (i.e. x', y' and z' of Figure 8.3) relative to $\boldsymbol{B}_0$. This orientation is different for the two mechanisms.

The anisotropy of the spin–lattice relaxation for the powder sample also confirms the presence of a jump mechanism. For PPV-d_2, the pattern is that of a jump model, similar to Figure 8.6(b) [89]. Also, the T_1 temperature dependence is large, indicating a thermally activated, large amplitude motion in the non-extreme narrowing region. A diffusional mechanism would lead to a smaller temperature dependence. Notably, the difference between the relaxation data for PPV-d_2 and PPV-d_4 can be attributed entirely to the different angles between the jump sites, $2 \times 72^\circ$ for PPV-d_2 versus 120° for PPV-d_4. Therefore, it was concluded that the distribution of rates must be similar.

Schadt *et al.* [28] have examined chain motion of poly(*p*-phenylene terephthalamide) (PPTA) another sp^2 hybridized chain. They find that 20–25% of N–D bonds undergo twofold jump motion about the 1,4 axis of the terephthalamide phenylene ring, but not the 1,4 axis of the diamine ring. The result suggested that dynamic heterogeneity accompanies the structural heterogeneity of PPTA. These studies of PPV and PPTA have revealed that considerable motion is possible in an sp^2 hybridized chain.

8.4.3 Segmental dynamics of Nylon

English and co-workers have carried out a careful study of the dynamics of Nylon-6,6 using full set of methylene deuterated samples [92–97]. These experiments provide a good recent example of the use of deuteron NMR to characterize the dynamics of a polymethylene chain. Nylon-6,6 is a semi-crystalline polymer, containing amorphous domains and crystalline domains. The signals from these two regions are clearly separable by means of saturation recovery, T_1 selection methods (Figure 8.2(d)). Hirschinger *et al.* [95] carried out a study of the crystalline domains and Miura [94] carried out a study of the amorphous domains.

At high temperature, the crystalline domains of Nylon are known to undergo a 'Brill' transition, a gradual change of the crystalline unit cell from monoclinic to pseudo-hexagonal. At issue was the change in segmental dynamics which accompanied the Brill transition. Figure 8.8 shows a set of deuterium spectra obtained by Hirschinger *et al.* [95] for one of the labels (at the 2,2 and 5,5 positions of the diamine segment). Spectra were also obtained from all other positions and were qualitatively similar, although there was variation in the linewidth at high temperature.

In Figure 8.8, a temperature-dependent change is evident from change from a static spectrum with $\eta = 0.0$ at low temperature to a motionally averaged spectrum with an asymmetry parameter of $\eta' \approx 1.0$ at high temperature. A spectrum with $\eta' = 1.0$ spectrum can result from two-site rotational jumps on a tetrahedral lattice. This spectrum was previously observed by Jelinski *et al.* [109] for the labelled butylene segments of PBT and was shown to provide evidence for a $tg+tg-t$ kink propagation mechanism for the methylene chain. Such kinks provide a simple mechanism for *trans-gauche* isomerization in a polymethylene chain whose ends are pinned. Deuteron lineshapes of an

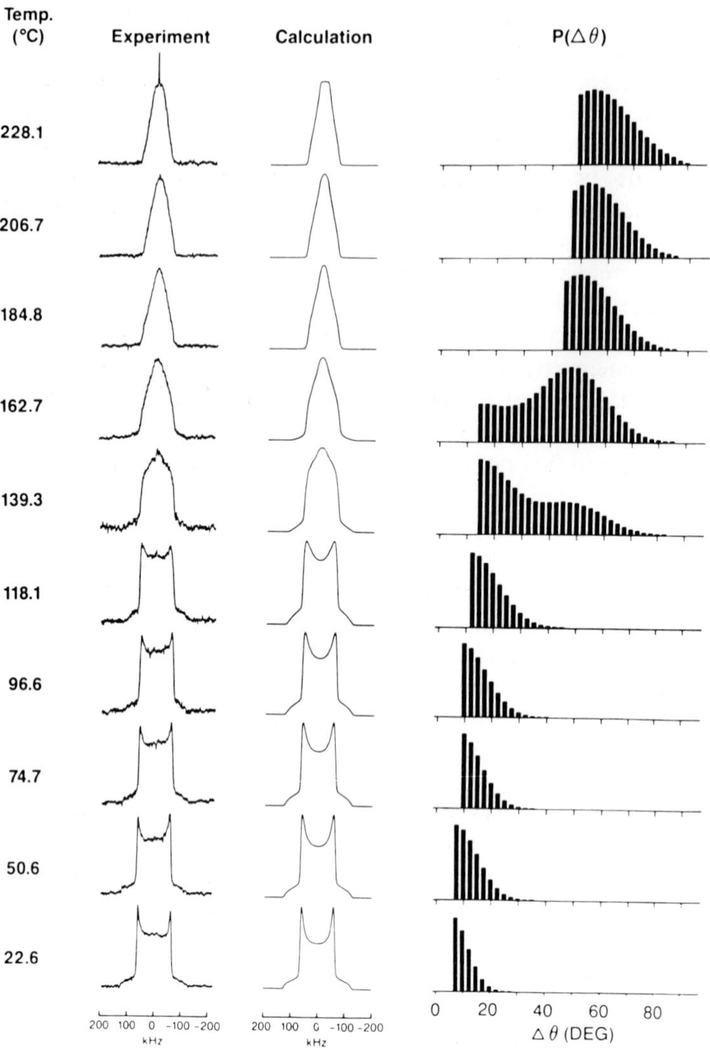

Figure 8.8 Deuteron spectra of Nylon-6,6. Simulations are for wide-amplitude libration of the diamine segment and the distributions of angle θ are shown. (Reprinted from J. Hirschinger, H. Miura, K.H. Gardner and A.D. English, *Macromolecules* **23** (1990) 2153–2169 by permission of the publishers, American Chemical Society.)

amide-deuterated Nylon-6,6 showed the N–D bonds to be static, probably due to the hydrogen bonding between chains. The diamine and diacid segments of Nylon are therefore similarly pinned and one might expect this type of motion. An alternative explanation of the spectra was that the chains undergo small angle continuous libration or diffusion in a potential well without jumps. For crystalline Nylon, the latter librational mechanism was shown to be present, and jumps were absent.

Spin–lattice relaxation measurements were obtained for deuterated Nylon over the temperature range of Figure 8.8. T_1 values varied from about 25 s at $-150°C$ to about 1 s at 200°C. This change (about one order of magnitude) is too small to be attributed to a jump mechanism (see Figure 8.6), although the jump mechanism does predict the negative slope. For a jump mechanism, the change from a static spectrum, to a fully averaged $\eta' = 1.0$ spectrum would require a change in jump rate of about 3–4 orders of magnitude and lead to a similar change in T_1. Instead, small angle libration, with a constant rate but a variable amplitude could explain the data. It is the widening of the libration amplitude that is the cause of the decrease in T_1 with increasing temperature, although the correlation time for the libration (about 60 ps) is in the extreme narrowing region.

Hirschinger *et al.* [95] also used the anisotropy of spin–lattice relaxation to distinguish the two mechanisms, in this case the relaxation time associated with quadrupolar order, T_{1Q}. A pulse sequence similar to Figure 8.2(b) was used. Quadrupolar order for the amorphous regions was shown to decay quickly, and for longer values of DE, only crystalline quadrupole order remained. The spectra could be closely simulated by a diffusion model and correlation time of 100 ps. Calculated T_{1Q} jump lineshapes possess a central peak that is not present in the experimental data.

The Brill transition is explained as the result of a widening of the amplitude of libration. The angular values are large at high temperature, greater than the typical thermal factors of a crystalline lattice, and the libration mechanism must be complex. Wendoloski *et al.* [97] have carried out a molecular dynamics simulation of Nylon-6,6 in order to explain the origin of the NMR data. Simulations successfully predict the amplitude of the libration, and the Nylon NMR data provide substantial experimental confirmation of a dynamics calculation for this complex motion, including some precise details. In particular, it was found that the amplitude of libration associated with the 4-methylene diacid segments was wider than of the 6-methylene diamine segments. The dynamics calculation based upon the crystal structure showed that the diacid segment occupies a region of lower density than the diamine segment and wider amplitude librational motions are possible.

The $\eta' = 1.0$ quadrupole echo spectra of the amorphous domains below T_g, obtained by Miura *et al.* [94] are qualitatively similar to the spectra of the crystalline domains of Figure 8.8. However, the lineshapes of the amorphous domains result predominantly from jumps or from a related wide-amplitude motion, and not from libration. Spectra from the amorphous domains possess a strong dependence upon the quadrupole echo pulse spacing, τ. Two signals were evident, a first with a strong τ-dependence and $\eta' = 1.0$, and a second equilibrium signal with a small τ dependence and a Pake doublet lineshape. The former was attributed to methylene groups undergoing *trans-gauche* isomerization and the latter to a population of more static groups that underwent only small-angle diffusion, perhaps at the crystallite boundaries. A strong temperature dependence of the spin–lattice relaxation time, T_1, was partially consistent with the jump model, although it was noted that explanation of the magnitude required a second mechanism to be present, probably small angle diffusion or slow libration near the Larmour frequency. The activation energy for the jump motion was determined to be $E_a = 8$ kcal/mol and it was suggested that the motion could be related to the γ-dielectric transition, which possesses a similar activation energy.

8.5 Multi-dimensional experiments

The 2-D exchange experiment [135] has played a central role in the area of multi-dimensional spectroscopy for a variety of different purposes, including the study of cross-relaxation, chemical exchange in liquids, and for the characterization of physical exchange processes and spin diffusion in solids (see chapter 6). A unifying theme of 2-D spectroscopy is the use of the isotropic chemical shift for selection of individual chemical species, which undergo exchange or cross-relaxation. The deuteron 2-D powder-exchange experiment instead makes use of the deuteron quadrupole splitting to label individual C–D bond orientations, which reorient to new positions during the mixing period. The reorientation is evident as the appearance of off-diagonal cross-peaks at the frequencies of the new positions. The 2-D exchange spectrum is in fact a direct frequency representation of the two-time correlation function for slow motion [55–57]. Exchange spectroscopy is necessarily applicable only to the study of slow reorientation ($\tau_c < 10^{-3}$ s); the quadrupole splittings of the exchanging reorientations must be distinguishable in the 1-D lineshape but exchange during the mixing period.

The 2-D deuteron exchange experiment [52–61] is closely related to the method of spin-alignment echoes (Figure 8.2(b)) and might be considered to be a 2-D version of this experiment. However, the 2-D powder-exchange method is also a conceptually different approach to the study of a reorientation mechanism. Data for which one works hard to simulate with 1-D spectroscopy are directly available in the 2-D experiment. For a jump mechanism, the appearance of elliptical ridges in the 2-D exchange pattern reveals the jump angle in an unambiguous or 'model-free' manner. A diffusional (or complex) mechanism is distinguished by the absence of ridges and may be characterized by simulation. Most important, elaboration of the experiment to three or more dimensions reveals the rotational correlation function at three or more times [52]. Data of this type can reveal information about the correlation of motions involving separate pairs of sites or different time-scales. A disadvantage of the deuteron exchange experiment is that slow motions are often accompanied by a long T_1 and impractical data collection times. Use of proton cross-polarization for preparation can avoid this disadvantage but practical use of cross-polarization has been limited to the spin 1/2 nuclei, hence the popularity of analogous experiments with ^{13}C and ^{31}P.

Figure 8.9(a) shows the pulse sequence used to obtain the deuteron exchange powder pattern [59]. The sequence differs from the spin 1/2 version (chapter 6) owing to spin 1 dynamics of the deuteron and the need for quadrupole echo detection to overcome probe ringing. The first two pulses, which bracket the evolution period, provide frequency selection of individual quadrupole splittings. As with the spin 1/2 experiment, in-phase and out-of-phase pulse pairs select the sine and cosine components of the evolving magnetization to yield quadrature detection in the F1 dimension. One can use either the hyper-

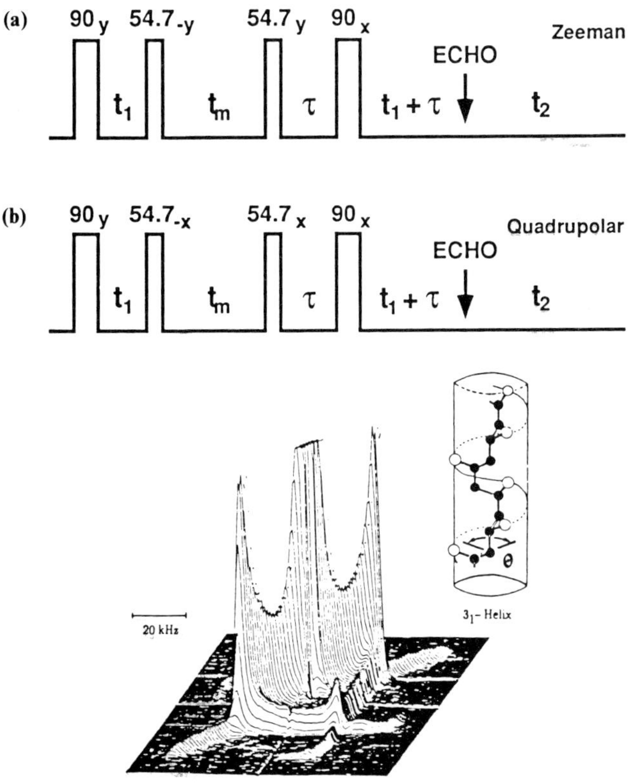

Figure 8.9 (a) The pulse sequence for the 2-D deuteron exchange experiment where t_1 is the incremented period for the first-time domain, t_2 is the aquisition period for the second time domain, τ is the quadrupole echo spacing and t_m is the mixing time period. (b) A 2-D exchange spectrum of crystalline i–PP. This confirms the 120° jump motion for the helix shown in the chain model. (Reprinted from D. Schaefer, H.W. Spiess, U.W. Sutter and W.W. Fleming, *Macromolecules* **23** (1992) 3431–3439 by permission of the publishers, American Chemical Society.)

complex Fourier transformation [59] or time proportional phase incrementation (TPPI) [136] to obtain a pure absorption phase spectrum. This process is somewhat more complicated for deuterium. As with the spin 1/2 experiment, in-phase pulses produce Zeeman order (which is maximum for the 90° pulse), but out-of-phase pulse pulses produce spin alignment or quadrupolar order; which is a maximum for a 45° pulse (see Figure 8.2(b) and the discussion of the vector model in section 8.2.2). The unusual 54.7° 'magic angle' pulse length is required to equalize the signals of the sine and cosine components. In practice, an empirical scale factor is usually also needed to correct for errors in pulse length. As with the spin 1/2 experiment, reorientation during the mixing period yields the off-diagonal intensity. The remaining two pulses provide quadrupole echo detection, and as usual phase cycling is used to remove unwanted coherences; these can appear in the deuteron experiment as echoes. Intensity errors result in the F1 dimension due to the lack of a point for $t_1 = 0.0$, and it is necessary to predict the first point. Alternatively, a reasonably accurate first point can be obtained from a normal quadrupole echo (Figure 8.2(a)) [136].

8.5.1 Exchange spectra of isotactic polypropylene

Figure 8.9(b) shows a 2-D exchange spectrum of the CD_3 groups of isotactic polypropylene, *i*-PP obtained by Schaefer *et al.* [11] at 114°C and with a mixing period of 150 ms. The usual CD_3 powder spectrum is on the diagonal. Note the elliptical ridges, which are apparent in the foreground of the spectrum. These ridges unambiguously reveal that the CD_3 vector of crystalline *i*-PP undergoes jumps with an angle of 113° with a correlation time τ_c in the range of 150 ms or less. The conformation of *i*-PP is a 3_1 helix, and a reorientation angle of 113° is consistent with a 120° jump of the helix (with a translation) about its axis.

That the data are unambiguous can be understood by selecting a horizontal slice in the foreground in line with the lower-left, parallel edge of the diagonal spectrum. This slice is the result of CD_3 vectors initially parallel to the field with a diagonal frequency splitting of $\Delta\nu = \Delta\nu_Q(3\cos^2(0°) - 1)$. The intersection of the ellipse with the slice is just a cross-peak with the frequency $\Delta\nu = \Delta\nu_Q(3\cos^2(113°) - 1)$. A jump of another angle would yield a different frequency, and this absence is experimentally evident in the spectrum. For other slices, the frequencies are not unique but the elliptical ridges result from singularities associated with the frequency distribution. Diffusion in small steps (without jumps) would lead to a distribution of angles centred about the diagonal frequency and the slice would contain a recognizably broad spectrum.

The elaboration of this experiment to three dimensions can be readily seen by imagining an additional mixing time and a vertical slice perpendicular to the cross-peak, noted above. The peaks in this imaginary slice would result only from CD_3 vectors first at 0° and then at 113°. Were the i-PP helix to rotate backwards to the original site a peak would be found in this slice at only $\Delta\nu = \Delta\nu_Q(3\cos^2(0°) - 1)$. Were the helix to also translate forward, a third frequency would be present in this slice corresponding to the orientation of the third site. Spiess and Schmidt-Rohr [52] have performed the three-dimensional experiment for a helix of poly(ethylene oxide) (PEO) with ^{13}C NMR. They found that jumps back to the original site were more probable, though up to three steps could be observed. To our knowledge, deuteron NMR has not yet been used with three dimensions.

8.5.2 Chain motion of poly(vinylidene fluoride)

The 2-D exchange experiment has also been used by Hirschinger *et al.* [46] to examine the chain dynamics of crystalline PVF_2. These results provide a good example of how a mechanism of motion can be used to determine a crystalline conformation. Ellipses were present in 2-D exchange spectra of PVF_2 which indicated reorientation angles of precisely 67° and $113 \pm 3°$. Previous X-ray data had established unit cell parameters and the conformation of crystalline PVF_2 as, tg^+tg^-, however, the exact bond angles associated with the conformation were unknown. Three X-ray structures were available for PVF_2, but only one of these structures was consistent with the reorientation angles implied by the NMR data. Therefore with NMR data, it was possible to select this particular structure from among the three.

Eight possible isomerizations could occur for the pinned tg^+tg^- conformation. Four of these isomerizations involved reorientation angles of about 67° or 113° and were consistent with the NMR data. However, each of the four motions implied a different electric dipole moment and only one of these (tg^+tg^- to g^-tg^+) was consistent with dielectric data. Therefore, from a combination of NMR data and dielectric data it was possible to uniquely determine the particular mechanism for reorientation of PVF_2 although neither of these methods would determine the reorientation mechanism alone.

8.5.3 Diffusive motion of atactic polypropylene near T_g

Deuteron NMR provides an excellent, albeit expensive, method for the determination of the glass transition, T_g. The quadrupole echo spectrum disappears (due to the echo distortion caused by isotropic motion) when the correlation time for chain motion approaches the quadrupole coupling frequency, typically about 30° above the calorimetric T_g. At higher temperatures, a narrow line appears due to fast isotropic motion. Early on, deuteron NMR and the spin alignment experiment (Figure 8.2(b)) were used to characterize slow motions of polystyrene [3]. More recently, deuteron NMR has been used to characterize

T_g in the presence of plasticizers [19] or to study the motion of individual components of polymer blends [20]. A goal of these experiments has been to provide a direct approach to confirm the applicability of WLF theory, and to the present WLF theory, has been useful in its extension to NMR data (see also chapter 6).

Recently the multi-dimensional exchange experiment has been applied to the question of chain motion near T_g. Figure 8.10 shows a pair of 2-D deuteron exchange spectra which show the diffusive motion of *a*-PP near T_g [10, 11]. In these spectra ridges are absent and the diagonal signal simply broadens at high temperature. The DE dependence of these spectra was well simulated by assuming a log-Gaussian distribution of correlation times. A more detailed and motivating question for these experiments has been whether the distribution of correlation times τ_c, associated with T_g or other mechanical transitions, is the result of isolated sites (1), or whether the correlation function itself is multi-exponential (2). A related question, associated with the study of miscible polymer blends and mixtures, is how the motions of the two components are coupled. Answers to both of these questions at present are preliminary and multi-dimensional experiments (including those with deuterons) will play a role in the answers. At present the multi-dimensional (a reduced 4-D experiment) ^{13}C exchange experiment has been used by Spiess and Schmidt-

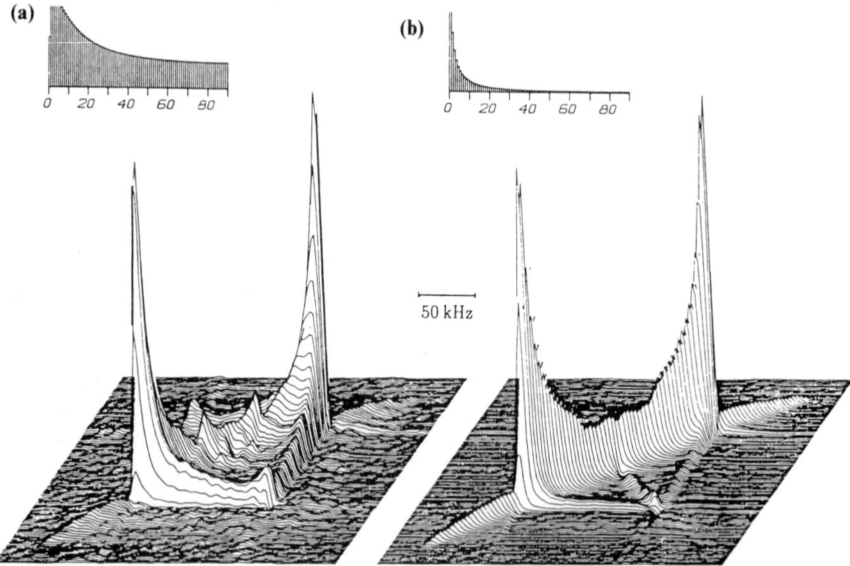

Figure 8.10 2-D deuteron exchange spectra of a-PP at two temperatures close to the glass-transition. (The simulated angular distributions of diffusive motion are shown for (a) 264 K and (b) 258 K. $t_m = 300$ ms. (Reprinted from D. Schaefer, H.W. Spiess, U.W. Sutter and W.W. Fleming, *Macromolecules* **23** (1992) 3431–3439 by permission of the publishers, American Chemical Society.)

Rohr [52] to examine the correlation function for *a*-PP. They find evidence that the correlation function itself is non-exponential (case 2).

In related experiments Inglefield and co-workers [21, 22] have used 2-D exchange spectroscopy to examine the motion of a ^{31}P probe molecule in a polymer blend, near T_g. They find that the probe molecule undergoes a complex motion with at least two correlation times. Also, 1-D deuteron NMR has shown that ring flips of PS-d_5 are not strongly affected by blending with PPO [20, 137]. A similar result has been obtained with a polyether-sulfone-d_4/polyimide blend [137]. The implication is that the phenylene-ring motion is local and is not coupled to the presence or absence of free volume. Also, the NMR T_g of PS-d_5 in a blend with PPO appears to be consistent with the calorimetric T_g and WLF theory, although more subtle experiments obtained through multi-dimensional NMR will be of value here. In contrast, an early study of the motions of a small molecular weight toluene-d_3 plasticizer in PS-d_3 appeared to be fast well below T_g [19].

It is important to also note the application of deuteron NMR to multi-dimensional experiments involving dipolar interactions and imaging [139]. Schmidt *et al.* [138] have recently presented a ^{13}C–^{2}H dipolar measurement with MAS which could be used in a 2-D experiment. In general, combination dipolar-exchange experiments have the potential to spatially correlate motions. Blümich *et al.* [139] have incorporated the deuteron experiment into an imaging method which allows imaging of molecular order and motion. Both of these experiments suggest future uses of deuteron NMR.

8.6 Conclusions

In the past 5 years, the technology associated with the deuteron NMR experiment has become readily available. In a sense, the deuteron has become just another nucleus for NMR characterization. However, the experiment is still thought to be 'elegant' or difficult. One reason is that the interpretation of deuteron NMR data (with rare exceptions) has always been a modelling process. A conceptual obstacle has been the lack of generally available calculation packages for the modelling of deuteron lineshapes. There is no obstacle to such programs other than general demand for them, and comparable software is certainly available for the calculation and modelling of diffraction data. In the future, areas which will be exploited involve the application of off-resonance irradiation and the use of magic-angle spinning or rotational echo methods. Deuteron spectroscopy will continue to be incorporated as a component of new multi-dimensional experiments, perhaps involving distance measurements, although at present spin 1/2 methods seem to provide an equally powerful and often more practical approach in this area. The value of deuteron experiments and other solid-state NMR methods has been the direct knowledge they have given about structure and mechanism. Many mechanisms remain to be solved.

Acknowledgements

I would like to thank Professor Frank E. Karasz for providing the resources for preparation of this chapter. I would also like to acknowledge a helpful conversation with Professor H.G. Zachmann and access to his work with oriented PET before publication, as well as useful comments by B. Blümich and Alan D. English.

References

1. H.W. Spiess, *Colloid Polym. Sci.* **261** (1983) 193–209.
2. H.W. Spiess, *J. Mol. Struct.* **111** (1983) 119–133.
3. H.W. Spiess, *Adv. Polym. Sci.* **66** (1985) 23–58.
4. H.W. Spiess, in *Developments in Oriented Polymers 1*, ed. I.M. Ward, Elsevier Applied Science, London (1982), pp. 47–48.
5. T.M. Alan and G.P. Drobney, *Chem. Rev.* **91** (1991) 1545–1590.
6. D.A. Torchia, *Ann. Rev. Biophys. Bioeng.* **13** (1984) 125–144.
7. J.H. Davis, *Chem. Phys. Lipids* **40** (1986) 223–258.
8. L.W. Jelinski, in *High Resolution NMR Spectroscopy of Synthetic Polymers in Bulk*, ed. R.A. Komoroski, VCH, Weinheim, Germany (1986), pp. 335–363.
9. R.G. Griffin, K. Beshah, R. Ebelhauser, T.H. Huang, E.T. Olejniczak, D.M. Rice, D.J. Siminovitch and R.J. Wittebort, in *The Time Domain in Surface and Structural Dynamics*, eds. G.J. Long and F. Grandjean, Kluwer, New York, (1988), pp. 81–105.
10. A. Hagemeyer, D. Schaefer, K. Schmidt-Rohr, B. Blümich and H.W. Spiess, *Mol. Cryst. Liquid Cryst.* **187** (1990) 223–230.
11. D. Schaefer, H.W. Spiess, U.W. Sutter and W.W. Fleming, *Macromolecules* **23** (1990) 3431–3439.
12. H.W. Spiess, *Lect. Notes. Phys.* **293** (1987) 247–249.
13. P.M. Henrichs, V.A. Nicely and D.R. Fagerburg, *Macromolecules* **24** (1991) 4033–4037.
14. P.M. Henrichs and T.E. Long, *Macromolecules* **24** (1991) 55–60.
15. B. Westermark and H.W. Spiess, *Makromol. Chem.* **189** (1988) 2367–2379.
16. B. Westermark and H.W. Spiess, *Makromol. Chem.* **189** (1988) 666–670.
17. D. Schaefer, M. Hansen, B. Blümich and H.W. Spiess, *J. Non-Cryst. Solids* **131–133** (1991) 777–780.
18. D. Schaefer, R.R. Rietz, W.H. Meyer and H.W. Spiess, *Ber. Bunsen-Ges. Phys. Chem.* **95** (1991) 1071–1076.
19. E. Rossler, H. Sillescu and H.W. Spiess, *Polymer* **26** (1985) 203–207.
20. M. Aurelio de Araujo, D. Oelfin, R. Stadtler and M. Moeller, *Makromol. Chem. Rapid Commun.* **10** (1989) 259–264.
21. P.T. Inglefield, A.A. Jones, Y.H. Chin and C. Zhang, *Polym. Preprints (Am. Chem. Soc. Div. Polym. Chem.)* **33**(1) (1992) 76–77.
22. C. Zhang, P. Wang, A.A. Jones, P.T. Inglefield and R.P. Kambour, *Macromolecules* **24** (1991) 338–340.
23. G. Stoppelmann, W. Gronski and A. Blume, *Polymer* **31** (1990) 1838–1853.
24. F.D. Blum, B.R. Sinha and F.C. Schwab, *Polym. Mater. Sci. Eng.* **59** (1988) 302–305.
25. K. Matsumura, K. Hayamizu, T. Nakane, H. Yangashita and O. Yamamoto, *J. Polym. Sci. B, Polym. Phys.* **25** (1987) 2149–2163.
26. P.B. Smith and D.J. Moll, *Macromolecules* **23** (1990) 3250–3256.
27. H. Deckman, M. Moeller, L. Covaert and P.J. Lemstra, *Int. Fund. Polym. Sci. Technol.* **5** (1991) 276–290.
28. R.J. Schadt, E.J. Cain, K.H. Gardner, V. Gabara, S.R. Allen and A.D. English, *Polym. Preprints (Am. Chem. Soc. Div. Polym. Chem.)* **32**(3) (1991) 253–254.
29. A.D. English, K.H. Gardner, R.J. Schadt, E.J. Cain, V. Gabara and S.R. Allen, *Polym. Preprints (Am. Chem. Soc. Div. Polym. Chem.)* **33**(1) (1992) 82–83.

30. E.J. Cain, K.H. Gardner, V. Gabara, S.R. Allen and A.D. English, *Polym. Preprints* (*Am. Chem. Soc. Div. Polym. Chem.*) **31**(1) (1990) 518–519.
31. S. Rober and H.G. Zachmann, *Polymer* **10** (1992) 2061–2075.
32. P. Sotta and B. Deloche, *Macromolecules* **23** (1990) 1999–2007.
33. P. Sotta and B. Deloche, *Macromol. Chem. Macromol. Symp.* **23** (1989) 183–190.
34. B. Deloche and P. Sotta, *Polym. Preprints* (*Am. Chem. Soc. Div. Polym. Chem.*) **29**(1) (1988) 424–425.
35. B. Deloche and P. Sotta, *Polymer* **29** (1988) 1171–1178.
36. P. Sotta, B. Deloche, J. Herz, A. Lapp, D. Durand and D.J.C. Rabadeux, *Macromolecules* **20** (1987) 2769–2774.
37. A. Dubault, B. Deloche and J. Herz, *Macromolecules* **20** (1987) 2096–2099.
38. C.D. Poon and E.T. Samulski, *J. Non-Cryst. Solids* **131–133** (1991) 509–515.
39. C.D. Poon, E.T. Samulski and A.I. Nakatani, *Makromol. Chem. Macromol. Symp.* **40** (1990) 109–120.
40. D. Yang, F. Li and Z. Qiu, *J. Macromol. Sci. Chem.* A27 (1990) 149–165.
41. H. Bauer, K. Muller, G. Kothe and W.D. Funke, *Angew. Makromol. Chem.* **185** (1991) 61–73.
42. G. Simon, *Polym. Bull.* **25** (1991) 365–371.
43. F. Li, S. Yue and Z.W. Qui, *J. Macromol. Sci. Chem.* **A24** (1987) 167–181.
44. M.I. Lifshits, *Polymer* **28** (1987) 454–485.
45. W. Gronski and G. Stoeppelmann, *Teubner-Texte Phys.* **9** (1986) 148–157.
46. J. Hirschinger, D. Schaefer, H.W. Spiess and A.J. Lovinger, *Macromolecules* **245** (1991) 2428–2433.
47. M.A. Doverspike, M.S. Conradi, A.S. DeReggi and R.E. Cais, *J. Appl. Phys.* **65** (1989) 541–547.
48. P.M. Henrichs and H.R. Luss, *Macromolecules* **21** (1988) 860–862.
49. P.P Smith, R.A. Bubeck and S.A. Bales, *Macromolecules* **21** (1988) 2058–2063.
50. M. Wehrle, G.P. Hellman and H.W. Spiess, *Coll. Poly. Sci.* **265** (1987) 815–822.
51. K.L. Ngai, R.W. Rendell, A.F. Yee and D.J. Plazek, *Macromolecules* **24** (1991) 61–67.
52. H.W. Spiess and K. Schmidt-Rohr, *Polym. Preprints* (*Am. Chem. Soc. Div. Polym. Chem.*) **32**(1) (1992) 68–69.
53. H.W. Spiess, *Chem. Rev.* **91** (1991) 1321–1328.
54. H.W. Spiess, *Polym. Preprints* (*Am. Chem. Soc. Div. Polym. Chem.*) **31**(**1**) (1990) 103–104.
55. S. Wefing and H.W. Spiess, *J. Chem. Phys.* **89** (1988) 1219–1233.
56. S. Wefing, S. Kaufmann and H.W. Spiess, *J. Chem. Phys.* **89** (1988) 1234–1244.
57. S. Kaufmann, S. Wefing, D. Schaefer and H.W. Spiess, *J. Chem. Phys.* **93** (1990) 197–214.
58. B. Blümich and H.W. Spiess, *Agnew. Chem.* **100** (1988) 1716–1734.
59. C. Schmidt, B. Blümich and H.W. Spiess, *J. Magn. Reson.* **79** (1988) 269–290.
60. C. Schmidt, S. Wefing, B. Blümich and H.W. Spiess, *Chem. Phys. Lett.* **130**(1986) 84–90.
61. P.A. Mirau, G. Koegler and G.E. Johnson, *Polym. Preprints* (*Am. Chem. Soc. Div. Polym. Chem.*) **33**(1) (1992) 84–85.
62. J. Hirschinger, W. Kranig and H.W. Spiess, *Colloid Polym. Sci.* **269** (1991) 993–1002.
63. W. Kranig, C. Boeffel and H.W. Spiess, *Macromolecules* **23** (1990) 4061–4067.
64. B. Huser, T. Pakula and H.W. Spiess, *Macromolecules* **22** (1989) 1960–1963.
65. B. Huser and H.W. Spiess, *Makromol. Chem. Rapid. Commun.* **9** (1988) 337–343.
66. A. Abe, S. Tabata and N. Kimura, *Polym. J.* **23** (1991) 69–72.
67. A. Abe and T. Yamazaki, *Macromolecules* **22** (1989) 2138–2145.
68. A. Abe and H. Furuya, *Macromolecules* **22** (1989) 2982–2987.
69. H. Furuya and A. Abe, *Polym. Bull.* **20** (1988) 467–470.
70. E. Meirovitch, E.T. Samulski, A. Leed, H.A. Scheraga, S. Ranavare, G. Nemethy and J.H. Freed, *J. Phys. Chem.* **91** (1987) 4840–4851.
71. P. Esnault, D. Galland, F. Volino and R.B. Blumstein *Macromolecules* **22** (1989) 3734–3741.
72. J.A. Ratto, F. Volino and R.B. Blumstein, *Macromolecules* **24** (1991) 2862–2867.
73. B.J. Fahie, C.A. Fyfe, G.A. Facey, A. Muhlebach, N. Niessner, J. Economy and J.R. Lyerla, *Mol. Cryst. Liquid Cryst.* **203** (1991) 127–135.
74. K. Kohlhammer, G. Kothe, B. Reck and H. Ringsdorf, *Ber. Bunsen-Ges. Phys. Chem.* **93** (1989) 1323–1325.
75. F. Schilling and P. Sozzani, *Polym. Preprints* (*Am. Chem. Soc., Div. Polym. Chem.*) **33**(1) (1992) 80–81.

76. R. Ebelhauser and H.W. Spiess, *Makromol. Chem.* **188** (1987) 2935–2949.
77. K. Mueller, A. Schleicher, E. Ohmens, A. Ferrarini and G. Kothe, *Macromolecules* **20** (1987) 2761–2768.
78. U. Falk, B. Westermark, C. Boeffel and H.W. Spiess, *Mol. Cryst. Liquid Cryst.* **153** (1987) 199–206.
79. M. Mauzac, H. Richard and L. Latie, *Macromolecules* **23** (1990) 753–758.
80. U. Pschorn, H.W. Spiess, B. Hisgen and H. Ringsdorf, *Makromol. Chem.* **187** (1986) 2711–2723.
81. E.T. Samulski, D.J. Photinos and H. Toriumi, *Polym. Preprints (Am. Chem. Soc., Div. Polym. Chem.)* **33**(1) (1992) 88–89.
82. J. Pluyter and E.T. Samulski, *Polym. Preprints (Am. Chem. Soc., Div. Polym. Chem.)* **33**(1) (1991) 140–141.
83. R. Stannarius, J.P. Crawford, L.C. Chien and J.W. Doane, *J. Appl. Phys.* **70** (1991) 135–143.
84. A. Golemme, S. Zumer, J.W. Doane and M.E. Neubert, *Phys. Rev. A* **37** (1988) 559–569.
85. J.H. Simpson, N. Egger, M.A. Masse, D.M. Rice and F.E. Karasz, *J. Polym. Sci. B. Polym. Phys.* **28** (1990) 1859–1869.
86. J.H. Simpson, D.M. Rice and F.E. Karasz, *Polymer* **32** (1991) 2340–2344.
87. J.H. Simpson, D.M. Rice and F.E. Karasz, *Macromolecules* **25** (1992) 2099–2016.
88. J.H. Simpson, D.M. Rice and F.E. Karasz, *J. Polym. Sci. B. Polym. Phys. Ed.* **30** (1992) 11–18.
89. J.H. Simpson, W. Wang, D.M. Rice and F.E. Karasz, *Macromolecules* **25** (1992) 3068–3074.
90. J.J. Dumais, L.W. Jelinski, M.E. Galvin, C. Dybowsk, C.E. Brown and P. Kovacic, *Macromolecules* **22** (1989) 612–617.
91. S. Kaplan, E.M. Conwell, A.F. Richter and A.G. Macdiramid, *Macromolecules* **22** (1989) 1669–1675.
92. J. Hirschinger, H. Miura and A.D. English, *Polym. Preprints (Am. Chem. Soc. Div. Polym. Chem.)* **30**(1) (1990) 312–314.
93. H. Miura and A.D. English, *Macromolecules* **21** (1988) 1544–1546.
94. H. Miura, J. Hirschinger and A.D. English, *Macromolecules* **23** (1990) 2169–2182.
95. J. Hirschinger, H. Miura, K.H. Gardner and A.D. English, *Macromolecules* **23** (1990) 2153–2169.
96. A.D. English, J.J. Wendoloski, K.H. Gardner, J. Hirschinger and H. Miura, *Polym. Preprints (Am. Chem. Soc. Div. Polym. Chem.)* **31**(1) (1990) 105–106.
97. J.J. Wendoloski, K.H. Gardner, J. Hirschinger, H. Miura and A.D. English, *Science* **247** (1990) 431–436.
98. L.J. Mathais and R.F. Colletti, *Polym. Preprints (Am. Chem. Soc., Div. Polym. Chem.)* **31**(1) (1990) 521–522.
99. R.F. Colletti, M. Jeno and L.J. Mathias, *Polym. Commun.* **32** (1991) 332–335.
100. L.J. Mathias and R.F. Colletti, *Macromolecules* **24** (1991) 5515–5521.
101. L.J. Mathias, R.F. Colletti and H.W. Spiess, *Polym. Preprints (Am. Chem. Soc. Div. Polym. Chem.)* **30**(1) (1990) 304–305.
102. T. Thomsen, H.G. Zachmann and H.R. Kricheldorf, *J. Macromol. Sci. Phys.* **30** (1991) 87–99.
103. A.D. Meltzer, H.W. Spiess, C.D. Eisenbach and H. Hayen, *Makromol. Chem. Rapid. Commun.* **12** (1991) 261–268.
104. A.D. Meltzer, H.W. Spiess, C.D. Eisenbach and H. Hayen, *Macromolecules* **25** (1992) 1993–1995.
105. J.A. Kornfield, H.W. Spiess, H. Nefzeger, H. Hayen and C.D. Eisenbach, *Macromolecules* **24** (1991) 4787–4795.
106. M. Mehring, *High Resolution NMR of Solids*, 2nd edition, Springer-Verlag, Berlin (1983).
107. H.W. Spiess and H.J. Sillescu, *J. Magn. Reson.* **42** (1981) 381–389.
108. D.M. Rice, R.J. Wittebort, R.G. Griffin, E. Meirovitch, E.R. Stimson, Y.C. Meinwald, J.H. Freed and H.A. Scheraga, *J. Am. Chem. Soc.* **103** (1981) 7707–7710.
109. L.W. Jelinski, J.J. Dumais and A.K. Engle, *Macromolecules* **16** (1983) 492–496.
110. J.H. Davis, K.R. Jeffrey, M. Bloom, M.I. Valic and T.P. Higgs, *Chem. Phys. Lett.* **42** (1976) 390–394.
111. H.W. Spiess, *J. Chem. Phys.* **72** (1980) 6755–6762.
112. M. Lausch and H.W. Spiess, *J. Magn. Reson.* **54** (1983) 466–479.
113. R.J. Wittebort, E.T. Olejniczak and R.G. Griffin, *J. Chem. Phys.* **86** (1987) 5411–5420.
114. A. Schleicher, K. Mueller and G. Kothe, *J. Chem. Phys.* **92** (1992) 6432–6440.

115. D.M. Rice, Y.C. Meinwald, H.A. Scheraga and R.G. Griffin, *J. Am. Chem. Soc.* **109** (1987) 1636–1640.
116. T. Lin and R.R. Vold, *J. Phys. Chem.* **95** (1991) 9032–9034.
117. S. Vega and A. Pines, *J. Chem. Phys.* **66** (1977) 5624–5644.
118. A.D. Ronemus, R.L. Vold and R.R. Vold, *J. Magn. Reson.* **70** (1986) 416–426.
119. P.M. Henrichs, J.M. Hewitt and M. Linder, *J. Magn. Reson.* **60** (1984) 280–298.
120. T.N. Barbara, *J. Magn. Reson.* **67** (1986) 491–500.
121. M. Bloom, J.H. Davis and M.I. Vallc, *Can. J. Phys.* **48** (1980) 1510–1517.
122. T.A. Early, *J. Magn. Reson.* **74** (1987) 337–343.
123. M.J. Folkes and I.M Ward, in *Structure and Properties of Oriented Polymers*, ed. I.M Ward, John Wiley and Sons, New York (1975), pp. 219–241.
124. B.E. Read, in *Structure and Properties of Oriented Polymers*, ed. I.M. Ward, John Wiley and Sons, New York (1975), pp. 150–186.
125. A.J. Brandolini and C. Dybowski, in *High Resolution NMR Spectroscopy of Synthetic Polymers in Bulk*, ed. R.A. Komoroski, VCH, Wienheim, Germany (1986), pp. 283–306.
126. G.S. Harbison, V. Veit-Dieter and H.W. Spiess, *J. Chem. Phys.* **86** (1987) 1206–1218.
127. D.L. Tzou, P. Desai, A.S. Abhiraman and T.L. Huang, *Polym. Preprints (Am. Chem. Soc. Div. Polym. Chem.)* **31**(1) (1990) 157–158.
128. J.J. Dumais, A. Choli, L.W. Jelinski, J.L. Hedrick and J.E. McGrath, *Macromolecules* **19** (1986) 1884–1889.
129. A.L. Choli, J.J. Dumais, A.K. Engle and L.W. Jelinski, *Macromolecules* **17** (1984) 2399–2404.
130. E. Fischer, *J. Mol. Struct.* **84** (1982) 219–226.
131. M.A. Masse, D.C. Martin, J.H. Petermann, E.L. Thomas and F.E. Karasz, *J. Mater. Sci.* **25** (1990) 311–320.
132. R.H. Boyd, *Polymer* **26** (1985) 1123–1133.
133. A. Jones, *Polym. Preprints (Am. Chem. Soc. Div. Polym. Chem.)* **33**(1) (1992) 63–64.
134. D.A. Torchia and A. Szabo, *J. Magn. Reson.* **49** (1982) 107–121.
135. J. Jeener, B.H. Meier, P. Bachman and R.R. Ernst, *J. Chem. Phys.* **71** (1979) 4546–4553.
136. M. Ziliox, Bruker Instruments Inc., personal communication.
137. S. Li, A. Wang, D.M. Rice and F.E. Karasz, unpublished results.
138. A. Schmidt, R.A. McKay and J. Schaefer *J. Magn. Reson.* **96** (1992) 644–650.
139. B. Blümich, P. Blumler, E. Guenther, G. Schauss and H.W. Spiess, *Makromol. Chem. Macromol. Symp.* **44** (1991) 37–45.

9 NMR in polymers using magnetic field gradients: imaging, diffusion and flow

P.T. CALLAGHAN

9.1 Introduction

High-resolution nuclear magnetic resonance spectroscopy relies on the existence of a uniform magnetic field across the sample volume, a result achieved by careful adjustment of currents in the 'shim' coils which produce a variety of correction terms representing different spatial harmonics of the field. A key element in the art of the spectroscopist lies in such optimisation. Because of the undesirable influence of magnetic field variations across the sample, their influence was an early consideration in the historical development of NMR. For example, the use of the spin echo in refocusing phase spreading associated with magnetic field inhomogeneity was developed by Hahn [1], and Carr and Purcell [2] in the early 1950s. In seeking to compensate for the effects of such field variations, it was clear to Hahn that these effects could also be put to specific use. For example, where the non-uniformity comprised a gradient in the magnetic field along some well-specified axis direction, the spin echo became sensitive to molecular translational motion along that axis. This understanding led, during the 1960s, to the development of the pulsed gradient spin echo (PGSE) method [3, 4] for the measurement of self-diffusion and its extensive application to the study of Brownian motion in polymers over the next two decades. As the technology associated with such measurement has improved and the variety of applications to molecular dynamics has widened, the study of more complex motion has been envisaged. This has called for a description of the PGSE method which encompasses more than the rudimentary case of unrestricted self-diffusion, and has led to a formalism in which PGSE NMR may be seen as closely related to inelastic neutron scattering [5, 6]. This approach has recently resulted in the measurement of internal polymer motions in high molar mass random coils [7], a result made possible by the use of magnetic field gradients sufficiently large to probe a distance scale smaller than the molecular dimensions. The scattering depiction has also resulted in the development of a method whereby pore morphology in polymeric structures can be deduced from the restricted diffusion of small molecules in an interpenetrating fluid [8].

While the use of magnetic field gradients to study molecular translational motion is almost as old as NMR itself, the realisation that these gradients could also be used to reconstruct an image of the molecular distributions came much later [9,10] in the 1970s. In common with the Hahn echo effect, NMR imaging relies on the fact that the spectrum of Larmor frequencies in the presence of a magnetic field gradient carries a spatial signature. This means that after a fixed period of precession, the angle of precession (i.e. the phase) of a nuclear spin will depend on the position of that spin in the magnetic field gradient. Consequently, the evolution of the spins over a period of such gradient application results in a spatially characteristic phase spectrum. Whereas in the PGSE experiment, phase differences are used to detect motion, in imaging, absolute phase is used to detect absolute position. The two experiments are therefore similar in principle and it is remarkable that such a long time elapsed before the tomographic applications of NMR were established.

Quite naturally, the initial development of magnetic resonance imaging (MRI) focused on its potential uses in medical tomography. In the late 1980s, however, MRI became a field of major application using the laboratory NMR spectrometer, with the development of small-scale and micro-imaging systems [11–14]. Furthermore, both the technology and theoretical formalism became inextricably linked with NMR spectroscopy for several reasons. First, the magnetic field gradient and soft-RF pulse equipment necessary for imaging became vital ingredients in gradient-accelerated two-dimensional NMR spectroscopy, in three-dimensional NMR, in solvent signal suppression methods, in chemical-selective excitation methods and in enabling PGSE NMR measurements which had previously been a feature only of specialised instruments. Second, the Fourier methods associated with imaging were identical to those associated with 2-D spectroscopy and lent themselves to a new depiction of the PGSE experiment in terms of an imaging process involving the average propagator of the molecular motion, a key element in the understanding of PGSE studies involving restricted diffusion [6, 15]. Finally, the amalgamation of imaging and PGSE methods has led to the development of dynamic NMR microscopy in which velocity and diffusion mapping can be carried out on systems exhibiting spatially heterogeneous motion. This has proved of considerable significance in the study of polymeric liquids under flow [16–18].

In this chapter, we review the various applications of magnetic field gradient methods in NMR spectroscopy of synthetic polymers, applications which encompass both PGSE NMR and NMR imaging. In doing so, we shall take advantage of hindsight and utilise an imaging perspective from the beginning.

9.2 Theory

9.2.1 Magnetic field gradients and NMR imaging

In general, the field gradient tensor describes the variation of three Cartesian components of magnetic field along the three independent Cartesian axes. Because in NMR applications such gradients result in additional magnetic fields much smaller than the polarising field, the Larmor frequency is affected solely by any components of these fields parallel to the polarising field axis, since orthogonal components have only the effect of slightly tilting the net field direction. In what follows, therefore, we shall be concerned only with the variation of the polarising field magnitude, B_0, so that the gradient, $\boldsymbol{G}$, is a vector with components given by grad B_0. Hence we define the local Larmor frequency as

$$\omega(\boldsymbol{r}) = \gamma B_0 + \gamma \boldsymbol{G} \cdot \boldsymbol{r} \tag{9.1}$$

where γ is the nuclear gyromagnetic ratio. This simple linear relation between the Larmor frequency and the nuclear spin coordinates, $\boldsymbol{r}$, contains the essence of the imaging principle. Figure 9.1(a) shows a simple pulse sequence involving the application of magnetic field gradients which allow a reconstruction of the nuclear spin density to be obtained. Since each isochromat of spins at position $\boldsymbol{r}$ precesses at local offset frequency $\gamma \boldsymbol{G} \cdot \boldsymbol{r}$, the net signal ($S$) in the heterodyne detection frame may be written in complex number formalism [6, 19]

$$S(\boldsymbol{k}) = \int \rho(\boldsymbol{r}) \exp(i 2\pi \boldsymbol{k} \cdot \boldsymbol{r}) \, \mathrm{d}\boldsymbol{r} \tag{9.2}$$

where $\rho(\boldsymbol{r})$ is the nuclear spin density and $\boldsymbol{k}$ is the reciprocal space dimension conjugate to $\boldsymbol{r}$ and given by $(1/2\pi)\gamma \boldsymbol{G} t$, t being the evolution time. This $\boldsymbol{k}$-space dimension is the domain in which the NMR signal is acquired. For example, where a fixed duration gradient pulse is applied, the phase angle acquired by a spin at position $\boldsymbol{r}$ will be given by $2\pi \boldsymbol{k} \cdot \boldsymbol{r}$. Such a gradient pulse is shown for the G_y direction in Figure 9.1(a) and is known as the phase gradient. By contrast, if a fixed amplitude gradient pulse is applied and the nuclear precession signal acquired at successive time intervals, a successively incremented spin phase is recorded in what is known as a read gradient. This is shown for the G_x direction in Figure 9.1(a). Successive application of these phase and read gradients allow one to sample the signal while evolving over orthogonal directions in $\boldsymbol{k}$-space.

The nuclear spin density can then be reconstructed by acquiring $S(\boldsymbol{k})$ over some appropriate volume of $\boldsymbol{k}$ space and performing an inverse Fourier transformation. Generally, reconstruction is carried out in two dimensions using a plane of spins prepared by a frequency-selective excitation. This latter process is known as slice selection and is a standard procedure in NMR imaging [6, 19–22]. The pulse sequence shown contains a frequency-selective RF pulse which has an approximately rectangular bandwidth and, when applied in the presence of a magnetic field gradient, excites a slice of rectangular slice of spins normal to the gradient axis. The two-dimensional reconstruction process used in such an imaging experiment bears a close resemblance to the 2-D Fourier methods used in standard 2-D NMR [23]. In the example shown in Figure 9.1(a) the read gradient corresponds to the f_2 dimension while the phase gradient corresponds to the f_1 dimension, the difference being that in this latter dimension, instead of the evolution time being stepped successively, the phase gradient amplitude is stepped [24, 25].

Figure 9.1(b) shows the receiver and gradient coil assembly used for NMR imaging in a typical widebore (89mm) superconducting magnet along with a block diagram indicating the ancillary apparatus required for NMR imaging using a laboratory spectrometer. Given a suitably optimised system, it is possible to reconstruct an image from hydrogen nuclei (the most sensitive spins) with a voxel resolution below $(100\,\mu m)^3$. Such imaging is known as NMR microscopy [6, 26]. An example of a micrograph [27] obtained from a cylindrical sample of natural rubber is shown in Figure 9.1(c)

The non-invasive character of NMR imaging is especially useful in medical applications but clearly less relevant in materials science. Furthermore, the spatial resolution of NMR microscopy is poor in comparison with optical or electron microscopy [28–30]. Consequently the value of the NMR imaging method must arise from an ability to provide unique, spatially resolved information at the molecular level, a feature that is generally referred to as imaging contrast. Specifically, contrast concerns the mapping of NMR parameters such as the local magnetic susceptibility, chemical shift, dipolar, quadrupolar and scalar couplings, relaxation times, and phase shifts associated with molecular translational motion. In addition, because the NMR signal gives a direct measure of the numbers of resonant nuclei, NMR imaging is highly quantitative and can be used to monitor slow changes in nuclear spin distribution as well as in the NMR parameters to which the image is sensitive.

The incorporation of contrast in equation (9.2) is achieved by defining a position-dependent modulation term in the integrand. Where the contrast is applied prior to the acquisition of the signal under the read gradient, this term will generally take the form of a simple attenuation/phase shift factor, represented by a complex function $C(\mathbf{r})$. Examples of this effect include the evolution due to spin relaxation prior to signal acquisition, chemical shift precession and attenuation/phase shifts arising from diffusion and flow. Where the contrast is present during the acquisition, its effect will be imposed as an additional time-dependent evolution due to the presence of the appropriate Hamiltonian term during the read acquisition. Two approaches are possible in dealing with such effects. Either the read gradient can be made sufficiently large to dominate these additional Hamiltonian terms, in which case they may be ignored, or, all spatial information can be imposed by phase encoding, in which case the signal is acquired in the absence of a read gradient so that the full effect of the additional Hamiltonian term may be observed. This latter approach is known as four-dimensional imaging in which the relevant NMR spectrum provides the extra dimension. Examples include chemical shift imaging (CSI) [31] and dipolar or quadrupolar spectrum imaging [32].

(a)

90_x
180_y
RF
time
slice
G_z
t_y
phase
G_y
k_x=0
read
G_x
t_x
begin acquisition
signal
T_E

(b)

B_0
z
y
x
B_1
computer
gradient coils
wave-form memory
RF modulator
current supplies
G_z
G_x
G_y

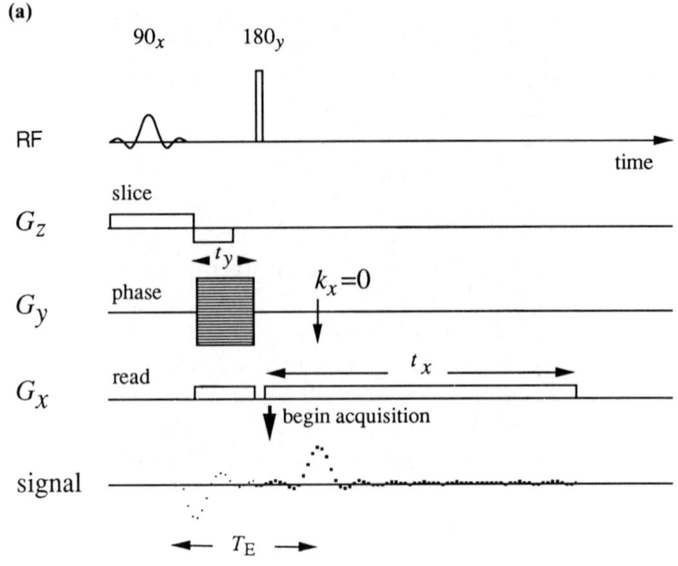

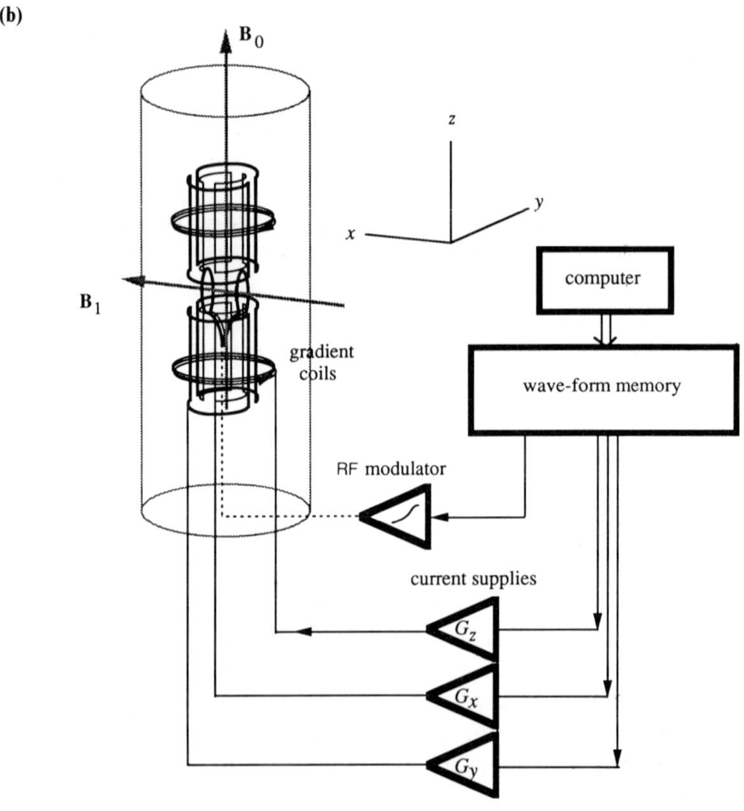

(c)

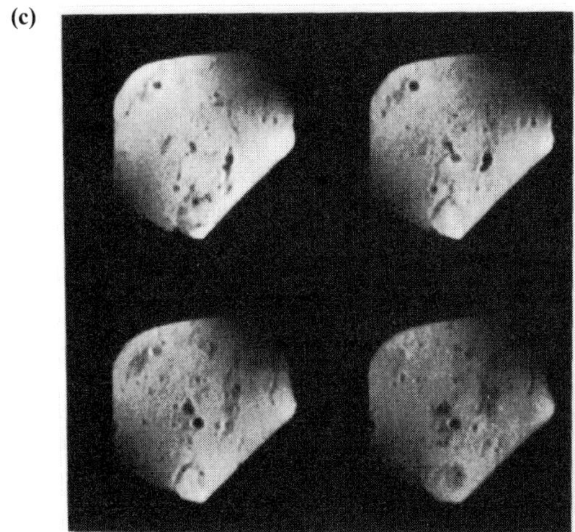

Figure 9.1 (a) Gradient and RF pulse sequence used for spin echo Fourier imaging. The initial (soft) RF pulse excites a rectangular layer of spins normal to the slice gradient (G_z) axis. T_E is the spin echo time which determines the degree of T_2 decay that occurs after the slice excitation. The three time lines labelled G_z, G_y and G_x respectively, indicate the magnetic field gradient pulses applied in the slice, phase and read directions. The combination of the sinc-modulated RF pulse with the slice gradient results in excitation of a slice of spins normal to the z-axis. Subsequent application of the G_y gradient results in phase-encoding of the spins along the y-axis of the slice plane. The NMR signal is then read under application of an orthogonal gradient applied along the x-axis. By stepping the amplitude of the phase gradient in successive experiments, the entire plane of k-space normal to z can be sampled. Subsequent Fourier transformation produces the image. (b) Schematic diagram showing hardware accessories needed for imaging and PGSE experiments using a laboratory NMR spectrometer based on a superconducting magnet. (c) 300 MHz ^{1}H NMR spin echo images obtained from natural rubber. The adjacent slices shown have a thickness of 500 μm while the pixel dimension is (45 μm)^2. (From W. Kuhn and M.A. Mattingley [27] and reproduced by permission of Bruker Analytische Messtechnik.)

9.2.2 Pulsed gradient spin echo NMR

By contrast with NMR imaging, the PGSE experiment [3–6, 33], shown in Figure 9.2(a), uses two narrow gradient pulses of amplitude $\boldsymbol{g}$, duration δ and separation Δ. These pulses effectively define the starting and finishing point of molecular translational motion over the well-defined timescale, Δ. Note that the clarity of that definition depends on the assumption that the spins move an insignificant distance during the gradient pulse itself, or in other words, $\delta \ll \Delta$. This narrow gradient pulse aproximation allows us to use a propagator formalism to describe the result of the PGSE experiment. Using the conditional probability $P_s(\boldsymbol{r}|\boldsymbol{r}', \Delta)$ [34] that a molecule starting at position $\boldsymbol{r}$ will move to position $\boldsymbol{r}'$ over the time Δ, we may write the echo attenuation

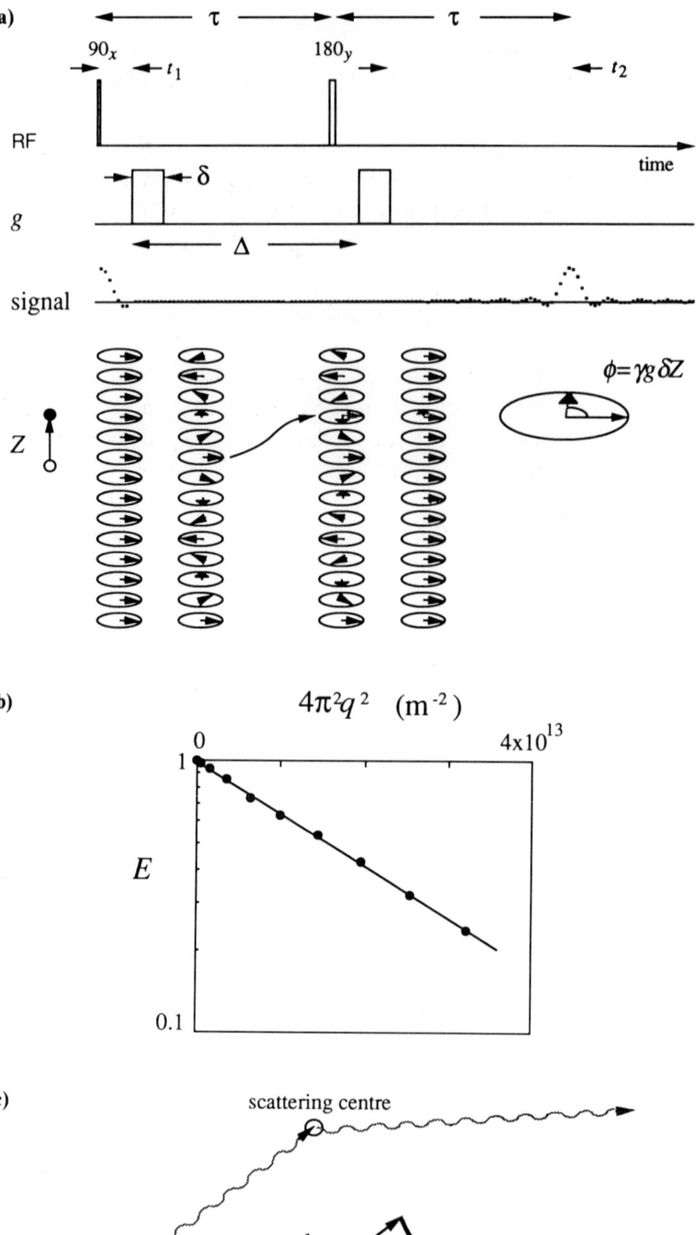
(a)
τ
τ
90_x
t_1
180_y
t_2
RF
time
δ
g
Δ
signal
Z
$\phi = \gamma g \delta Z$
(b)
$4\pi^2 q^2$ (m^{-2})
0
$4x10^{13}$
1
E
0.1
(c)
scattering centre
k_{in}
q
k_{out}

as [6, 35]

$$E_{\Delta}(\boldsymbol{g}) = \int\int \rho(\boldsymbol{r}) P_s(\boldsymbol{r}|\boldsymbol{r}', \Delta) \exp[i\gamma\delta \boldsymbol{g}\cdot(\boldsymbol{r}' - \boldsymbol{r})]\,\mathrm{d}\boldsymbol{r}\,\mathrm{d}\boldsymbol{r}' \tag{9.3}$$

where $\rho(\boldsymbol{r})$ is the starting spin density, the quantity mapped in conventional NMR imaging, $\int \rho(\boldsymbol{r})P_s(\boldsymbol{r}|\boldsymbol{r}', \Delta)\,\mathrm{d}\boldsymbol{r}$ defines the average propagator [15], $\overline{P}_s(\boldsymbol{R}, \Delta)$, the probability that a molecule at any starting position is displaced by $\boldsymbol{R} = \boldsymbol{r}' - \boldsymbol{r}$ over time Δ. Consequently equation (9.3) may be written [6].

$$E_{\Delta}(\boldsymbol{q}) = \int \overline{P}_s(\boldsymbol{R}, \Delta) \exp[i2\pi \boldsymbol{q}\cdot\boldsymbol{R}]\,\mathrm{d}\boldsymbol{R} \tag{9.4}$$

where the reciprocal space vector $\boldsymbol{q}$ is given by $(2\pi)^{-1}\gamma g\delta$. Just as inverse Fourier transformation of $S(\boldsymbol{k})$ in equation (9.2) returns an image of the spin density $\rho(\boldsymbol{r})$, so inverse Fourier transformation of $E_{\Delta}(\boldsymbol{q})$ with respect to $\boldsymbol{q}$ returns an image of the average propagator, $\overline{P}_s$. Note that the pulse sequence shown employs a spin echo although the measurement may equally be carried out using a stimulated echo, the advantage of the latter sequence being that the relaxation over the observation time Δ depends on T_1 rather than T_2. For many polymer molecules $T_1 \gg T_2$.

It is of interest to consider equation (9.4) in the case of two simple examples of motion of importance in this chapter, namely self-diffusion and flow. We write down the result where the motions are superposed.

$$\overline{P}_s(\boldsymbol{R}, \Delta) = (4\pi D_s \Delta)^{-3/2} \exp\left(-\frac{(\boldsymbol{R} - \boldsymbol{v}\Delta)^2}{4D_s\Delta}\right) \tag{9.5a}$$

$$E_{\Delta}(\boldsymbol{q}) = \exp(i2\pi \boldsymbol{q}\cdot\boldsymbol{v}\Delta - 4\pi^2 q^2 D_s \Delta) \tag{9.5b}$$

where $\boldsymbol{v}$ is the constant velocity, D_s is the molecular self-diffusion coefficient and $q = |\boldsymbol{q}|$. It is apparent that the phase spreading associated with diffusion causes echo attenuation whereas velocity is responsible for engendering a phase shift in the echo. It is also apparent that the measurement of velocity using this method will require phase-sensitive detection.

The idea that the echo attenuation function could be inverted to yield an image of the average propagator is relatively new and of special value in

Figure 9.2 (a) Gradient and RF pulse sequence for pulsed gradient spin echo NMR using a spin echo. The lower part of the diagram shows the phase twist produced along the direction of the magnetic field gradient when the pulse of duration δ is applied. The second pulse causes perfect 'unwinding' of this twist provided that molecules containing spins do not move. Any displacement by Z along the gradient direction results in a residual phase shift $\gamma g\delta Z$ or $2\pi qZ$ where $q = (2\pi)^{-1}\gamma g\delta$. (b) Semi-logarithmic (Stejskal–Tanner) spin echo attenuation plot for ^{1}H PGSE NMR on 10% (w/v) solution fo 200 kDa polystyrene in CCl_4. Note that q^2 is conjugate to the mean squared distance travelled by the polymer molecules over the observation time $\Delta_r = 27$ ms. (c) Schematic diagram indicating the neutron scattering analogy with PGSE NMR in the limit of small δ. $\boldsymbol{q}$ is equivalent to the product $(2\pi)^{-1}\gamma\delta g$.

applications where the motion is complex. Until recently the predominant use of the PGSE NMR method was in the study of polymer self-diffusion. Here the Gaussian form of equation (9.5b) is utilised in the data analysis since, in the absence of flow, this relation reduces to the well-known Stejskal–Tanner relation [4]

$$E_\Delta = \exp(-\gamma^2 g^2 \delta^2 D_s \Delta_r) \tag{9.6}$$

where for finite gradient pulse durations Δ_r is given by $(\Delta - \delta/3)$. Thus the polymer self-diffusion coefficient can be obtained from the slope of semi-logarithmic plots of E_Δ versus $\gamma^2 g^2 \delta^2 (\Delta - \delta/3)$. The PGSE experiment has been widely used to measure self-diffusion in polymer melts and solutions [36–53], and an example of a semi-logarithmic plot for random coil polystyrene in CCl_4 solution is shown in Figure 9.2(b).

It is important to note that the structure function described in equation (9.4) bears a formal resemblance to that which applies in neutron scattering where $\boldsymbol{q}$ is the scattering wave vector. PGSE NMR and neutron scattering are closely analogous as illustrated in Figure 9.2(c), the main differences being the detection of $E_\Delta(\boldsymbol{q})$ in the time domain of Δ in the case of PGSE and in the frequency domain in the case of neutron scattering, and the differing scale of time and distance regimes to which the respective methods are sensitive. Neutron scattering [54–56] is confined to measuring rms displacement below 50 Å on a timescale shorter than a few tens of microseconds whilst, until recently, PGSE NMR has been confined to distances greater than 500 Å. The time regime for PGSE NMR is constrained by polymer T_2 and T_1 relaxation times and available gradient strength to $1\ \text{ms} \lesssim \Delta \lesssim 1\ \text{s}$.

Equation (9.4) is equivalent to the neutron inelastic scattering function for the incoherent fraction,

$$S_{\text{incoherent}} = \overline{N^{-1} \sum_i \exp[i2\pi \boldsymbol{q} \cdot (\boldsymbol{r}_i(t) - \boldsymbol{r}_i(0))]} \tag{9.7}$$

where t corresponds to the observation time Δ and the sum is taken over all scattering centres, typically the monomer units of the polymer. By contrast the coherent inelastic neutron scattering and quasi-elastic light scattering is described by

$$S_{\text{coherent}} = \overline{N^{-2} \sum_i \sum_j \exp[i2\pi \boldsymbol{q} \cdot (\boldsymbol{r}_j(t) - \boldsymbol{r}_i(0))]} \tag{9.8}$$

This sensitivity to relative motion, $(\boldsymbol{r}_j(t) - \boldsymbol{r}_i(0))$, makes the interpretation of these methods considerably more difficult, the direct measurement of self-motion being a major advantage in PGSE NMR.

The representation of $E_\Delta(\boldsymbol{q})$ by the average phase shift $\overline{\exp[i2\pi \boldsymbol{q} \cdot \boldsymbol{R}]}$ leads to a useful Taylor expansion

$$E_\Delta(q) \approx 1 - (1/2!)(2\pi q)^2 \overline{Z^2} + (1/4!)(2\pi q)^4 \overline{Z^4} + \cdots \tag{9.9}$$

where Z is the component of displacement along the gradient direction defined by $\boldsymbol{q}$. Equation (9.9) is especially helpful in the case of non-Brownian or restricted motion since it tells us that the initial decay of $E_\Delta(q)$ with respect to q will always yield the ensemble-averaged mean squared displacement, $\overline{Z^2}$.

The additional NMR apparatus required for pulsed gradient spin echo NMR is a subset of that required for NMR microscopy and illustrated in Figure 9.1(b). Normally, shaped RF pulses are unnecessary and only a single gradient coil is required, although it is essential that this be actively screened if a superconducting magnet environment is used, to avoid the possibility of induced eddy currents or quenching. For diffusion measurements above $10^{-11}\,\mathrm{m^2\,s^{-1}}$, no special precautions need to be taken apart from ensuring that the sample is held firmly in place and that the gradient pulse areas are matched to within 0.1%. In the range 10^{-11}–$10^{-13}\,\mathrm{m^2 s^{-1}}$, special care must be taken to reduce noise and ripple in the gradient current supply and to provide clean switching of the gradient pulses [57, 58]. The distance scale $(2\pi q)^{-1} \sim 500\,\text{Å}$ represents the lower reliable limit of the method, corresponding to lower diffusion limit of around $10^{-14}\,\mathrm{m^2 s^{-1}}$ over an echo time of around 100 ms and is largely determined by the need to precisely balance the areas of the two gradient pulses and prevent movement in the sample if spurious phase artefacts are to be avoided. Such artefacts always lead to enhanced echo attenuation and hence anomalously high D_s values.

Recent variants [59, 60] of the PGSE NMR method have provided access to the distance scale between 50 and 500 Å thus effectively bridging the gap with neutron scattering. One approach is to utilise the enormous magnetic field gradients that exist in the stray fields of superconducting magnets, a method first suggested for solid state NMR imaging and known by the acronym STRAFI (for STRAy Field Imaging) [61]. The STRAFI method produces uniform and constant magnetic field gradients in excess of $20\,\mathrm{T\,m^{-1}}$ and in its PGSE NMR application [59] is used in conjunction with a stimulated echo as illustrated in Figure 9.3(a). In this technique, the time between the first two 90° RF pulses defines the duration of gradient-induced precession (i.e. δ) in the transverse plane. In consequence, any problems associated with clean switching of large gradient pulses does not arise. Although the magnetic field gradient is applied continuously, its effect during the z-storage period τ_2 is simply a homospoiling of residual transverse magnetisation. The disadvantage of the STRAFI method is twofold. First there is a loss of spectral information as the stimulated echo is sampled in the presence of the gradient. Second there is a severe loss of signal due to the very narrow regions of spins excited, since the finite bandwidth RF pulse is applied in the presence of a strong gradient, and there is a major increase in noise because of the need to sample the echo with large acquisition bandwidth. Despite these disadvantages, the method has proven especially effective in the case of polymer melts where diffusion coefficients below $10^{-14}\,\mathrm{m^2 s^{-1}}$ have been measured. An alternative approach termed Modulus Addition using Spatially Separated

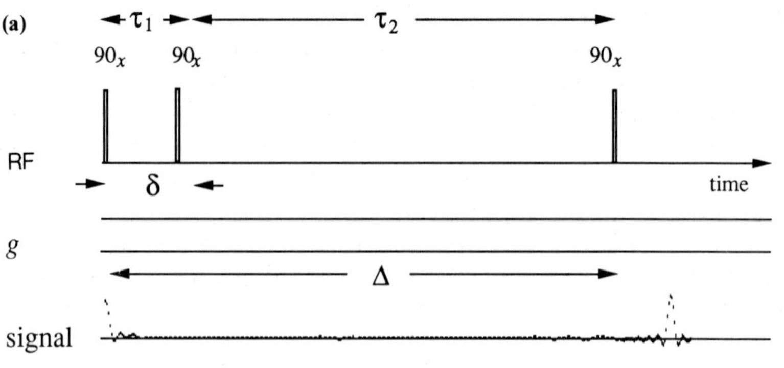
(a)
τ_1
τ_2
90_x
90_x
90_x
RF
δ
time
g
Δ
signal

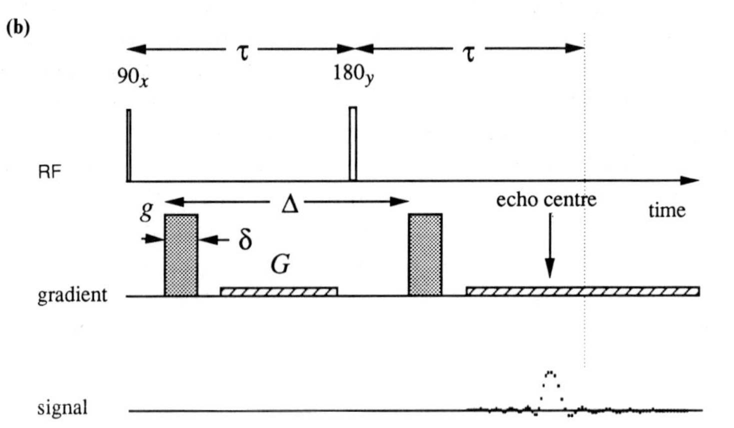
(b)
τ
τ
90_x
180_y
RF
g
Δ
echo centre
time
δ
G
gradient
signal

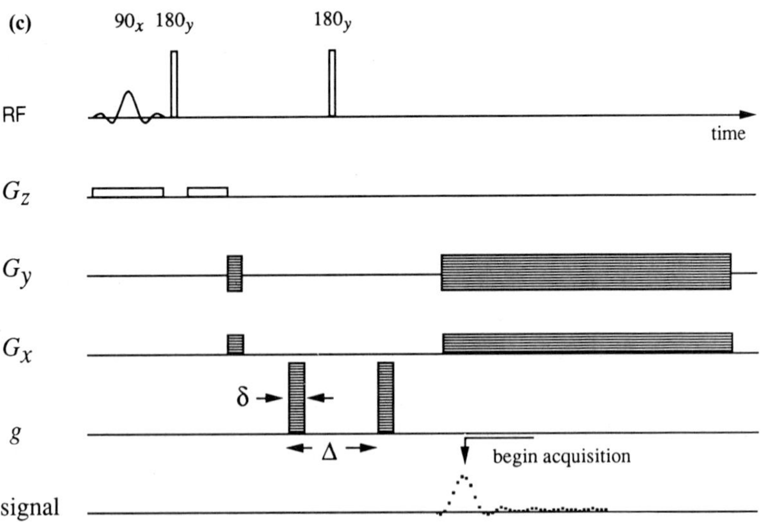
(c)
90_x 180_y
180_y
RF
time
G_z
G_y
G_x
δ
g
Δ
begin acquisition
signal

Echo spectroscopY (PGSE-MASSEY [60]), is illustrated in Figure 9.3(b). In this method, a small read gradient (G) is used and an image of the sample obtained along the gradient axis. This enables helical phase twists due to echo area imbalance to be resolved and these along with phase shift due to sample movement are removed by co-adding power spectra. This method suffers less signal-to-noise ratio loss than STRAFI and has proven effective in application to polymer solutions [7].

Because the primary aim of PGSE NMR is to gain information about molecular translation, a property common to all parts of a molecule, subtle features of the intramolecular structure which are so powerfully revealed by high resolution 1-D and 2-D ^{1}H and ^{13}C NMR spectroscopy are seldom required. Because the proton is ubiquitous and NMR-sensitive, it is almost universally used in PGSE NMR and in NMR imaging measurements using synthetic polymers. Where it is necessary to distinguish the signals from polymer and solvent or between different polymer species in the same system, it is usual to rely on per-deuteration to mask the signal from the unwanted component (for example the solvent) or to utilise rudimentary chemical shifts in the ^{1}H NMR spectrum to provide separate windows on the components of interest. Note, however, that the vastly different magnitudes of solvent and polymer D_s values generally means that solvent deuteration is unnecessary. This is because under these circumstances, equation (9.6) comprises a component signal sum of two exponentials with widely differing decay rates so that the resolution of the diffusion coefficients of the two components is straightforward.

9.2.3 *Dynamic NMR imaging*

Figure 9.3(c) shows a pulse sequence in which the PGSE method is amalgamated with NMR imaging [6, 16, 62, 63]. As a consequence, the echo attenuation factor, $E_\Delta(\boldsymbol{q})$ becomes a contrast factor as described in section 9.2.1 so that the signal acquired is effectively modulated both in $\boldsymbol{k}$- and $\boldsymbol{q}$-space. Thus

$$\begin{aligned} S(\boldsymbol{k},\boldsymbol{q}) &= \int \rho(\boldsymbol{r}) E_\Delta(\boldsymbol{q}) \exp(i2\pi\boldsymbol{k}\cdot\boldsymbol{r})\,\mathrm{d}\boldsymbol{r} \\ &= \int \rho(\boldsymbol{r}) \left\{ \int \overline{P}_s(\boldsymbol{R},\Delta) \exp[i2\pi\boldsymbol{q}\cdot\boldsymbol{R}]\,\mathrm{d}\boldsymbol{R} \right\} \exp(i2\pi\boldsymbol{k}\cdot\boldsymbol{r})\,\mathrm{d}\boldsymbol{r} \end{aligned} \qquad (9.10)$$

←

Figure 9.3 (a) Gradient and RF pulse sequence for STRAFI pulsed gradient spin echo NMR using a stimulated echo to define the period δ over which the transverse magnetisation is subject to precession in the constant gradient. The period τ_1 effectively defines the duration δ of the gradient pulse applied to the transverse magnetisation. (b) Gradient and RF pulse sequence for PGSE MASSEY in which a small read gradient (G) is used to resolve the phase twist associated with gradient pulse mismatch. Each echo is Fourier transformed to produce a 1-D 'image' and subsequent signal averaging occurs by adding power spectra. (c) Gradient and RF pulse sequence for dynamic NMR imaging. The imaging gradients are preceded by a PGSE pulse pair which are successively stepped to phase encode for motion.

where implicitly $\overline{P}_s(\boldsymbol{R}, \Delta)$ is the average propagator at each pixel $\boldsymbol{r}$ of the image. Double inverse Fourier transformation of $S(\boldsymbol{k}, \boldsymbol{q})$ with respect to both $\boldsymbol{k}$ and $\boldsymbol{q}$ returns $\rho(\boldsymbol{r})\overline{P}_s(\boldsymbol{R}, \Delta)$. By normalising this function with the image density $\rho(\boldsymbol{r})$ acquired under zero PGSE gradient, one reconstructs $\overline{P}_s(\boldsymbol{R}, \Delta)$ for each pixel of the image. Generally, it would be too time-consuming to carry out a full analysis in the six-dimensions of $\boldsymbol{k}$ and $\boldsymbol{q}$. Instead, it is conventional to apply the $\boldsymbol{q}$ gradient (g) in a single direction in any given experiment, stepping its value from zero to some maximum number, n_D, of order 10 to 20 steps. In that sense the method is akin to multi-slice imaging. For each step (or 'q-slice)') a complex image is reconstructed. At the completion of acquisition, and zero-filling from n_D to N to improve digital resolution, the modulated image signal in each pixel is Fourier transformed along the q-direction to return $\overline{P}_s(Z, \Delta)$ for that pixel.

By appropriate processing of these average propagators, details of the local motion can be calculated. For example, the width of $\overline{P}_s(Z, \Delta)$ is determined by the rms Brownian motion $(2D_s\Delta)^{1/2}$ whilst the displacement of $\overline{P}_s(Z, \Delta)$ along the Z axis is determined by the flow displacement $v\Delta$ where v is the local molecular velocity. In his manner maps of $D_s(\boldsymbol{r})$ and $v(\boldsymbol{r})$ may be constructed. The method is described in detail elsewhere [63]. Dynamic NMR microscopy can provide detailed velocity maps for liquids moving inside a confining vessel and, at the same time, give access to molecular properties at differing locations in the velocity field. Consequently, it has great potential as a tool for investigating the molecular basis of complex rheological properties in non-Newtonian fluids.

9.3 Applications of NMR imaging in polymers

9.3.1 Solid-state imaging

NMR imaging in solids is complicated by the very short spin–spin relaxation times which prevail. The rapid decay of transverse magnetisation makes slice selection extremely difficult because of the finite duration of soft, narrowband RF pulses, and in addition requires that very large magnetic field gradients be used if the effects of dipolar broadening are to be overwhelmed. In consequence, a variety of highly specialised methods are required to achieve an NMR image from molecules in the solid state. A detailed review of solid state imaging is beyond the scope of this chapter and may be found elsewhere [6, 64].

Polymeric solids exhibit a wide spectrum of local molecular motion. In the classical subdivision into thermosets, rubbers and thermoplastics, both the thermosets and the thermoplastics below the glass transition temperature would exhibit transverse relaxation time on the sub-millisecond timescale, the so-called solid regime of NMR. It is these materials which require specialised solid state imaging methods rather than the elastomeric materials where

T_2 values may be considerably longer. Because of the critical role of sensitivity in these experiments, the use of the proton as the sensitive nucleus is prevalent. However, by the use of elegant coherence transfer methods, both ^{13}C and ^2H NMR imaging experiments have been performed [32, 65].

Three approaches to solid-state imaging have proven particularly successful. The first involves using the large magnetic field gradients in the superconducting magnets stray field. The STRAFI imaging methods, pioneered by Samoilenko *et al.* [61, 66], is based on a point-wise acquisition of rapidly decaying signal as the sample is moved stepwise through the sensitive region. The necessary use of wide acquisition bandwidth inherently reduces the signal-to-noise ratio by comparison with imaging in the liquid state. While the method is consequentially slow, it has resulted in high quality images in rigid polymeric solids, an example of which is shown in Figure 9.4(a).

A second method involves the use of multiple RF pulse sequences developed to prolong the magnetisation in the face of dipolar broadening effects [67, 68]. This approach requires that intense gradient pulses be applied in specified 'windows' of a line-narrowing sequence such as MREV-8 (see chapter 7). While the method is quite effective, it is highly demanding both on the equipment and on the experimenter. A variation of this multiple pulse method involves the use of oscillating gradients in conjunction with a multiple solid echo or Hahn echo train [69, 70]. This approach does not attempt to remove the effective dipolar Hamiltonian but to recycle the magnetisation in a manner that makes it available for multiple acquisition, thus compensating for the signal-to-noise loss associated with broad bandwidth acquisition. The method is less technically demanding than line-narrowing and has been applied successfully to heavily crosslinked rubber for which $T_2 \approx 900\,\mu s$ [70].

A third approach to solid-state imaging is to use magic-angle spinning coupled with synchronously rotating gradients to reduce the dipolar linewidth so that field gradient broadening will dominate [71–74]. The method of gradient rotation is ingenious and involves the phase-locked application of oscillating currents in quadrature in orthogonal gradient coils. While MAS techniques at up to 10 kHz rotation speed are effective in removing dipolar interactions in ^{13}C, it is considerably more difficult to reduce the larger proton dipolar width so that MAS imaging is most effective in polymers for which local motional-narrowing has already narrowed the line. For more rigid solids, a combination of rotation and multiple pulse line-narrowing (CRAMPS) has proven effective [72, 73].

One exciting new mode of solid-state imaging involves the use of pure phase encoding of spatial information so that the FID is acquired in the absence of the magnetic field gradient and the Hamiltonian terms which result in spectral broadening such as the dipolar interaction, quadrupole interaction and anisotropic chemical shift can be observed as a fourth dimension of contrast. This requires that the transverse magnetisation be sustained sufficiently long for phase encoding to be applied, one method being the generation of multiple

(a)

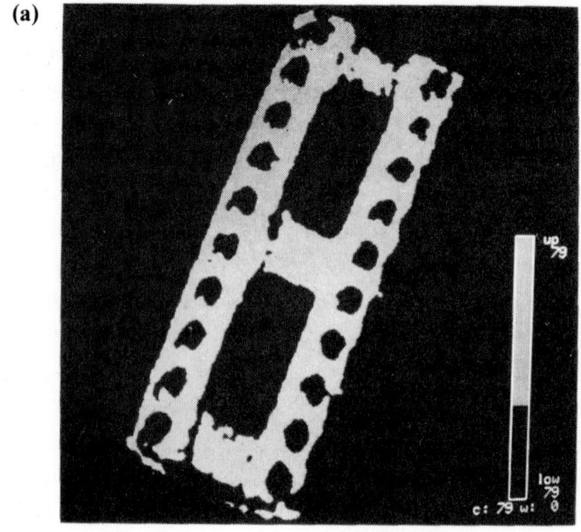

(b)

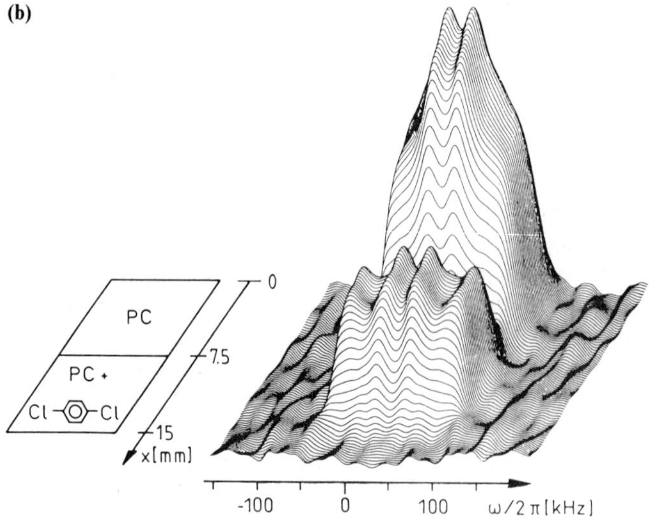

Figure 9.4 (a) ^{1}H NMR image of an integrated circuit socket obtained in the stray field of a superconducting magnet using the STRAFI method. The spatial resolution is better than 100 μm, clearly revealing the metal connectors on the socket. (From A.A. Samoilenko and K. Zick [66] and reproduced by permission of Bruker Analytische Messtechnik.) (b) Spectroscopic image of two adjacent pieces of polycarbonate, one with and the other without plasticiser. Note the orthogonal spatial and spectroscopic dimensions. This ^{2}H NMR image is obtained using phase-encoding on a state of two quantum coherence. Note that the narrow component in the pure polycarbonate arises from a local motion which is suppressed when the additive is present. (From B. Blümich *et al.* [76] and reproduced by permission of Hüthig and Wepf Verlag.)

quantum coherence. The method has proven most effective in deuteron NMR where the quadrupole interaction is used to indicate the local order parameter [32, 75, 76]. Blümich, Spiess and co-workers have combined deuterium NMR spectroscopy and imaging to examine the local molecular order in a number of solid polymer systems, including drawn deuterated polyethylene [32] and deuterated polycarbonate containing regions with and without plasticiser [77] as shown in Figure 9.4(b).

9.3.2 Elastomer imaging and solvent imaging

By contrast with rigid polymers and thermosets, amorphous thermoplastics and rubbers above the glass transition temperature may have T_2 values considerably longer than 1 ms because of the influence of significant re-orientational motion at the local level. A feature of these materials is the short T_2^* arising from chemical shift variation and from susceptibility artifacts causes by microscopic voids. Conventional Fourier methods, such as the pulse sequence illustrated in Figure 9.1(a), may be used to obtain images in materials with T_2 values of a few milliseconds. For shorter T_2 values of around 1 ms, significant signal loss may occur during selective slice excitation using narrow bandwidth soft RF pulses. For these systems, it is advantageous to use hard pulse excitation of the whole sample and 3-D imaging, as shown in Figure 9.5(a). Komoroski and co-workers [78–80] have used this approach to obtain high resolution images in tyre rubber as shown in Figure 9.5(b). This method has proven effective in revealing the distribution of carbon black in these tyre samples [79].

The effect of magnetic susceptibility variations around voids and impurities in polymeric solids may be nicely demonstrated by comparing images obtained using gradient and spin echo methods [6] where the respective attenuation of the images is due to T_2^* and T_2. A characteristic feature of susceptibility artefacts is the bright arrowhead distortion apparent in Figure 9.5(b).

Above and below the glass-transition polymer ^{1}H NMR relaxation times differ markedly. This effect has been used [81–83] to demonstrate the penetration of plasticising solvents into a glassy polymer at temperatures below T_g in which the diluent has the effect of reducing the local glass-transition temperature and consequently substantially increasing T_2. This leads to a significant proton NMR signal arising where previously the short T_2 had rendered the material invisible in the image. Mareci *et al.* [81] have used this method to examine the penetration of deuterated chloroform into PMMA as shown in Figure 9.5(c) ($T_g = 310\,K$) while Koenig and co-workers have used deuterated solvent swelling of crosslinked polybutadiene to render the polymer chains more visible to conventional NMR imaging [83]. This technique has enabled the determination of heterogeneity in the crosslink density in these materials.

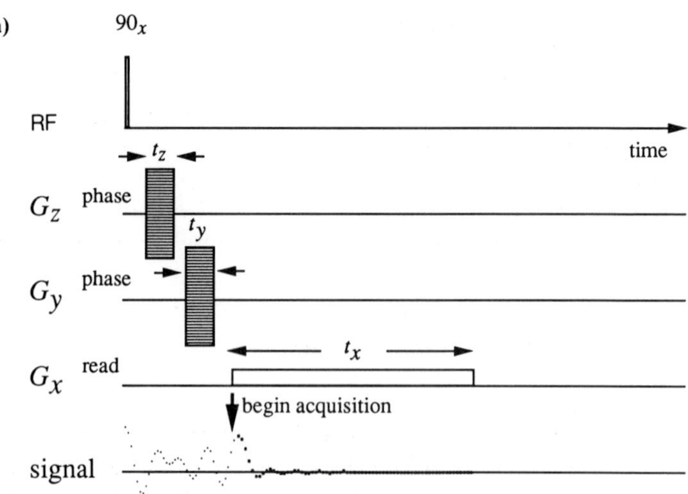
(a)
90x
RF
time
tz
Gz
phase
ty
Gy
phase
tx
Gx
read
begin acquisition
signal

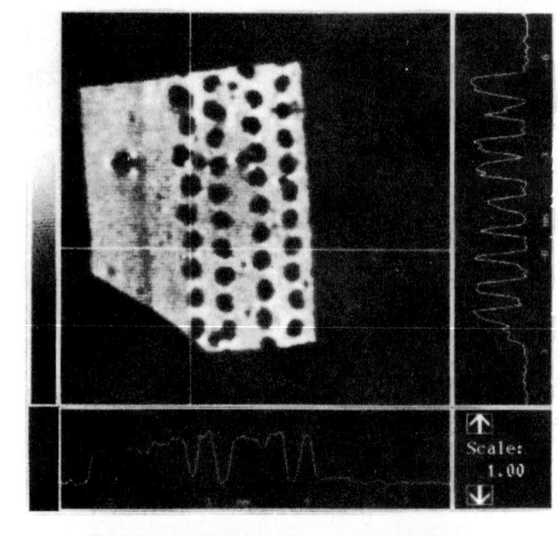
(b)
Scale:
1.00

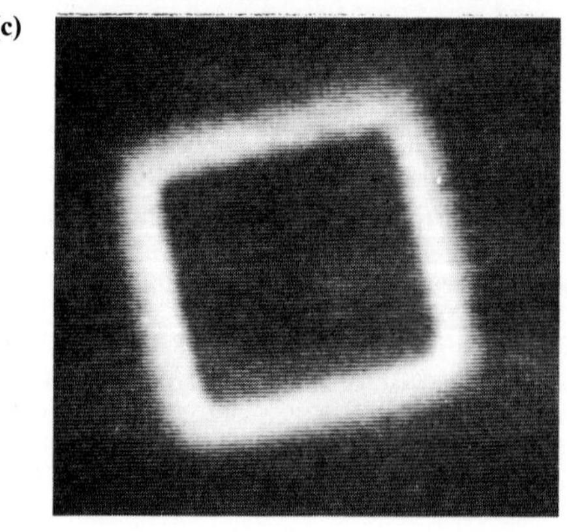
(c)

9.3.3 *Monomer and solvent imaging*

One of the very practical applications of NMR imaging in polymers concerns the observation of time variation in the physical state of polymer during polymerisation, crosslinking and solvent penetration. In the polymerisation process, the local T_2 values undergo a sharp transition as the monomer is incorporated into the chain, especially at temperatures below T_g, so that for an imaging pulse sequence involving a suitably long spin echo duration, the image signal disappears once local polymerisation occurs. This effect has been utilised by Hall *et al.* [84] to observe the polymerisation of poly-(methylmethacrylate)(PMMA) in a specially designed reaction cell which is placed inside the RF coil of the imaging system. The same approach has been used to observe crosslinking in epoxies and adhesive setting [85–87]. These studies have the potential to relate the heterogeneity of polymerisation rates to subsequent solid-state morphology, an important factor in improving bulk material properties.

The penetration of solvents into solid polymers may also be effectively studied using NMR imaging, leading to quantitative information concerning inter-diffusion mechanisms [88–94]. These experiments are performed by immersing a polymer rod or block in a bath of proton-bearing solvent and observing subsequent images of the solvent distribution as it penetrated the solid polymer. Some nice demonstrations of this application have been given by Weisenberger and Koenig [88, 89] with their study of solvent inter-diffusion in PMMA above and below T_g. Above the glass-transition, the solvent diffuses relatively freely exhibiting the Fickian property of a solvent front advancing in proportion to the square root of elapsed time. Below T_g, the solvent is hindered but has the effect of locally reducing the glass-transition temperature by swelling, thus resulting in a constant concentration behind the solvent front which in turn advances in direct proportion to time. This latter diffusion is termed Case II while Fickian diffusion Case I. An example of both types of diffusion for methanol in PMMA is shown in Figure 9.6. Grinstead *et al.* [94] have also used this approach to study the lowering of T_g which results when PMMA is exposed to water, thus gaining information about the effect of water on the craze resistance of this polymer.

←

Figure 9.5 (a) NMR imaging sequence in which 2-D phase encoding is employed to reduce the relaxation delay associated with soft-pulse excitation. (b) 200 MHz ^{1}H NMR image from a section of steel belted tyre obtained using the pulse sequence of Figure 9.4(c). The slice thickness is 500 μm and the pixel resolution is $(200\,\mu\text{m})^2$. Note the bright arrowhead distortion arising from a susceptibility artefact around the defect on the left of the material. (From S.N. Sarkar and R.A. Komoroski [79] and reproduced by permission of the American Chemical Society.) (c) ^{1}H NMR image from a bar of poly(methylmethacrylate) at room temperature which has been immersed in deuterated chloroform for 3 h. The solvent penetration has caused a substantial increase in T_2 thus making the polymer chains visible. (From T.H. Mareci, *et al.* [81] and reproduced by permission of Elsevier.)

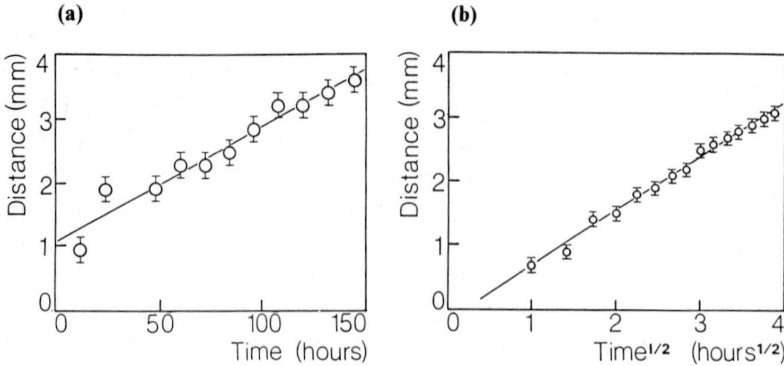

Figure 9.6 Diffusion of methanol onto poly(methylmethacrylate) rod: (a) shows Case II mutual diffusion at 30°C while (b) shows Case I (Fickian) diffusion at 60°C. (From L.A. Weisenberger and J.L. Koenig [88] and reproduced by permission of the Society for Applied Spectroscopy.)

9.4 Pulsed gradient spin echo diffusion studies in polymers

9.4.1 Centre of mass diffusion in polymer liquids

Macromolecular self-diffusion coeffcients in polymer solutions and melts are strongly influenced by concentration and molar mass. In solution the concentration affects the Brownian motion in a number of ways. Firstly, concentration influences the local friction of polymer segments, secondly, concentration influences the screening of the excluded volume interaction which in turn determines the polymer conformation; thirdly, there are concentration-dependent screening influences on the hydrodynamic interaction between segments via the solvent velocity field, finally, there is a concentration-dependence of topological constraints exerted by polymer entanglements. In both solution and melts, the molar mass determines both the total chain friction and the entanglement effect. Pulsed gradient spin echo NMR studies have been of considerable significance in elucidating these phenomena.

In practice the PGSE method is routinely capable of measuring self-diffusion coefficients down to a lower limit of $10^{-13}\,m^2s^{-1}$ with relative ease. The regime above $10^{-13}\,m^2s^{-1}$ covers the range of dilute solution centre-of-mass diffusion coefficients for all polymers. PGSE studies in such systems have shown good agreement with alternative methods, such as quasi-elastic light scattering, and have confirmed the infinite dilution scaling dependence on molar mass, $D_s \sim M^{-\nu}$, where ν ranges between 0.5 and 0.6 depending on solvent quality [95–97]. Studies of concentration dependence of the friction factor in dilute solution have also proven of value in revealing information about polymer conformation and shape [98].

Because of its sensitivity to self-motion of the 'labelled' chain, as indicated in equation (9.7), PGSE NMR has proven of especial value in studying the dense phase of polymer liquids. Where polymer chains strongly overlap, light scattering measurements are complicated by the inter-segment terms exhibited in equation (9.8) and thus lead to results that are dominated by collective motions of the system. By contrast, both PGSE NMR and the incoherent fraction of inelastic neutron scattering reveal the rms displacements of individual polymer segments. The complementarity of the distance scales to which the two methods are sensitive makes PGSE NMR the method of choice for studying centre of mass self-diffusion. Only forced Rayleigh scattering [99] can compete with PGSE NMR in the direct measurement of D_s in dense polymer liquids and this latter technique requires a special photochromic label to be attached to the polymer. Furthermore, the PGSE NMR method is ideally suited to the study of mixed phase systems and to polymer blends [100] where the chemical shift selectivity in the NMR spectrum enables one to independently determine D_s for different polymeric components. The use of the ^{13}C PGSE NMR has also proven effective in distinguishing molecular species in polymer solutions [101].

Topological effects (i.e. forbidden chain crossings) are considered to dominate the Brownian motion of random coil polymers when chain overlap effects become significant [96, 97]. For solutions, such overlap effects become important above the polymer volume fraction $\varphi^* = b^3 N/R_0^3$ where N is the number of Kuhn statistical segments of dimension b in the random coil of end-to-end length R_0. The regime $\varphi > \varphi^*$ is termed semi-dilute. In both melts and semi-dilute solutions, the onset of topological constraints is considered to take place when the molar mass exceeds a critical value, M_c. In the Doi–Edwards tube depiction [97, 102] (see chapter 4), the topological effects are accounted for in a mean field description in which the motion is characterised by one-dimensional Rouse-like curvilinear diffusion in an entanglement tube formed by the topological constraints of surrounding chains. This curvilinear motion, first proposed by de Gennes [103], is termed 'reptation'. The diameter of the Doi–Edwards tube, a, is an important parameter in describing the dynamics of these systems. This diameter is considered to correspond with the dimension of the random coil of molar mass M_c, and is concentration-dependent in the case of the semi-dilute solution.

Reptation theory predicts scaling laws for centre of mass self diffusion above M_c. For melts, the characteristic 'signature' for reptation is given by $D_s \sim M^{-2}$, a result that has been confirmed both by PGSE NMR and by other methods applicable in the molten state. Figure 9.7 shows the dependence of D_s on M for molten mono-disperse polyethylene and polyethylene oxide melts obtained using PGSE NMR. For semi-dilute solution, reptation theory [96, 97] gives $D_s \sim \varphi^{(2-\nu')/(1-3\nu)} M^{-2}$ where ν and ν' are the excluded volume exponents appropriate, respectively, to static and dynamic dimensions of the random coil section of length a. Both exponents undergo a transition from 0.6 to 0.5 as the displacement between chain monomers decreases [104], but at a markedly different rate, thus leading to a strong variation of the effective

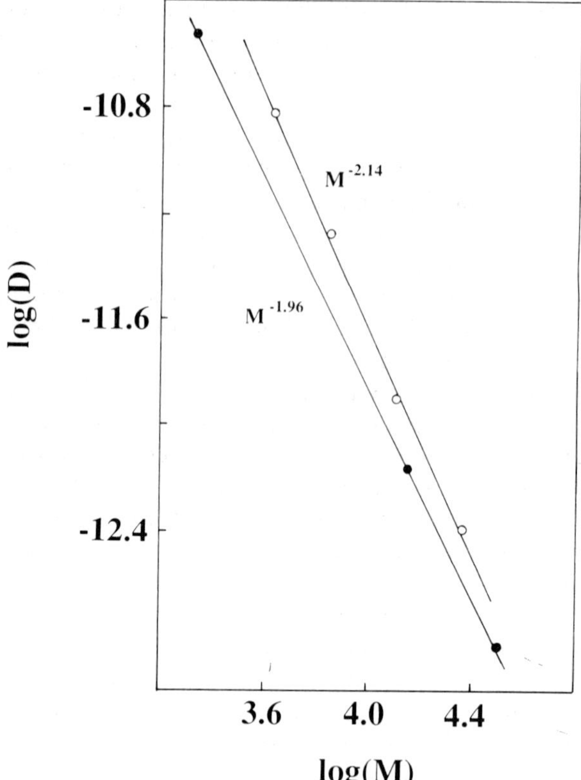

Figure 9.7 Dependence of D_s on M obtained from PGSE NMR measurements on molten, mono-disperse polyethylene (●) and poly(ethylene oxide) (○) at 150°C. (From T.M. Huirua *et al.* [53] and reproduced by permission of the American Chemical Society.)

exponent $(2 - v')/(1 - 3v)$ between -1.75 and -3.0 as φ increases [42]. This effect strongly undermines the concentration scaling hypothesis. Furthermore, the variation in monomer friction as φ increases further complicates the φ-dependence of D_s [105, 106]. Remarkably, however, once this monomer friction effect is approximately accounted for, the generalised D_s versus φ dependence of random coil polymers exhibits a universal behaviour as shown in Figure 9.8(a) [52, 105]. This graph contains data from polystyrene poly-(ethylene oxide), dextran, gelatin, cellulose acetate and poly(dimethylsiloxane), in a wide range of solvents. Common behaviour is observed despite the range of polymer glass transition temperatures and solvent quality applicable in each case. In these data, $D_s(\varphi)$ has been corrected for monomer friction effects by utilising local rotational correlation times obtained from nuclear spin relaxation measurements. The corrected $D_s'(\varphi)$ values are then normalised to $D_s(0)$ while concentrations are normalised to the overlap cross-over value,

φ^*. A noteworthy feature of Figure 9.8(a) is the asymptotic dependence on D_s' on φ^{-3}, irrespective of solvent quality. This provides strong evidence for chain exponents being predominantly Gaussian at short range, even in good solvents.

The molar mass scaling law in semi-dilute solution exhibits a similarly universal character as evident in Figure 9.8(b), with an asymptotic scaling of M^{-2}. However, it should be emphasised that the regimes of apparent scaling in both the concentration and molar mass plots are precariously small. In consequence, the contention that reptation is proven by these results has been seriously questioned [107–109]. Certainly other analytic dependences can be used to represent these data. Furthermore, other theories of dense polymer dynamics predict similar scaling laws. Finally and most damaging, PGSE NMR studies on semi-dilute solutions of star-branched polymers [110, 111] have shown a similar concentration dependence to that exhibited in Figure 9.8(a) despite the fact the reptation theory would predict severe attenuation of the Brownian dynamics in these systems. To conclude, while the predictions of reptation are consistent with the observed dependence of centre-of-mass diffusion in semi-dilute solutions, the data are by no means conclusive. Of critical importance is the existence of sufficient numbers of entanglements to justify a mean field description. Many studies of self-diffusion in semi-dilute solution have employed polymers with molar masses below 1×10^6 Da for which the tube diameter a is comparable with the polymer dimension $\bar{R}$. In order to clarify the role of reptation in determining centre-of-mass self-diffusion, it will be important to extend PGSE NMR measurements to polymers with $M > 10^6$ Da.

In many polymer systems, polydispersity is a feature that must be accounted for in analysing the results of PGSE NMR measurements. Allowing for both relaxation time and diffusion coefficient dependence on molar mass, the echo attenuation expression in equation (9.6) must be modified to read

$$E_\Delta = \int_0^\infty P(M)R(M)\exp[-\gamma^2 g^2 \delta^2 D_s(M)\Delta_r]\,\mathrm{d}M \qquad (9.11)$$

where $P(M)$ is the mass distribution of the polymer and $R(M)$ is a relaxation decay factor dependent on $T_1(M)$ or $T_2(M)$ decay, respectively, depending on whether a spin echo or stimulated echo PGSE sequence is employed. The influence of this factor can be subtle since T_1 and T_2 in polymers are principally influenced by the local segmental dynamics. In solutions, these relaxation rates are fairly insensitive to molar mass at lower concentrations. In melts, the polymers characteristically exhibit bi-exponential T_2 relaxation with the slow component (of dominant influence in PGSE NMR) arising from the chain ends and independent of molar mass. By contrast, D_s will generally depend strongly on M, especially where chain overlap effects are important so that we may consider its influence to be dominant. Provided

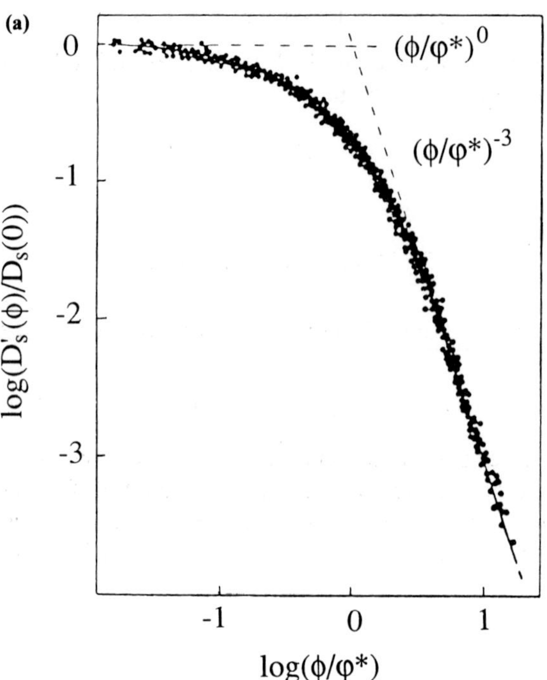
(a)
0
-1
-2
-3
$(\phi/\varphi^*)^0$
$(\phi/\varphi^*)^{-3}$
$\log(D'_s(\phi)/D_s(0))$
-1
0
1
$\log(\phi/\varphi^*)$

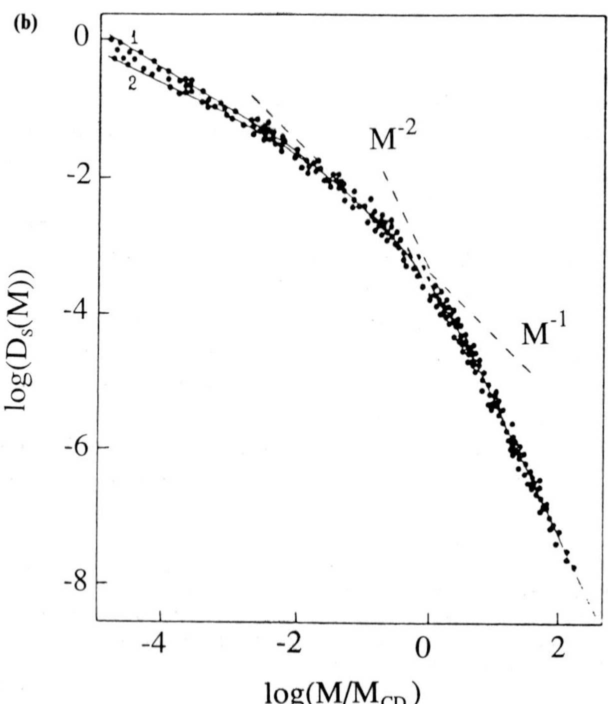
(b)
0
-2
-4
-6
-8
1
2
M^{-2}
M^{-1}
$\log(D_s(M))$
-4
-2
0
2
$\log(M/M_{CD})$

small echo attenuations are used ($E_\Delta > 0.5$), equation (9.11) reduces, in accordance with equation (9.9), to

$$E_\Delta \approx 1 - \gamma^2 g^2 \delta^2 \overline{D_s(M)} \Delta \tag{9.12}$$

where

$$\overline{D_s(M)} = \int_0^\infty P(M) D_s(M) \, dM \tag{9.13}$$

One particularly useful depiction of $P(M)$ is the log-normal distribution for which

$$P(M) = (2\pi\sigma^2)^{1/2} M^{-1} \exp[-(L - L_0)^2/2\sigma^2] \tag{9.14}$$

L being $\ln(M)$. L_0 and σ are the centre and standard deviations, respectively, of the log normal distribution. In this case, allowing that D_s scales with molar mass as $D_s \sim M^\alpha$, yields a simple power series evaluation of equation (9.11) whose leading term is determined by the mass-averaged diffusion coefficient to be [112]

$$\overline{D_s(M)} = D_s(M_w) \exp[(\alpha^2 - \alpha)\sigma^2/2] \tag{9.15}$$

where $D_s(M_w)$ is the self-diffusion coefficient of the polymer with mass-averaged molar mass, M_w.

9.4.2 Internal motions in high polymers

Rather than focusing on scaling laws for centre-of-mass self-diffusion, a more convincing test is to examine the entire motional regime of the polymer, and especially the internal modes over the distance regime 50–1000 Å to which access is given by the PGSE MASSEY and PGSE STRAFI methods. An indication that internal motions can be observed is given in the pioneering STRAFI measurements of Kimmich *et al.* [59] using poly(dimethylsiloxane) melts. Their D_s data exhibit a deviation from the usual M^{-2} scaling as the distance scale being detected drops below the polymer dimensions.

One useful approach to the description of internal motion is to follow the time dependence of the ensemble-average mean-squared displacement, $\overline{Z^2}$, along the laboratory frame gradient axis as the PGSE pulse separation time, Δ, is varied. The characteristic behaviour of $\overline{Z^2}$ as respectively dependent on $\Delta^{1/4}$, $\Delta^{1/2}$ and Δ^1 over the various time regimes represents a signature for

Figure 9.8 Dependence of (a) D_s' on ϕ/φ^* and (b) D_s on M/M_{CD} for a wide variety of polymer and solvent systems. Note that in (a) the monomer friction effect has been removed and the data follow a universal curve. In (b) D_s is taken from an equivalent concentration and is shown plotted in arbitrary units. M_{CD} corresponds to the intersection of the M^{-2} and M^{-1} asymptotes. (From A.I. Maklakov *et al.* [105] and reproduced by permission of the Gulf Publishing Company.)

reptative motion [97]. The transition from $\Delta^{1/4}$ to $\Delta^{1/2}$ occurs when the observation time matches the Rouse time of the polymer whilst the transition $\Delta^{1/2}$ to Δ^1 occurs at the tube disengagement time τ_d. At $\Delta \sim \tau_d$, $\overline{Z^2} \sim R_0^2$.

Figure 9.9 shows $\overline{Z^2}$ versus Δ for three different molar mass polystyrenes in 9% volume fraction CCl_4 solutions [7] in the regime $\Delta \sim \tau_d$. For those experiments performed with $\Delta \lesssim \tau_d$, $E_\Delta(q)$ is not the simple Gaussian associated with centre-of-mass Brownian motion but is given by the incoherent inelastic neutron scattering structure factor for the internal modes of the polymer [102]. This q-dependence can be used to examine the mode-structure and to observe local transverse motion in the tube [7]. However, irrespective of the dynamical behaviour, $\overline{Z^2}$ can be obtained from the low q dependence of $E_\Delta(q)$. A transition region from $\overline{Z^2} \sim \Delta^1$ to $\overline{Z^2} \sim \Delta^{1/2}$ is cleary visible for two of the three polymer systems as the observation time is reduced below 1 s. In both cases, the position of this transition closely agrees with the theoretical reptation prediction. For the 15×10^6 Da polystyrene at 10% concentration, no such transition is apparent, consistent with the calculated value of τ_d

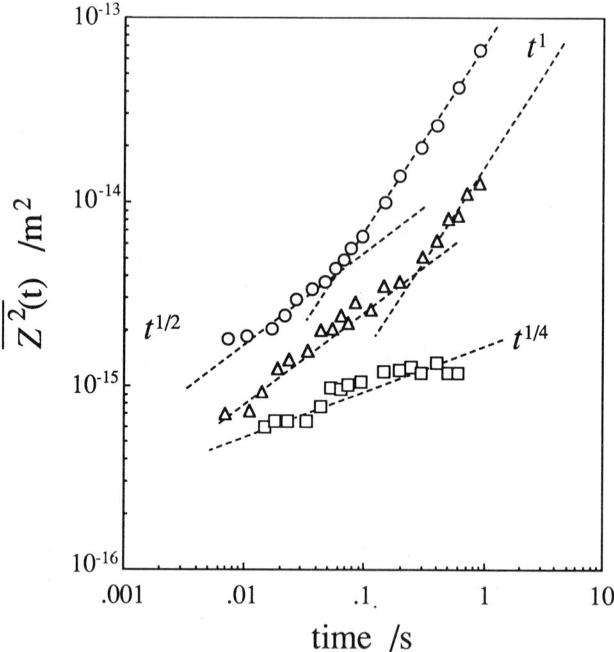

Figure 9.9 Ensemble-averaged mean squared displacement $\overline{Z^2}$ versus time for the statistical sub-units of polystyrene in 9% volume fraction semi-dilute solution with CCl_4, as obtained using ^1H PGSE NMR. (○) 1.8×10^6 Da; (△) 3.0×10^6 Da; (□) 15×10^6 Da. The data are compared with asymptotic lines for $t^{1/4}$, $t^{1/2}$ and t^1 scaling where t corresponds to the PGSE observation time △. The transition points are consistent with calculated τ_d values. (From P.T. Callaghan and A. Coy [7] and reproduced by permission of the American Institute of Physics.)

and a clear $\Delta^{1/4}$ region is apparent, albeit at a lower mean-squared displacement than suggested by Doi–Edwards theory. While such transitional behaviour in the internal motions of high polymers has been found in computer simulations [113–116], this PGSE NMR data provides a direct measurement of mean-squared displacement over the motional regimes long considered a signature for reptation.

9.4.3 Diffusion of small molecules in polymer systems

The self-diffusion of highly mobile small molecules in macromolecular solutions is very easily measured using PSGE NMR techniques because of the typically long T_2 values and rapid diffusion. Such studies can provide indirect information about polymer conformation via the solvent/macromolecular interactions. For example, the macromolecule presents an obstruction around which the small molecule must diffuse, thus attenuating its D_s value in a manner that depends characteristically on the polymer concentration. Another important interaction influencing the small molecule D_s concerns the attractive or bonding interactions with the polymer, sometimes referred to as the solvation effect. Obstruction factors have been formulated for oblate and prolate ellipsoids, two well-known descriptions being the Wang and Jonsson models [117, 118]. Both incorporate the additional effect of solvation. Indirect shape factors have been obtained for a wide variety of random coil, rigid rod and latex particle systems by careful analysis of the polymer concentration dependence of normalised solvent D_s values using both ^1H and ^2H PGSE NMR [98, 100, 119–121].

In dense polymers, the self-diffusion of small plasticising solvent molecules has been measured PGSE methods [105]. For polymers above the glass-transition temperature, it is common to model the solvent diffusion using the free volume theory as modified for polymer systems by Fujita [122] and by Vrentas and Duda [123]. For solvent diffusion in dense polymers below T_g, an alternative model has been given by Frisch and Stern [124].

9.5 Flow studies using dynamic NMR imaging

High polymer melts and solutions exhibit non-linear viscoelastic properties such as the non-Newtonian dependence of the shear stress on shear rate [125], commonly known as shear-thinning. A number of constitutive equations have been developed to describe the relationship between the shear stress, σ_{xy}, and the shear rate, $\dot{\gamma} = \partial v_x/\partial y$, in simple Couette flow, one of the better known being that of the 'power law' fluid [126]

$$\sigma_{xy} = K\dot{\gamma}^n \tag{9.16}$$

where K and n are constants for a particular fluid. Whence the non-linear viscosity η may be written

$$\eta(\dot{\gamma}) = K\dot{\gamma}^{n-1} \tag{9.17}$$

The power law exponent, n, is unity for a Newtonian fluid and less than unity for a shear-thinning fluid. One of the central questions of polymer physics concerns the molecular basis for the constitutive equations. Because NMR is so sensitive to molecular dynamical parameters, the simultaneous mapping of velocity profiles and molecular properties such as the polymer self-diffusion coefficient by means of the dynamic NMR microscopy technique offers an effective test of much molecular models.

One molecular explanation for shear thinning concerns breakdown in entanglement renewal as adjacent polymers separate [97]. Thus non-Newtonian properties are expected to arise when the shear rate is comparable with the slowest polymer relaxation process, the rate of tube disengagement, τ_d^{-1}. A particularly simple experiment involves forcing the polymer solution through a capillary. The shear rate will exhibit a variation from zero at the capillary centre to a maximum at the walls, this maximum value varying inversely as the capillary radius. By choosing capillaries that are sufficiently narrow, it is possible to achieve shear rates that exceed the characteristic conformational relaxation rates of the polymer molecule. Because flow velocities in NMR imaging experiments must be below $100\,\mathrm{mm\,s^{-1}}$ if the excited slice is to remain within the RF coil over the duration of the spin echo, this determines the shear rates accessible to the method given some chosen capillary diameter [63]. For example, to observe shear rates of order $100\,\mathrm{s^{-1}}$, it is necessary to use submillimetre diameter capillaries, thus confirming the need for microscopic resolution in these experiments.

Figure 9.10(a) compares velocity profiles obtained in a dynamic NMR microscopy experiment performed at 30°C using 1.6×10^6 Da poly(ethylene oxide) (PEO) dissolved in water at a range of concentrations between 0.5% and 4.5% (w/v) where the solution is forced through a Teflon capillary with internal diameter $700\,\mu\mathrm{m}$ [17, 18]. A transition from Newtonian to non-Newtonian behaviour is observed as the concentration increases, an effect that is consistent with a measured value of φ^* of around 0.5%. Pressure heads of up to 21 atm were used to drive the polymer solution from the header tank reservoir through the capillary. A power law fit to the high concentration velocity profiles yields an exponent of 0.4, similar to that found in laser Doppler anemometry experiments using polyethylene melts [127].

It should be noted that the dominant ^{1}H NMR signal for these data arises from the water solvent. While the velocity profiles of solvent and polymer solute will be the same, the Brownian motion of the two components will be markedly different. Figures 9.10(c), (d) show solvent velocity and self-diffusion maps obtained from the highest concentration solution for which some evidence of enhanced solvent diffusion is apparent in the high shear region of the capillary. These data are obtained at a pressure head of 2100 kPa

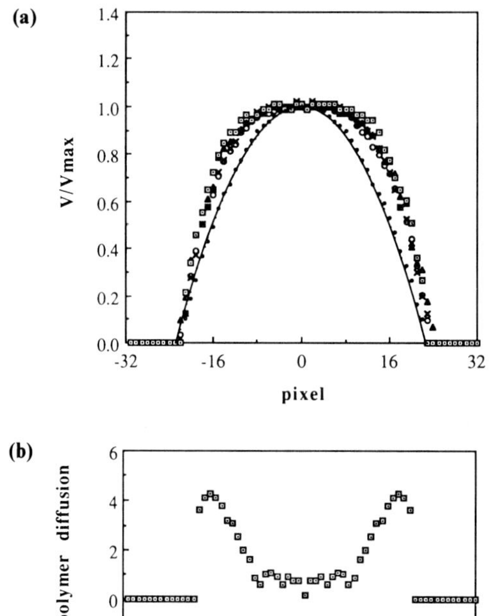

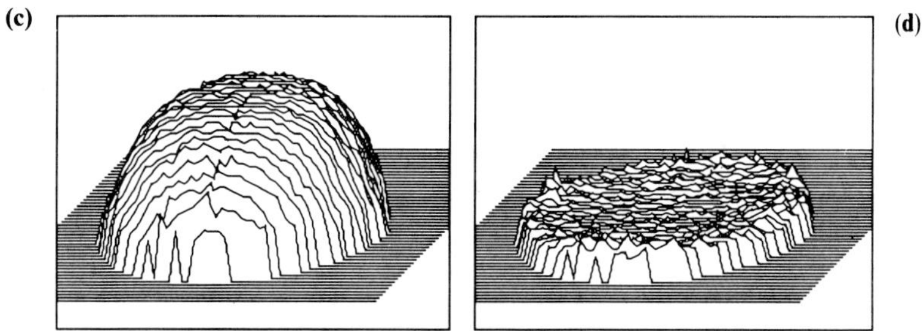

Figure 9.10 (a) Normalised velocity profiles for different concentration solutions of poly(ethylene oxide) in water obtained using dynamic NMR microscopy. The concentrations increase in equal steps from 0.5% (w/v) (·) to 4.5% (w/v) (⊡). (b) The polymer self-diffusion profile for the highest concentration solution in units of $10^{-13}\,m^2\,s^{-1}$. Note that this was obtained in a separate experiment so that the capillary wall does not fall at precisely the same pixel as in (a). (c) Water solvent velocity and (d) diffusion maps for the 4.5% (w/v) poly(ethylene oxide) solution. (From Y. Xia and P.T. Callaghan [18] and reproduced by permission of the American Chemical Society.)

and exhibit a central velocity maximum of 7.4 mm s^{-1} with maximum shear rate at the capillary radius of 41 s^{-1}. In order to obtain polymer self-diffusion coefficients, the experiments were repeated using PEO/D_2O solutions. Figure 9.10(b) shows the poly(ethylene oxide) self-diffusion profile obtained at 4.5% concentration. It is clear that above a critical radius, $r_c \approx 0.5R$, in the capillary flow, dramatic enhancement of the polymer Brownian motion occurs. Using the fitted power law velocity profile, this corresponds to a critical shear rate (0.08 s)$^{-1}$. By comparison, the tube renewal time may be calculated from our knowledge of the equilibrium self-diffusion coefficient since from the tube model, τ_d is given by $R_0^2/3\pi^2 D_s$, yielding $\tau_d \approx 0.07$ s. τ_d^{-1} is sufficiently close to the observed critical shear rate to suggest that the tube model may provide a useful approach in understanding the basis of shear thinning. It has also been shown [18] that the self-diffusion profile can be interpreted quantitatively using Doi–Edwards theory without the need for adjustable parameters.

It should be noted that the measurement of such small self-diffusion coefficients in the presence of a large local velocity shear requires the use of large magnetic field gradients (5 T m^{-1}) along with special pulse sequence which removes phase shifts due to coherent (i.e. stationary) flow but retains those due to stochastic motion. Such an effect is achieved by using a double echo variant of the PGSE pulse sequence [6, 8, 63].

9.6 *q*-Space imaging of morphology in colloidal and polymeric systems

The diffusive motion of molecules trapped inside a pore is strongly dependent on timescale. For example, except for a very small fraction in close proximity to the walls, molecules with self-diffusion coefficient D inside a spherical pore of size a will exhibit free Brownian motion for PGSE NMR observation times $\Delta \ll a^2/D$. By contrast, for $\Delta \gg a^2/D$, the mean-squared displacement $\overline{Z^2}$ will be independent of time and characteristic of the pore dimensions. In general, diffusive restrictions result in non-Gaussian conditional probabilities, $P_s(\boldsymbol{r}'|\boldsymbol{r}, \Delta)$, with time dependences strongly related to length scales and the local molecular self-diffusion coefficient. This behaviour can be used to probe the microstructure of the diffusive barriers [6]. While an exact description of the PGSE NMR experiment for all times Δ is tractable only for a few simple geometries [128, 129], a generalised formalism is available in the pore equilibration limit, $\Delta \gg a^2/D$ [6, 130, 131]. Here the conditional probabilities are independent of starting position so that $P_s(\boldsymbol{r}|\boldsymbol{r}', \Delta)$ reduces to $\rho(\boldsymbol{r}')$, the pore molecular density function. In consequence, the averaged propagator $\overline{P}_s(\boldsymbol{R}, \infty)$ becomes an autocorrelation function of $\rho(\boldsymbol{r}')$,

$$\overline{P}_s(\boldsymbol{R}, \infty) = \int \rho(\boldsymbol{r} + \boldsymbol{R})\rho(\boldsymbol{r})\,\mathrm{d}\boldsymbol{r} \tag{9.18}$$

and the echo attenuation function reduces to the power Fourier spectrum of $\rho(\boldsymbol{r}')$,

$$E_{\infty}(\boldsymbol{q}) = |S(\boldsymbol{q})|^2 \tag{9.19}$$

where $S(\boldsymbol{q})$ is analogous to the signal measured in conventional NMR imaging as defined in equation (9.2). To use the sphere example,

$$E_{\infty}(q) = |3(2\pi qa\cos(2\pi qa) - \sin(2\pi qa)]/(2\pi qa)^3|^2 \tag{9.20}$$

The behaviour when $q \ll a^{-1}$ is

$$\begin{aligned} E_{\infty}(q) &\approx 1 - \tfrac{1}{5}(2\pi qa)^2) \\ &\approx \exp(-\tfrac{1}{5}\gamma^2\delta^2 g^2 a^2) \end{aligned} \tag{9.21}$$

Equation (9.21) has ben used to determine droplet sizes in emulsions and disperse phases using the PGSE NMR method [132, 133].

Equation (9.20) has a node when $qa \approx 3/4$, reminiscent of optical diffraction and indeed the treatment lends itself to diffractive analogy. Unlike conventional NMR imaging where $S(\boldsymbol{k})$ is acquired directly, the loss of phase information inherent in equation (9.19) is formally similar to that suffered in X-ray or optical diffraction. Consequently, Fourier inversion does not yield $\rho(\boldsymbol{r})$. Structural features are more easily ascertained by taking advantage, for example, of characteristic features of the X-ray diffraction pattern, such as the nodes referred to.

The diffraction formalism carries over nicely to systems of interconnected pores, where molecular diffusion leads to successive incorporation of neighbouring pores in the effective 'diffraction grating' as time advances, in a manner akin to X-ray diffraction from a finite lattice of scattering centres. A detailed description of the physics is given elsewhere but the essence of the depiction is given by the relation [6, 8, 130, 131]

$$E_{\Delta}(q) \approx \overline{|S_0(q)|^2}\,[\mathscr{F}\{L(Z)\} \otimes \mathscr{F}\{d(Z,\Delta)\}] \tag{9.22}$$

where $S_0(q)$ is the local pore structure factor, $L(Z)$ describes the relative displacement along the magnetic field gradient of pores in the pore lattice, $d(Z,\Delta)$ is the diffusive envelope describing the migration of molecules from pore to pore, and $\mathscr{F}\{\ \}$ represents Fourier transformation. To pursue the X-ray analogy, $\overline{|S_0(q)|^2}$ plays the role of the form factor while $\mathscr{F}\{L(Z)\}$ is the reciprocal lattice convolved with $\mathscr{F}\{d(Z,\Delta)\}$, a broadening factor associated with the finite number of scattering centres. The most relevant evaluation of $E_{\Delta}(q)$ is for the 'pore glass' for which equation (9.22) can be shown to yield [131]

$$E_{\Delta}(q) \approx \overline{|S_0(q)|^2}\exp\left[-\left(\frac{6D_{\mathrm{p}}\Delta}{b^2+3\zeta^2}\right)\left(1-\exp(-2\pi^2q^2\zeta^2)\frac{\sin(2\pi qb)}{2\pi qb}\right)\right] \tag{9.23}$$

where b is the mean pore spacing with standard deviation ζ and D_{p} is the long-range self-diffusion coefficient relevant to the permeability of the porous structure.

Figure 9.11(a) shows the result of PGSE ^{1}H NMR measurements [8] of the echo signal from water molecules diffusing in the void spaces between a closed packed assembly of polystyrene latex spheres of mean sphere diameter 15.8 μm. Despite the random orientation of the pore 'lattice', a first-order coherence peak might be expected in analogy with X-ray diffraction from a powder. For diffusion times sufficiently large for water molecules to move to neighbouring pores ($\Delta > 50$ ms) such a peak is clearly visible at $q^{-1} \approx 16\,\mu$m,

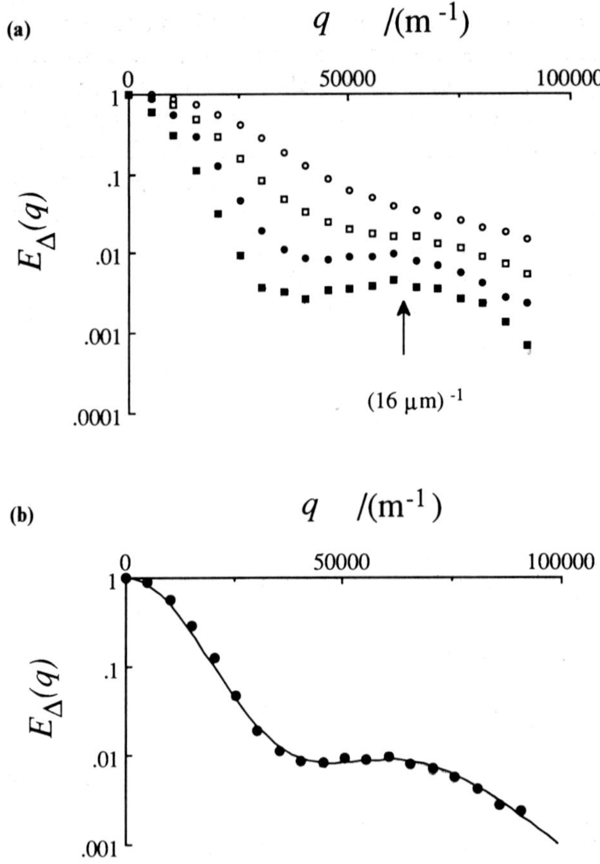

Figure 9.11 (a) The ^{1}H PGSE NMR echo attenuation, $E_\Delta(q)$, at a series of diffusion times Δ, for water diffusing in a packed array of latex spheres, diameter 15.8 μm. The feature at $q = 60\,000\,\text{m}^{-1}$ corresponds to a pore separation distance of 16 μm. The data correspond to $\Delta = 20$ ms (○), 40 ms (□), 70 ms (●), and 110 ms (■). (b) The 70 ms data, fitted using the pore glass expression of equation (9.23). (From P.T. Callaghan *et al.* [8] and reproduced by permission of Macmillan Magazines Ltd.)

in correspondence with the mean pore spacing. Figure 9.11(b) shows a quantitative fit to the 170 ms data using equation (9.23) and assuming a spherical pore shape. This yields a pore spacing $b = 16.1\ \mu m$, spacing standard deviation $\zeta = 2.9\ \mu m$ and pore diameter $a = 5.6\ \mu m$, structural parameters that correspond with the physical dimensions of the packed array.

9.7 Conclusions

Until very recently, the use of magnetic field gradients in NMR spectroscopy of polymers was restricted to a few specialist groups who have measured polymer and small molecule self-diffusion coefficients using specially developed apparatus. With the advent of NMR microscopy as a routine feature in modern NMR spectrometers, the use of field gradient methods, both in imaging applications, and in the study of molecular translational motion, may soon be widespread. The unity of the imaging and PGSE NMR methodologies has been stressed in this chapter. This linkage arises not only because of the common apparatus involved but because of the insights that are gained by utilising the Fourier depiction in understanding in PGSE method. This formalism, inherent in imaging, has led to an extension of the method beyond the study of polymer self-diffusion, to applications involving polymer internal dynamics, the imaging of heterogeneous flow and the indirect imaging of structural boundaries via restricted diffusion. These new NMR techniques offer exciting vehicles for exploring macromolecular motion and the morphology of heterogeneous polymer phases.

References

1. E.L. Hahn, *Phys. Rev.* **80** (1950) 580–594.
2. H.Y. Carr and E.M. Purcell, *Phys. Rev.* **94** (1954) 630–638.
3. D.W. McCall, D.C. Douglass and E.W. Anderson, *Ber. Busenges. Phys. Chem.* **67** (1963) 336–340.
4. E.O. Stejskal and J.E. Tanner, *J. Chem. Phys.* **42** (1965) 288–292.
5. P.T. Callaghan, *Aust. J. Phys.* **37** (1984) 359–387.
6. P.T. Callaghan, *Principles of Nuclear Magnetic Resonance Microscopy*, Oxford University Press, Oxford (1991).
7. P.T. Callaghan and A. Coy, *Phys. Rev. Lett.* **68** (1992) 3176–3179.
8. P.T. Callaghan, A. Coy, D. MacGowan, K.J. Packer and F.O. Zelaya, *Nature* **351** (1991) 467–469.
9. P.C. Lauterbur, *Nature* **242** (1973) 190–191.
10. P. Mansfield and P.K. Grannell, *J. Phys. C* **6** (1973) L422–L426.
11. J.B. Aguayo, S.J. Blackband, J. Shoeniger, M.A. Mattingley and M. Hinterman, *Nature* **322** (1986) 190–191.
12. C.D. Eccles and P.T. Callaghan, *J. Magn. Reson.* **68** (1986) 393–398.
13. P.C. Lauterbur, in *NMR in Biology and Medicine*, eds Shu Chien and Chien Ho, Raven Press, New York (1986).
14. H. Kamei and Y. Katayama, *IEEE 8th Annu. Conf. of the Engineering in Medicine and Biology Society*, IEEE (1986) Abstracts p. 1159.

15. J. Karger and W. Heink, *J. Magn. Reson.* **51** (1983) 1–7.
16. P.T. Callaghan, C.D. Eccles and Y. Xia, *J. Phys. E* **21** (1988) 820–822.
17. Y. Xia and P.T. Callaghan, *Makromol. Chem. Macromol. Symp.* **34** (1990) 277–286.
18. Y. Xia and P.T. Callaghan, *Macromolecules* **24** (1991) 4777–4786.
19. P. Mansfield and P.G. Morris, *NMR Imaging in Biomedicine*, Academic Press, New York (1982).
20. D.R. Bailes and D.J. Bryant, *Contemp. Phys.* **25** (1984) 441–475.
21. P. Mansfield, *Physics of NMR Spectroscopy in Biology and Medicine*, 345–369 Soc. Italiana di Fisica, Bologna, Italy (1988) pp. 345–369.
22. J. Frahm and W. Hanicke, *J. Magn. Reson.* **60** (1984) 320–332.
23. R.R. Ernst, G. Bodenhausen and A. Wokaum, *Principle of Nuclear Magnetic Resonance in One and Two Dimensions*, Clarendon Press, Oxford (1987).
24. W.A. Edelstein, J.M.S. Hutchison, G. Johnson and T.W. Redpath, *Phys. Med. Biol.* **25** (1980) 751–756.
25. G. Johnson, J.M.S. Hutchison, T.W. Redpath and L.M. Eastwood, *J. Magn. Reson.* **54** (1983) 374–384.
26. W. Kuhn, *Angew. Chem.* **29** (1990) 1–19.
27. W. Kuhn and M. Mattingley, *Special Bruker Report on NMR Microscopy*, Bruker GmbH, Karlsruhe (1988).
28. P.T. Callaghan and C.D. Eccles, *J. Magn. Reson.* **71** (1987) 426–445.
29. P.T. Callaghan and C.D. Eccles, *J. Magn. Reson.* **78** (1988) 1–8.
30. P.T. Callaghan, *J. Magn. Reson.* **87** (1990) 304–318.
31. A.A. Maudsley, S.K. Hilal, W.H. Perman and H.E. Simon, *J. Magn. Reson.* **51** (1983) 147–152.
32. E. Günther, B. Blümich and H.W. Spiess, *Mol. Phys.* **7** (1990) 477–489.
33. P. Stilbs, *Prog. Nucl. Magn. Reson. Spectrosc.* **19** (1987) 1–15.
34. P.A. Egelstaff, *An Introduction to the Liquid State*, Academic Press, London (1967).
35. E.O. Stejskal, *J. Chem. Phys.* **43** (1965) 3597–3603.
36. M. Tirrell, *Rubber Chem.* **57** (1984) 523–556.
37. E.D. von Meerwall, *Adv. Polym. Sci.* **54** (1983) 1–29.
38. J.E. Tanner *Macromolecules* **4** (1971) 748–750.
39. P.T. Callaghan and D.N. Pinder, *Macromolecules* **13** (1980) 1085–1091.
40. P.T. Callaghan and D.N. Pinder, *Macromolecules* **14** (1981) 1334–1340.
41. P.T. Callaghan and D.N. Pinder, *Macromolecules* **16** (1983) 968–972.
42. P.T. Callaghan and D.N. Pinder, *Macromolecules* **17** (1984) 431–437.
43. G. Fleischer, *Polym. Bull.* **9** (1983) 152–158.
44. R. Bachus and R. Kimmich, *Polymer* **24** (1983) 964–970.
45. W. Brown and P. Stilbs, *Polymer* **24** (1983) 188–192.
46. G. Fleischer, *Polym. Bull.* **11** (1984) 75–80.
47. G. Fleischer, *J. Colloid Polym. Sci.* **262** (1984) 919–926.
48. G. Fleischer, *Polymer* **26** (1985) 1677–1682.
49. G. Fleischer and E. Straube, *Polymer* **26** (1985) 241–246.
50. V.A. Sevrugin, V.D. Skirda and A.I. Maklakov, *Polymer* **27** (1986) 290–292.
51. F.A. Grinberg, V.D. Skirda, A.I. Maklakov, L.Z. Rogovina and I.P. Storogyk, *Polymer* **28** (1987) 1075–1078.
52. V.D. Skirda, V.I. Sundulov, A.I. Maklakov, I.R. Gafurov and G.I. Vasiljev, *Polymer* **29** (1988) 1294–1300.
53. T.M. Huirua, R. Wang and P.T. Callaghan, *Macromolecules* **23** (1990) 1658–1664.
54. W.M. Lomer and G.C. Low, in *Thermal Neutron Scattering*, ed. P.A. Egelstaff, Academic Press, New York (1965).
55. G.L. Squires, *Contemp. Phys.* **17** (1976) 411–441.
56. E. Balcar and S.W. Lovesey, *The Theory of Magnetic Neutron and Photon Scattering*, Oxford University Press, Oxford (1989).
57. P.T. Callaghan, C.M. Trotter and K.W. Jolley, *J. Magn. Reson.* **37** (1980) 247–259.
58. P.T. Callaghan, K.W. Jolley and C.M. Trotter, *J. Magn. Reson.* **39** (1980) 525–527.
59. R. Kimmich, W. Unrath, G. Schnur and E. Rommel, *J. Magn. Reson.* **91** (1991) 136–140.
60. P.T. Callaghan, *J. Magn. Reson.* **88** (1990) 493–500.
61. A.A. Samoilenko, D. Yu. Artemov and L.A. Sobeldina, *Bruker Rep.* **2** (1987) 30–31.

62. P.T. Callaghan, in *Flow Visualization V*, ed. R. Reznicek, Hemisphere, New York (1990), pp. 188–193.
63. P.T. Callaghan and Y. Xia, *J. Magn. Reson.* **91** (1991) 326–352.
64. J.J. Attard and P.J. McDonald, *Int. J. Imaging Systems Technol. (USA)* **2** (1990) 47–51.
65. A. Knüttel, K.H. Spohn and R. Kimmich, *J. Magn. Reson.* **86** (1990) 542–548.
66. A.A. Samoilenko and K. Zick, *Bruker Rep.* **1** (1990) 40–41.
67. J.B. Miller and A.N. Garroway, *J. Magn. Reson.* **77** (1988) 187–191.
68. D.G. Cory, J.B. Miller and A.N. Garroway, *J. Magn. Reson.* **90** (1990) 205–213.
69. S.P. Cottrell, M.R. Halse and J.H. Strange, *Meas. Sci. Technol.* **1** (1990) 624–629.
70. J.H. Strange, *Phil. Trans. R. Soc. London Ser. A* **333** (1990) 427–439.
71. D.G. Cory, J.W.M. Van Os and W.S. Veeman, *J. Magn. Reson.* **76** (1988) 543–547.
72. D.G. Cory, A.M. Reichwein, J.W.M. Van Os and W.S. Veeman, *Chem. Phys. Lett.* **143** (1988) 467–471.
73. D.G. Cory, J.C. de Boer and W.S. Veeman, *Macromolecules* **22** (1989) 1618–1621.
74. D.G. Cory and W.S. Veeman, *J. Magn. Reson.* **84** (1989) 392–397.
75. B. Blümich, *Adv. Mater.* **3** (1991) 237–244.
76. B. Blümich, P. Blümler, E. Günther, G. Schauss and H.W. Spiess, *Makromol. Chem. Macromol. Symp.* **44** (1991) 37–45.
77. B. Blümich, P. Blümler, E. Günther and G. Schauss, *Bruker Rep.* **2** (1990) 22–24.
78. C. Chang and R.A. Komoroski, *Macromolecules* **22** (1989) 600–607.
79. S.N. Sarkar and R.A. Komoroski, *Macromolecules* **25** (1992) 1420–1426.
80. R.A. Komoroski and S.N. Sarkar, *Mater. Res. Soc. Symp. Proc.* **217** (1991) 3–14.
81. T.H. Mareci, S. Donstrup and A. Rigamonti, *J. Mol. Liquids* **38** (1988) 185–206.
82. S.R. Smith and J.L. Koenig, *Macromolecules* **24** (1991) 3496–3504.
83. M.R. Krejsa and J.L. Koenig, *Rubber Chem. Technol.* **64** (1991) 635–640.
84. L.D. Hall *et al.*, to be published.
85. A.O.K. Nieminen and J.L. Koenig, *J. Adhesion Sci. Technol.* **2** (1988) 407–414.
86. A.O.K. Nieminen and J.L. Koenig, *Appl. Spectrosc.* **43** (1989) 1358–1362.
87. A.O.K. Nieminen J. Liu and J.L. Koenig, *J. Adhesion Sci. Technol.* **3** (1989) 455–462.
88. L.A. Weisenberger and J.L. Koenig, *Appl. Spectrosc.* **43** (1989) 1117–1126.
89. L.A. Weisenberger and J.L. Koenig, *J. Polym. Sci. Polym. Lett.* **27** (1989) 55–57.
90. B.C. Perry and J.L. Koenig, *J. Polym. Sci. Polym. Chem.* **27** (1989). 3429–438.
91. R.S. Clough and J.L. Koenig, *J. Polym. Sci. Polym. Lett.* **27** (1898) 451–454.
92. K.P. Hoh, B. Perry, G. Rotter, H. Ishida and J.L. Koenig, *J. Adhesion* **27** (1989) 245–249.
93. R.A. Grinstead and J.L. Koenig, *Macromolecules* **25** (1992) 1229–1234.
94. R.A. Grinstead, L. Clark and J.L. Koenig, *Macromolecules* **25** (1992). 1235–1241.
95. H. Yamakawa, *Modern Theory of Polymer Solutions*, Harper and Row, New York, (1971).
96. P.G. de Gennes, *Scaling Concepts in Polymer Physics*, Cornell University Press, Ithaca, NY (1979).
97. M. Doi and S.F. Edwards, *The Theory of Polymer Dynamics*, Oxford University Press, Oxford (1987).
98. P.T. Callaghan and J. Lelievre, *Anal. Chim. Acta* **189** (1986) 145–166.
99. H. Hervet, L. Leger and F. Rondelez, *Phys. Rev. Lett.* **42** (1979) 1681–1684.
100. P.J. Daivis, D.N. Pinder and P.T. Callaghan, *Macromolecules* **25** (1992) 170–178.
101. M.E. Moseley and P. Stilbs, *Chem. Scr.* **16** (1980) 114–115.
102. M. Doi and S.F. Edwards, *J. Chem. Soc. Faraday Trans. 2* **74** (1978) 1789–1701.
103. P.G. de Gennes, *J. Chem. Phys.* **55** (1971) 572–579.
104. G. Weill and J. des Cloizeaux, *J. Phys. (Paris)* **40** (1979) 99–105.
105. A.I. Maklakov, V.D. Skirdaand and N.F. Fatkullin, in *Encyclopedia of Fluid Mechanics*, ed. N.P. Cheremiisinoff, Gulf, Houston (1991).
106. D.N. Pinder, *Macromolecules* **23** (1990) 1724–1729.
107. G.D.J. Phillies, *Macromolecules* **19** (1986) 2367–2376.
108. T.P. Lodge, N.A. Rotstein and S.Prager, *Adv. Chem. Phys.* **79** (1990) 1–132.
109. N. Nemoto, T. Kojima, T. Inoue, M. Kishine, T. Hirayama and M. Kurata, *Macromolecules* **22** (1989) 3793–3798.
110. T.P. Lodge, P. Markland and L.M. Wheeler, *Macromolecules* **22** (1989) 3409–3418.
111. E.D. von Meerwall, D.H. Tomich, J. Grigsby, R.W. Pennisi, L.J. Fetters and N. Hadjichristidis, *Macromolecules* **16** (1983) 1715–1722.

112. P.T. Callaghan and D.N. Pinder, *Macromolecules* **18** (1985) 373–379.
113. K.E. Evans and S.E. Edwards, *J. Chem. Soc. Faraday Trans., 2* **77** (1981) 1891–1912.
114. J.M. Deutsch, *Phys. Rev. Lett.* **49** (1982) 926–929.
115. A Baumgartner, *Annu. Rev. Phys. Chem.* **35** (1984) 419–435.
116. K. Kremer, G.S. Grest and I. Carmesin, *Phys. Rev. Lett.* **61** (1988) 566–569.
117. J.H. Wang, *J. Am. Chem. Soc.* **76** (1954) 4755–4765.
118. B. Jonsson, H. Wennerstrom, P.G. Nilsson and P. Linse, *Colloid Polym. Sci.* **264** (1986) 77–88.
119. M. Litowska, *Colloid Polym. Sci.* **262** (1984) 461–465.
120. T.M. Garver and P.T. Callaghan, *Macromolecules* **24** (1991) 420–430.
121. W. Miller and R. Olayo, *Macromolecules*, in press.
122. H.L. Frisch and S.A. Stern, *CRC Crit. Rev. Solid State Matter Sci.* **11** (1983) 123.
123. J.S. Vrentas and J.L. Duda, *Macromolecules* **9** (1976) 785–790.
124. H.L. Frisch and S.A. Stern, *CRC Crit. Rev. Solid State Matter Sci.* **11** (1983) 123–187.
125. J.D. Ferry, *Viscoelastic Properties of Polymers*, 3rd edition, Wiley, New York (1980).
126. A.H.P. Skelland, *Non-Newtonian Flow and Heat Transfer*, J. Wiley, New York (1967).
127. M.R. Mackley and I.P.T. Moore, *J. Non-Newtonian Fluid Mech.* **21** (1986) 337–358.
128. J.E. Tanner and E.O. Stejskal, *J. Chem. Phys.* **49** (1968) 1768–1777.
129. J.E. Tanner, *J. Chem. Phys.* **69** (1978) 1748–1754.
130. P.T. Callaghan, D. McGowan, K.J. Packer and F. Zelaya, *J. Magn. Reson.* **90** (1990) 177–182.
131. P.T. Callaghan, A. Coy. T.P.J. Halpin, D. MacGowan, K.J. Packer and F.O. Zelaya, *J. Chem. Phys.* **97** (1992) 651–662.
132. K.J. Packer and C. Rees, *J. Colloid Interface Sci.* **40** (1976) 206–218.
133. P.T. Callaghan, K.W. Jolley and R.S. Humphrey, *J. Colloid Interface Sci.* **93** (1983) 521–529.

Index